AF361882

Advances in Lipidomics

Advances in Lipidomics: Biomedicine, Nutrients and Methodology

Editors

Olimpio Montero
David Balgoma
Luis Gil-de-Gómez

MDPI • Basel • Beijing • Wuhan • Barcelona • Belgrade • Manchester • Tokyo • Cluj • Tianjin

Editors
Olimpio Montero
Spanish National Research Council (CSIC)
Spain

David Balgoma
University of Uppsala
Sweden

Luis Gil-de-Gómez
Children's Hospital of Philadelphia
USA

Editorial Office
MDPI
St. Alban-Anlage 66 4052 Basel, Switzerland

This is a reprint of articles from the Special Issue published online in the open access journal *Metabolites* (ISSN 2218-1989) (available at: https://www.mdpi.com/journal/metabolites/special_issues/advances_lipidomics).

For citation purposes, cite each article independently as indicated on the article page online and as indicated below:

LastName, A.A.; LastName, B.B.; LastName, C.C. Article Title. *Journal Name* **Year**, *Volume Number*, Page Range.

ISBN 978-3-0365-1186-3 (Hbk)
ISBN 978-3-0365-1187-0 (PDF)

Contents

About the Editors

Olimpio Montero's research is focused in mass spectrometry-based metabolomics with a special focus in lipidomics. He has worked in diverse research centers and institutes, as well as collaborated with groups specializing in different areas of mass spectrometry, metabolomics, and lipidomics. Dr. Montero has published about 60 scientific documents, including journal articles and book chapters. Currently member of the Metabolomics Society, he has reviewed more than 70 papers for about 20 different scientific journals.

David Balgoma is a researcher at Uppsala University. His areas of expertise are mass spectrometry and lipidomics with a focus on data pre-treatment and statistical analysis.

Luis Gil-de-Gómez has developed a research portfolio in the study of how metabolism shapes the innate and adaptive immune response in diverse diseases, including Alzheimer's disease, malnutrition, and cancer. He is experienced in chromatography, mass spectrometry, and flow cytometry techniques. Based on these platforms, the impact of cellular metabolic state has been studied with different approaches: lipidomic profiling of the inflammation process, diagnosis method based on alterations of plasma metabolites, the interaction of the gut microbiota with the immune response, and the metabolic reprogramming on the immunity activation and immune tolerance balance.

Preface to "Advances in Lipidomics: Biomedicine, Nutrients and Methodology"

Lipidomics, a primary branch of metabolomics, have rapidly developed into a novel research discipline. This feature has arisen from robust evidence of lipid reprogramming in metabolic disorders, cancer, or cardiovascular diseases. The importance of the lipid composition in a variety of pathologies has been elucidated from alternative perspectives. This book contains studies regarding lipidomics methodology, the involvement of lipids in metabolism and biomedicine, and advances in nutrition.

Regarding advances in methodology, Panzenobeck et. al. automated the annotation of glycosylinositolphosphoceramides in plants. Jenkins et al. presented a high-throughput method of lipid extraction. Magny et al. combined the prediction of molecular behavior with mass spectrometric data in order to annotate lipids. Regarding advances in lipid metabolism in biomedicine, Saito et al. reported the lipid profile associated to liver injury induced by drugs. Franco et al. investigated the sexual dimorphism in mouse epidermis of phospholipids, cholesteryl esters, acylcarnitines, and sphingolipids. Azbukina et al. profiled oxylipins in plasma from patients with Wilson disease. Miehle et al. studied the remodeling of lipids during adipogenesis. Korczyńska et al. profiled fatty acids in serum from patients with chronic kidney disease. Balgoma et al. reviewed the involvement of lipids in human positive ssRNA virus infection. Gil-de-Gómez et al. reviewed the potential of lipids to modulate the immune response in cancer. Tomczyk et al. reviewed the lipidome profile in models of cardiovascular disease. Finally, regarding nutrition and biomedicine, Ferreri et al. reviewed the relationship between the lipidome, nutrition, and signaling metabolic pathways in cancer.

The reader interested in a broad perspective of lipidomics and the involvement of the lipidome in different diseases will find this book interesting in order to obtain state-of-the-art applications and discoveries.

Olimpio Montero, David Balgoma, Luis Gil-de-Gómez
Editors

 *metabolites*

Article

Chasing the Major Sphingolipids on Earth: Automated Annotation of Plant Glycosyl Inositol Phospho Ceramides by Glycolipidomics

Lisa Panzenboeck [1], Nina Troppmair [1], Sara Schlachter [1], Gunda Koellensperger [1,2,3], Jürgen Hartler [4] and Evelyn Rampler [1,2,3,*]

[1] Department of Analytical Chemistry, Faculty of Chemistry, University of Vienna, Waehringer Str. 38, 1090 Vienna, Austria; lisa.panzenboeck@univie.ac.at (L.P.); a01556591@unet.univie.ac.at (N.T.); a11774156@unet.univie.ac.at (S.S.); gunda.koellensperger@univie.ac.at (G.K.)
[2] Vienna Metabolomics Center (VIME), University of Vienna, Althanstraße 14, 1090 Vienna, Austria
[3] Chemistry Meets Microbiology, University of Vienna, Althanstraße 14, 1090 Vienna, Austria
[4] Institute of Pharmaceutical Sciences, University of Graz, Universitätsplatz 1/I, 8010 Graz, Austria; juergen.hartler@uni-graz.at
* Correspondence: evelyn.rampler@univie.ac.at; Tel.: +43-1-4277-52381

Received: 20 August 2020; Accepted: 16 September 2020; Published: 19 September 2020

Abstract: Glycosyl inositol phospho ceramides (GIPCs) are the major sphingolipids on earth, as they account for a considerable fraction of the total lipids in plants and fungi, which in turn represent a large portion of the biomass on earth. Despite their obvious importance, GIPC analysis remains challenging due to the lack of commercial standards and automated annotation software. In this work, we introduce a novel GIPC glycolipidomics workflow based on reversed-phase ultra-high pressure liquid chromatography coupled to high-resolution mass spectrometry. For the first time, automated GIPC assignment was performed using the open-source software Lipid Data Analyzer (LDA), based on platform-independent decision rules. Four different plant samples (salad, spinach, raspberry, and strawberry) were analyzed and the results revealed 64 GIPCs based on accurate mass, characteristic MS2 fragments and matching retention times. Relative quantification using lactosyl ceramide for internal standardization revealed GIPC t18:1/h24:0 as the most abundant species in all plants. Depending on the plant sample, GIPCs contained mainly amine, N-acetylamine or hydroxyl residues. Most GIPCs revealed a Hex-HexA-IPC core and contained a ceramide part with a trihydroxylated t18:0 or a t18:1 long chain base and hydroxylated fatty acid chains ranging from 16 to 26 carbon atoms in length (h16:0–h26:0). Interestingly, four GIPCs containing t18:2 were observed in the raspberry sample, which was not reported so far. The presented workflow supports the characterization of different plant samples by automatic GIPC assignment, potentially leading to the identification of new GIPCs. For the first time, automated high-throughput profiling of these complex glycolipids is possible by liquid chromatography-high-resolution tandem mass spectrometry and subsequent automated glycolipid annotation based on decision rules.

Keywords: glycolipidomics; GIPC; glycosyl inositol phospho ceramides; Lipid Data Analyzer; lipidomics; sphingolipids; ultra-high pressure liquid chromatography; high-resolution mass spectrometry; LC-MS; automated annotation

---

1. Introduction

The sphingolipidome of plants contains glycosyl inositol phospho ceramides (GIPCs), glycosylceramides and ceramides, whereas sphingomyelin, globosides, sulfatides or gangliosides are absent. GIPCs were characterized as the major sphingolipid on earth due to their high abundance in plants and fungi, which comprise a large portion of the biomass of the biosphere [1]. GIPCs were first described more than 60 years ago as "phytoglycolipids" [2]. The total plant lipid content can consist of up to 40% GIPCs [3]. The structure of these plant sphingolipids has three major subunits: (1) a polar inositol containing part, (2) the sphingoid backbone with a long-chain base (amino-alcohol) linked by an amide bond to a (3) fatty acyl chain moiety [2,4]. The terms d, t and q refer to the hydroxylation state of the whole ceramide or long-chain base (LCB) moiety, ranging from two (d) to four (q) hydroxy groups. The term h denotes a hydroxylation of the fatty acyl group (i.e., the ceramide moiety q40:1 can correspond to a t18:1 LCB connected to a h22:0 fatty acyl). Di- and trihydroxylation of LCBs with t18:0, t18:1(8Z and 8E) (the main sphingoid base in some species), and d18:0, d18:1(8Z and 8E), d18:2 (4E/8Z and 4E/8E) and fatty acid components varying in chain-length, saturation and hydroxylation state (h16:0–h26:1, 20:0 to 28:0) have been reported in plant GIPCs [5,6]. Different GIPC core structures were determined from higher plants ranging from simple high-abundant A-series species with Hex-HexA-IPC and HexN(Ac)-HexA-IPC (Hex = hexose, HexA = hexuronic acid, IPC = inositol phospho ceramide, HexN = hexosamine, and HexNAc = N-acetyl hexosamine) to low abundant F-series species containing several arabinoses and hexoses [3,7]. Despite the fact that GIPCs are an integral part of the plant plasma membrane, there is still little knowledge concerning its molecular organization and the way this organization is involved in signaling processes necessary for cellular adaptation [1]. To understand the interplay of GIPCs with different enzymes and their detailed function in the plasma membrane in plants, comprehensive structural information provided by observation tools such as NMR or MS are necessary.

Even though GIPCs were discovered 60 years ago, their analysis remains challenging due to the lack of available standards, automated annotation software and reference databases. For example, CHEBI [8] does not provide any GIPCs and the comprehensive LIPID MAPS Structure Database (LMSD) contains only one GIPC (A-NH$_2$-t18:1/h24:0) [9]. As GIPCs consist of a sugar head group linked to a lipid subunit causing amphiphilic properties, they are neither well covered by common glycomics nor lipidomics workflows. Consequently, specialized glycolipidomics analysis strategies are required, e.g., applying a mixture of 2-propanol (IPA), hexane and water [10]. The combination of liquid chromatography and mass spectrometry (LC-MS) has been used due to its unprecedented potential to annotate GIPCs by *m/z*, retention time and fragmentation pattern [7,11]. Unambiguous GIPC identification requires both retention time evaluation and detection of structural subunits by tandem mass spectrometry (MS2), due to the absence of commercial standards. Most GIPC LC-MS-based analysis workflows were performed almost a decade ago by electrospray ionization followed by analysis with low resolution mass spectrometers (QQQ, QTRAP) [7,11]. Meanwhile, high-resolution mass spectrometers (such as TOF, orbitrap, FTICR) have been established with up to 1,000,000 resolution enabling GIPC analysis by accurate mass [12]. Additionally, ultra-high pressure liquid chromatography (up to 1500 bar) with sub 2-μm particles provides high chromatographic resolution and excellent sensitivity. Up to now, GIPC analysis has been performed by tedious manual annotation and curation [1,7,12,13] and expert knowledge was necessary to interpret glycosphingolipid tandem mass spectrometry fragmentation patterns [14–16]. The instrumentation advancements of the recent years paved the way for automated high-throughput GIPC analysis. In this work, a variety of plants, i.e., iceberg lettuce (*Lactuca sativa var. capitata nidus tenerimma*), deep frozen spinach (*Spinacia oleracea*), raspberries (*Rubus idaeus*), and strawberries (*Fragaria*) were analyzed by the combination of reversed-phase (RP) ultra-high pressure liquid chromatography (UHPLC) and high-resolution mass spectrometry (HRMS). For the first time, automated GIPC annotation will be performed using the Lipid Data Analyzer (LDA) and platform-independent decision rules [17].

2. Results

Here we describe a novel workflow by RP-HRMS/MS using the open-source program LDA [17] for automated GIPC assignment. Method development considerations and guidelines for the automated structural analysis of GIPCs are provided. Finally, we test the developed glycolipidomics workflow for different plant samples, leading to a reference database of GIPCs, including fragmentation and retention time information.

2.1. Method Development for Automated GIPC Assignment

GIPCs were extracted by a mixture of IPA, n-hexane and water [18]. So far, most LC-MS-based GIPC chromatographic separations relied on the use of tetrahydrofuran (THF) containing solvents [7,11–13]. However, the usage of THF in the eluent system has some drawbacks: (1) it is aprotic and cannot donate a proton; thus, for ionization, pairing with a protic solvent (usually water) is necessary; (2) it can attack tubing (especially PEEK tubing); (3) it tends to polymerize (usually in APCI mode); and (4) it is highly flammable. In order to avoid the use of THF, we developed a novel GIPC method based on RP-HRMS/MS, facilitating a 30 min isopropanol gradient (detailed information can be found in the Materials and Methods Section 4.3). GIPC detection was performed using both negative and positive electrospray ionization and high-resolution Orbitrap MS (see Materials and Methods Section 4.4). Importantly, GIPC analysis requires relatively high RF voltages (S-lens RF level of 45) to ensure efficient transport of medium size glycolipids in the mass spectrometer. Figure 1 shows the extracted ion chromatogram of GIPCs in salad samples analyzed by RP-HRMS, based on data-dependent MS2 (ddMS2) in positive and negative ion modes. The GIPCs displayed in Figure 1 belong to the A-series (Hex(R1)-HexA-IPC) with R1 being a hydroxyl group and the ceramide portion consisting of a hydroxylated saturated fatty acyl chain attached to a t18:1 long chain base.

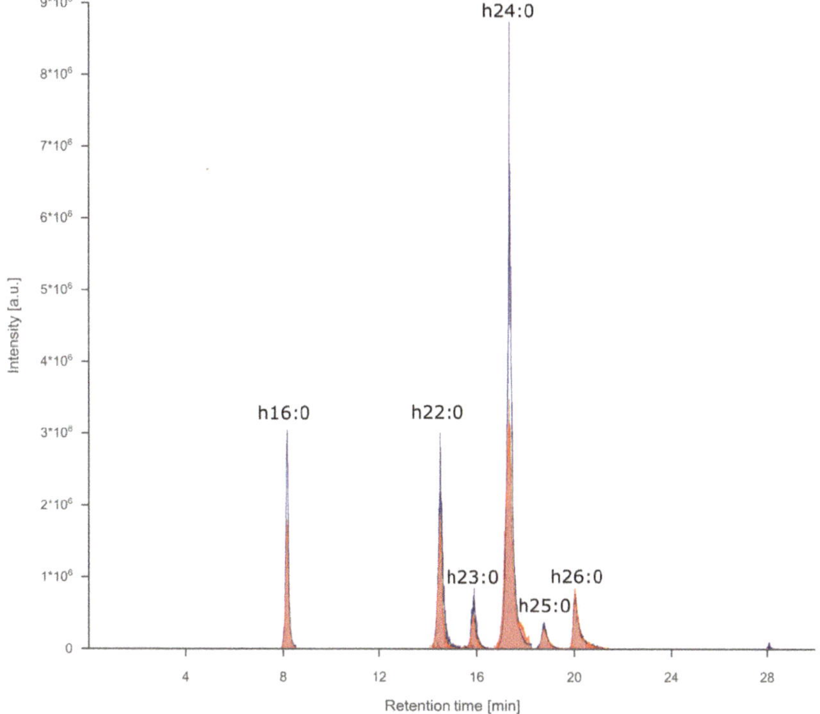

Figure 1. Extracted ion chromatogram of glycosyl inositol phospho ceramides (GIPCs) in salad samples analyzed by RP-HRMS/MS analysis using ddMS2 in positive (red) and negative (blue) ion modes. Assigned GIPCs belong to the Hex-HexA-IPC series with a t18:1 long-chain base (LCB) and varying chain length of the hydroxylated saturated fatty acids. Retention times coincided in positive and negative ion modes. Increasing carbon numbers result in belated elution.

As no commercial standards are available, GIPC assignment has to be conducted with caution. In such a situation, the use of the equivalent carbon number model (ECN) is required [19,20]. The ECN model originates from state of the art lipidomics workflows and is based on elution orders observed in RP columns: (1) longer fatty acid chains will increase the retention time (see Figure 1) and (2) more double bonds will decrease the retention time [21] (see Table S1). To increase the level of confidence in GIPC annotation, we accepted only GIPCs that: (1) were detectable by accurate mass (±5 ppm) in MS1 at the same retention time in both positive and negative ion modes (Figure 1); (2) showed MS2 spectra with characteristic fragments for the ceramide and sugar part in at least one ion mode and; (3) fulfilled the ECN model.

2.2. Structural Elucidation and GIPC Annotation Based on MS2 Information

In this work, we introduce the first automated GIPC annotation workflow based on structural information provided by acquired MS2 spectra. Structural analysis and automated GIPC annotation was performed based on a set of in-house developed decision rules for the freely available software LDA [17,22]. As no standards were available, blank extractions (no GIPC annotations found) and GIPC annotations in salad [13] and spinach [12] reported in the literature were used to validate GIPC assignments (Figure 1, Table A1 and Table S2). Various LCBs (d18:0, d18:1, d18:2, t18:0, and t18:1) and fatty acids (FAs) (16–26) with or without hydroxylation have been reported [5,13]. Moreover, R1 in Figure 2A can either be a hydroxyl (OH), an amine (NH_2) or an N-acetylamine (NAc) group, increasing the number of putative GIPCs even within a single series.

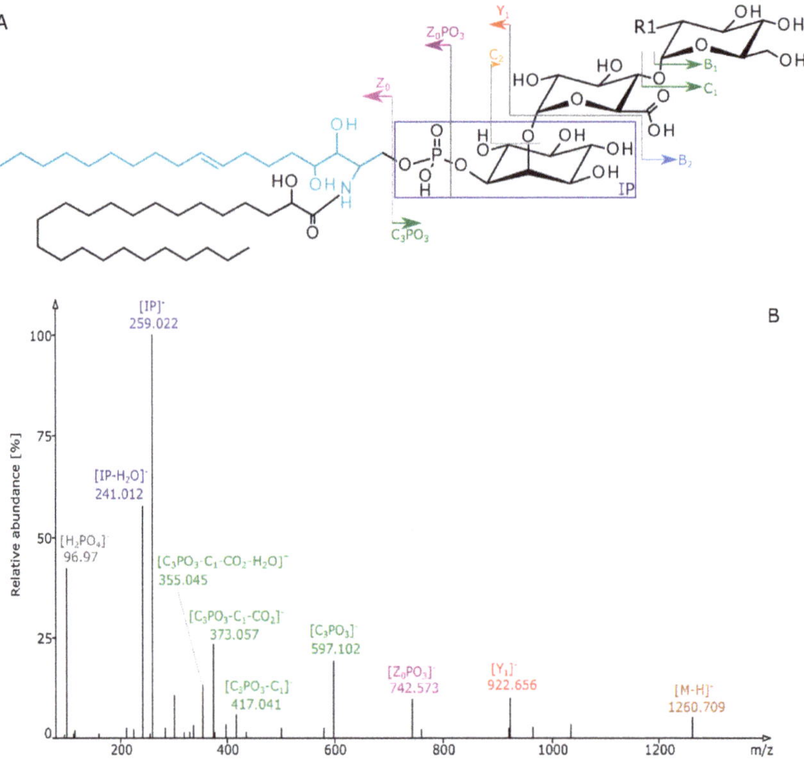

Figure 2. *Cont.*

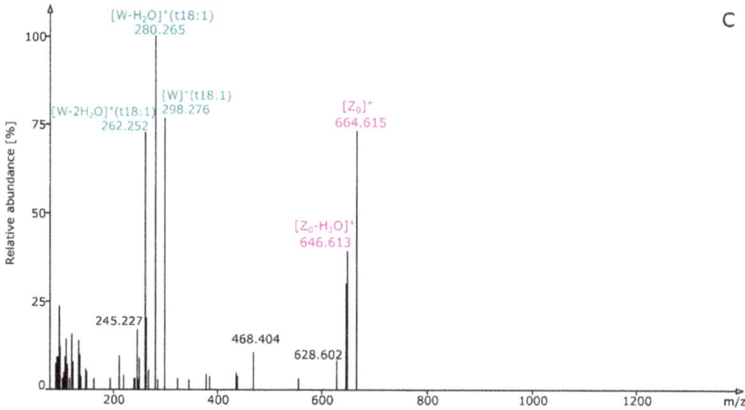

Figure 2. Overview of the GIPC fragmentation for the example of GIPC A-OH-t18:1/h24:0 in salad: (**A**) The fragment assignment of GIPC A-OH-t18:1/h24:0 (adapted from [23]). The W fragment is shown in a light blue color. Please note that a full structural characterization is not possible by RP-HRMS/MS, (**B**) The product ion spectrum in negative ion mode at m/z 1260.7237, showing characteristic fragments m/z 241 and 259, 355, 373 and 417. The sugar head group was confirmed by the $[C_3PO_3]^-$ fragment (m/z 597, R1 = OH). $[Z_0PO_3]^-$ and $[Y_1\text{-}H]^-$ fragments prove the ceramide moiety. (**C**) The positive ion mode ddMS2 spectrum of the $[M + H]^+$ precursor, exhibiting the $[W]^+$, $[W\text{-}H_2O]^+$ and $[W\text{-}2H_2O]^+$ fragments at m/z 298, 280 and 262, which are characteristic for the t18:1 LCB.

The final decision rule set was based on well-defined fragments (fragment rules) and their intensity relationships (intensity rules) (Folder S1). The characteristic fragments $[IP]^-$ (m/z 259) and $[IP\text{-}H_2O]^-$ (m/z 241) are mandatory in negative ion mode (e.g.: Figure 2B). However, these fragments are not specific, since they are produced by other phosphoinositol-containing lipids too. Thus, for a confident identification, negative or positive ion mode fragments indicating the sugar or ceramide part have to be detected.

In the majority of cases (see level 2 annotations, Table A1 and Table S2), MS2 spectra with GIPC fragmentation patterns were detected in both negative and positive mode. Depending on the fragmentation pattern and the level of confidence [24] of the structural elucidation, GIPCs are assigned as either: (1) series-R1-hydroxylation stage-carbon number (LCB + FA)-number of double bonds (LCB + FA) if the exact ceramide composition is not known or (2) series-R1-LCB/FA. Figure 2B displays an exemplary ddMS2 spectrum of A-OH-q42:1 with m/z 1260.7237, in salad recorded in negative ion mode. The positive ion mode fragmentation pattern of the $[M + H]^+$ precursor (m/z 1262.7389, Figure 2C) revealed further structural details, based on the identification of $[W]^+$, $[W\text{-}H_2O]^+$ and $[W\text{-}2H_2O]^+$ fragments, indicating an A-OH-t18:1/h24:0 GIPC. Additional GIPC confirmation is possible by Z_0 fragments ($[Z_0]^+$, $[Z_0\text{-}H_2O]^+$) of the $[M + H]^+$ precursor and by the sodium adduct $[M + Na]^+$ (Figure A1), where sugar fragments are readily observable. GIPCs were annotated based on single ionization information only if (1) in negative ion mode in addition to the apparent $[IP]^-/[IP\text{-}H_2O]^-/[H_2PO_4]^-$ fragments at m/z 259, 241 and 97, other characteristic fragments were detectable e.g., $[C_3PO_3]^-$ (m/z 596 – R1 = NH_2, m/z 597 – R1 = OH, m/z 638 – R1 = NAc), $[C_3PO_3\text{-}C_1\text{-}CO_2]^-$ (m/z 373) or $[C_3PO_3\text{-}C_1\text{-}CO_2\text{-}H_2O]^-$ (m/z 355) or (2) in positive ion mode the $[IP]^+$ (m/z 261)/$[IP + Na]^+$ (m/z 283) and fragments indicating the ceramide moiety (e.g., Z_0) were identified by LDA. The detailed fragment information used for GIPC annotation can be found in Table S2.

GIPC annotation can be hampered by the presence of isobaric masses for qX:Y NH_2 and t(X − 2):(Y − 1) NAc (where X refers to the carbon number (LCB + FA) and Y refers to the number of double bonds (LCB + FA), respectively). This may result in false positive GIPC identifications, because these classes share the same characteristic fragments *m/z* 241, 259, 355, 373 and 417. The correct structural elucidation is possible if additional fragments such as $[C_3PO_3]^-$ (R1 = OH − *m/z* 597, R1 = NH_2 − *m/z* 596, R1 = NAc − *m/z* 638) in negative ion mode or if LCBs in positive ion mode can be identified based on $[W]^+$, $[W-H_2O]^+$ and $[W-2H_2O]^+$ fragments. In the ddMS2 spectra of the $[M + H]^+$-precursor, trihydroxylated LCBs are characterized by the presence of three W fragments ($[W]^+$, $[W-H_2O]^+$ and $[W-2H_2O]^+$), such as t18:0 (*m/z* 300, 282 and 264) and t18:1 (*m/z* 298, 280, 262), while dihydroxylated species miss the $[W-2H_2O]^+$ fragment, e.g., d18:0 (*m/z* 284, 266), d18:1 (*m/z* 282, 264) and d18:2 (*m/z* 280, 262). As such, both LCB hydroxylation levels can be clearly distinguished.

2.3. Analysis of Different Plant GIPCs by UHPLC-HRMS Suggesting t18:2 LCB

The novel RP-HRMS/MS and GIPC annotation workflow was used to analyze different plant samples, namely salad (*Lactuca sativa var. capitata nidus tenerimma*), deep frozen spinach (*Spinacia oleracea*), raspberries (*Rubus idaeus*) and strawberries (*Fragaria*). As glycosphingolipid analysis is not negatively impacted by alkaline hydrolysis [10], alkaline hydrolysis was performed to simplify lipid profiles by removing the phospholipid background in the unknown plant samples (strawberry and raspberry, detailed information can be found in the Materials and Methods Section 4.2.2). Figure A2 shows the RP-HRMS/MS GIPC profile for the five most abundant GIPCs determined in spinach, strawberry and raspberry samples. For the sake of clarity, the five most abundant GIPCs in salad (A-NAc-t18:1/h24:0, A-NH₂-t18:1/h24:0, A-OH-t18:1 h22:0 and h24:0, A-OH-t18:0/h24:0) are not displayed in Figure A2. Irrespective of the plant sample, the species group A-R1-t18:1/h24:0 was always the most abundant one. While in spinach R1 was always N-acetylamine (A-NAc-t18:1 h22:0 to h26:0) for the five dominating GIPCs, in strawberries the major GIPCs contained a hydroxyl group as R1 (A-OH-t18:1 h23:0 to h26:0 and A-OH-t18:0/h24:0). In contrast to that, raspberries had an amine group as R1 for four out of five shown GIPCs (A-NH₂-t18:0/h24:0, A-NH₂-t18:1 h22:0 and h24:0, A-NH₂-t18:2/h24:0 and A-OH-t18:1/h24:0), emphasizing the structural diversity of GIPCs in different plants.

By analyzing different GIPCs, the NAc, NH₂ and OH-species from the A series could be detected (Figure 3A–C) with high confidence by (1) accurate determination of mass, (2) matching retention times of ion modes, (3) characteristic fragments and (4) the ECN model. We recommend checking isotopic patterns to avoid false positive hits. For a comprehensive overview of the annotated GIPCs see Table A1.

Due to the absence of commercially available GIPC standards, relative quantification of the individual species was performed using C16 lactosyl(ß) ceramide (d18:1/16:0) as the internal standard. This compound is similar in structure (sugar and ceramide moiety) and retention time (14 min). Even though lactosyl ceramide (d18:1/16:0) may be present in plants, we could not detect it in our samples, thus, making it suitable as the internal standard in our workflow. Normalization by the internal standard (area ratio) and dry weight was performed for MS1-based relative quantification by Skyline [25] (Figure 3A–C). Estimated concentrations in the nmol to μmol range per gram dry weight were observed, which is consistent with the literature [12,18].

In summary, 64 GIPCs in salad (19), spinach (8), strawberry (10) and raspberry (27) were annotated (Table A1). Ranking of the GIPC annotations was performed according to the guidelines of the metabolomics society [24,26], leading to 48 level 2 (matching accurate masses and MS2 in negative and positive mode) GIPCs, 13 level 3 (MS2 in one ion mode with matching accurate masses in both ion modes) GIPCs and 3 level 3** (matching accurate masses in both ion modes, MS2 in one ion mode but lacking information on IP fragments in positive ion mode or lacking sugar information in negative ion mode) GIPCs. The annotations found in spinach and salad are in accordance with literature [12,13]. To the best of our knowledge, this is the first report on GIPCs in strawberries and raspberries.

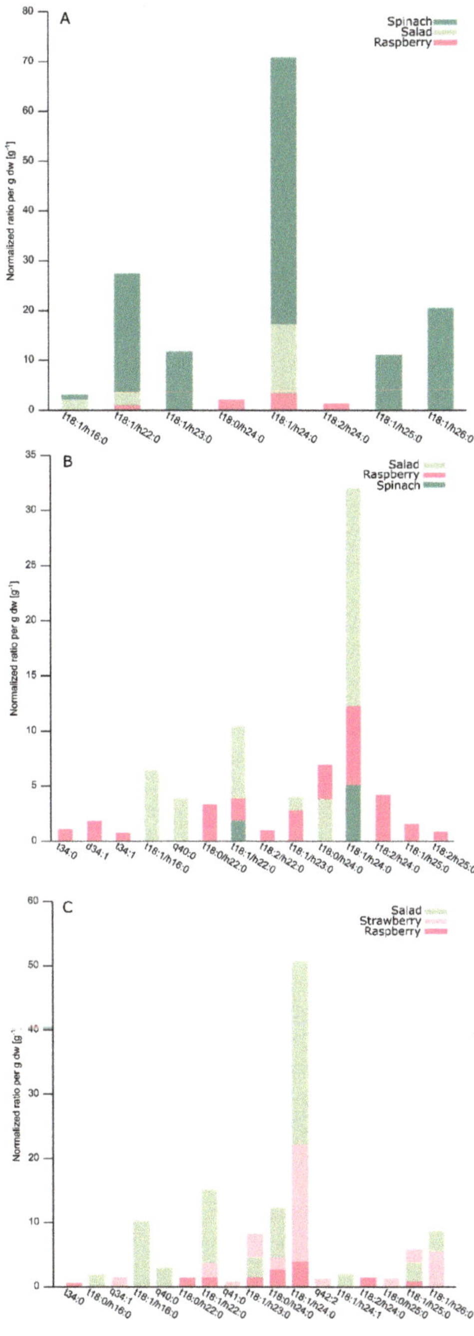

Figure 3. The normalized ratio per gram dry weight for annotated GIPCs in salad (light-green), spinach (green), strawberries (rose) and raspberries (dark-red), by using different substituents for the functional group R1: (**A**) NAc, (**B**) NH$_2$, and (**C**) OH (more detailed information can be found in Table A1).

Within the annotated GIPCs with structural information on LCB and fatty acyl composition, t18:1 followed by t18:0 and t18:2 were the most prominent LCBs in the analyzed plants (Table A1). For GIPCs containing N-acetylamine residues t18:1 was the most abundant LCB with regard to normalized ratios per gram dry weight (Figure 3A). The same holds true for the amine or hydroxyl group containing GIPCs, with additional high abundance of t18:0 LCBs (Figure 3B,C). While in spinach solely t18:1 LCBs were detected, salad, strawberries and raspberries show more variation in terms of LCB composition with presence of both t18:1 and t18:0 (Table A1).

Interestingly, besides the expected t18:0 and t18:1 LCBs (R1 = NAc, NH_2, OH), we additionally annotated four t18:2 (R1 = NAc, NH_2, OH) species in raspberries (Figure 3A–C). These annotations are verified by coinciding retention times in positive ion mode (Figure A3A), detection of characteristic fragments in MS2 spectra (Figure A3B,C) and conformity with the ECN model (Table S1). However, we could not find any report in the literature of t18:2 species, which can be explained as up to now no automated GIPC annotation was possible and t18:2 GIPC species were only detected in raspberries. As no standards are available, it is difficult to prove the presence of this species and further investigation is needed. A confirmed t18:2 LCB would indicate a much higher diversity in sphingolipids than anticipated in the past. Another hint for the complex nature of GIPCs in raspberries is the additional annotation of GIPCs with di- and trihydroxylated variants compared to all other analyzed plants.

Concerning LCB and fatty acyl combinations, t18:1/h22:0 and t18:1/h24:0 showed equal annotation numbers for N-acetylamine or amine containing GIPCs (Figure 3A,B). Independent of the NH_2, OH or NAc functional group, only two odd chain fatty acids (h23:0, h25:0) were detected and no fatty acids with a length from 17 to 21 carbon atoms were found. For the hydroxyl group containing GIPC variants, the combinations t18:1/h22:0, t18:1/h23:0, t18:0/h24:0, t18:1/h24:0 and t18:1/h25:0 were found in equal annotation numbers. Overall, plant GIPCs with a combination of t18:1 LCB and a h24:0 fatty acyl moiety were the most abundant ones in terms of normalized ratios per gram dry weight (Figure 3A–C, Table A1).

3. Discussion

GIPCs are the major sphingolipids on earth [1]. Hence, it is important to understand their function and distribution in plants and fungi. However, GIPC analysis remains extremely challenging, as tailored extraction strategies for this glycolipid class are necessary. GIPC analysis is in its infancy due to the lack of standards and databases. In this work, we present the first automated high-throughput GIPC annotation workflow which is based on RP-HRMS/MS. By using a novel 30 min gradient based on isopropanol with a reversed-phase column, packed with sub 2-μm particles, fast GIPC analysis was possible at the same time avoiding standard eluent use of tetrahydrofuran. Four different plant samples were analyzed. For salad and spinach, literature information has been available [12,13], while for raspberry and strawberry, GIPC profiles were completely uncharacterized. Using strict filtering by (1) accurate mass determination (±5 ppm) with matching retention times for both ion modes in MS1, (2) MS2 spectra with characteristic fragments and (3) expected retention time series, we produced a database of 64 GIPCs (Table A1). As no GIPC standards are available, only GIPC annotation hits with level 2 and 3 confidence [24] were possible. The most prominent MS2 fragments for GIPCs are [IP] fragments in both ion modes ([H]$^-$: *m/z* 241, 259; [H]$^+$: *m/z* 261; [Na]$^+$: *m/z* 283). However, additional sugar or ceramide fragments are essential for correct GIPC annotation. The high MS2 mass range coverage (*m/z* 65 to 2500) provided by the Orbitrap was beneficial to determine GIPC low mass fragments such as *m/z* 79 [PO_3]$^-$ or 97 [H_2PO_4]$^-$, besides high mass precursors such as 1261 [M − H]$^-$ (Figure 2).

Relative quantification with the internal standard lactosyl ceramide revealed GIPC t18:1/h24:0 as the most abundant species, independent of the plant sample. Depending on the plant sample, GIPCs contained mainly amine, N-acetyl or hydroxyl residues. Most GIPCs showed a Hex-HexA-IPC core with a trihydroxylated t18:0 or t18:1 long-chain base ceramide part a and hydroxylated fatty acid chains ranging from h16:0 to h26:0. Interestingly, in raspberry, four GIPCs contained t18:2, which was not reported so far. This finding would suggest the existence of more complex sphingolipid species in nature than previously anticipated. Further analysis by orthogonal methods such as NMR, GC-MS or IMS and available GIPC standards would be necessary to confirm the presence of the t18:2 GIPC group. Different analytical strategies could also resolve potential isomeric species and provide comprehensive details on the sugar moiety present in GIPCs. Nevertheless, this example shows the power of this workflow to detect promising novel GIPC candidates in an automated fashion. In order to support LC-MS-based GIPC analysis in general, we provide the mass lists for GIPCs in positive and negative ion modes (Tables S3 and S4), as well as the fragmentation rules (Folder S1) for setting up the automated GIPC analysis by Lipid Data Analyzer. Even though we confirmed the GIPCs exclusively from the A-series, the presented strategy is also suitable to determine less or more complex GIPC series, such as 0, B, C, D, E and F. However, extended analytical workflows (e.g., multi-stage fragmentation/MSn) and additional software method development might be necessary. Precursor mass lists for positive ([M + H]$^+$) and negative ([M − H]$^−$) ion modes comprising series 0–F, LCBs d18:0, d18:1, d18:2, t18:0 and t18:1 and fatty acyls h15:0–h26:0, h15:1–h26:1 and n20:0–n28:0 (n = non-hydroxylated), as reported in the literature [5,13], can be found in Tables S5 and S6. In general, we believe that LC-HRMS/MSn combined with automated annotation based on decision rules will pave the way for more complex glycolipidomics profiling.

4. Materials and Methods

4.1. Material

The plant material used was derived from salad (*Lactuca sativa var. capitata nidus tenerimma*), deep frozen spinach (*Spinacia oleracea*), raspberries (*Rubus idaeus*) and strawberries (*Fragaria*). (A more detailed description of plant samples can be found in Table A2.)

All chemicals were of LC-MS grade. Acetonitrile (ACN), methanol (MeOH), IPA and water were bought from Honeywell (Offenbach, Germany) and n-hexane was bought from VWR (Vienna, Austria). Butylated hydroxytoluene (BHT) was purchased from Sigma-Aldrich (Vienna, Austria), ammonium formate (AF) from Sigma-Aldrich (Vienna, Austria) and formic acid from VWR (Vienna, Austria). C16 Lactosyl(ß) Ceramide (d18:1/16:0) (D-lactosyl-ß-1,1' N-palmitoyl-D-erythro-sphingosine) was purchased from Avanti Polar Lipids, Inc. (Alabaster, Albama, USA), was used as internal standard (IS) and dissolved in an appropriate amount of IPA to achieve a concentration of 100 µM.

4.2. Sample Preparation

Salad was manually cut into small pieces before being weighed into falcon tubes (50 mL, VWR, Vienna, Austria) using a CPA225D balance (Sartorius, Vienna, Austria). Raspberries and strawberries (whole fruits) were homogenized with a hand blender (Tefal/SEB, Ecully, France). Raspberries, strawberries and deep-frozen homogenized spinach were directly weighed into 10 mL glass vials (more details can be found in Table A2). In order to prevent potential oxidation of lipids, 3 mL of an approximately 0.01% BHT solution in IPA were added and samples were mixed. Subsequently 30 µL IS were spiked into all samples except for one replicate (to test for potential IS presence in plants). Salad samples were homogenized using an ultra-turax (miccra d-1, Heitersheim, Germany) which was cleaned with 70% IPA and dried between the samples. In order to inhibit lipase activity, all samples were incubated at 75 °C for 30 min under constant shaking [27]. The warm salad samples were subsequently transferred into glass vials. The following sections provide a detailed overview of the extraction strategies that were applied.

4.2.1. One-Phase Extraction

The extraction of GIPCs from salad and spinach was performed as previously reported [18] using a mixture of IPA, n-hexane and water. Amounts of 3.47 mL IPA, 0.6 mL n-hexane and 1.93 mL water were added to the salad and spinach samples. In order to ensure sufficient accessibility of the plant material, samples were vortexed and manually shaken prior to incubation at 60 °C for 15 min under constant shaking.

4.2.2. One-Phase Extraction Combined with Alkaline Hydrolysis

To avoid the occurrence of glycerophospholipids, which might reduce GIPC ionization efficiency and lead to potential false identifications, alkaline hydrolysis was applied for the raspberry and strawberry samples, using an adapted workflow [28]. After incubating the plant material with the BHT solution for 30 min at 75 °C under constant shaking, 3.47 mL IPA and 0.6 mL n-hexane were added. Samples were vortexed and put on a shaker for 15 min at 60 °C. As soon as the samples had reached room temperature 707 μL 1 M KOH in MeOH was added and the solution was vortexed. After shaking the samples for 2 h at 37 °C, they were left at room temperature. Subsequently 100% formic acid was added until a pH of ~6–7 was reached and 1.93 mL water was added before repeating the incubation step.

4.2.3. Centrifugation, Drying and Reconstitution

Irrespective of the extraction strategy, the warm samples were centrifuged at 1000 rpm for 10 min at 4 °C and the supernatant was transferred into a separate glass vial. The solvent was evaporated to dryness overnight in a Genevac EZ-2 Series Personal Evaporator (SP Scientific, Ipswich, UK) and the dried residue was reconstituted in 2 mL IPA:H_2O (65:35) [13]. Samples were vortexed prior and after ultrasonication at 30 °C for 15 min. Subsequently, 500 μL of this solution was filtered directly into HPLC vials through a ClariStep filter (Sartorius, Vienna, Austria). Pools were prepared separately for each plant by pipetting 50 μL of each biological replicate into a separate HPLC vial. A quality control pool was prepared by combining 30 μL of the pooled samples.

4.3. Reversed-Phase Chromatography

Liquid chromatography was performed using a C18 Acquity UHPLC HSS T3 reversed phase column (2.1 × 150 mm, 100 Å, 1.8 μm, Waters, Vienna, Austria) equipped with a VanGuard Pre-column (2.1 × 5 mm, 100 Å, 1.8 μm, Waters, Vienna, Austria) at a column temperature of 40 °C. The flow rate was 0.25 mL/min and the backpressure was 460 bar at the starting conditions. Gradient elution with a total runtime of 30 min was performed using the solvent A: ACN:H_2O (3:2, *v/v*) and the solvent B: IPA:ACN (9:1, *v/v*), both of which contained 0.1% formic acid and 10 mM ammonium formate.

The gradient can be described as follows: 0–2 min 30% B, 2–3 min ramp to 55% B, 3–17 min ramp to 67% B, 17–22 min ramp to 100% B, 22–26 min 100% B, followed by an equilibration step from 26 to 30 min using 30% B. A Vanquish Duo UHPLC system (Thermo Fisher Scientific, Germering, Germany) was used and injections were performed with an autosampler. An injection volume of 10 μL was chosen and the injector needle was flushed with 75% IPA and 1% formic acid in between the injections.

4.4. High-Resolution Mass Spectrometry

The LC system was coupled to a Q Exactive HF (Thermo Fisher Scientific, Bremen, Germany) high resolution mass spectrometer, applying a HESI ion source with an S-lens RF level of 45. Measurements were carried out in positive and negative modes using different parameters. The following settings were applied in positive mode: spray voltage: 3.5 kV, capillary temperature 220 °C, sheath gas flow rate: 30, and auxiliary flow rate: 5. In negative mode parameters were adapted as follows: spray voltage: 2.8 kV, capillary temperature 250 °C, sheath gas flow rate: 35 (a.u.), and auxiliary flow rate: 10 (a.u.). The top 10 data-dependent MS2 spectra were obtained at a scan range of 500 to 3000 *m/z* with HCD using normalized collision energies of 35 (+35 in positive ion mode, −35 in negative mode), an MS1 resolution of 15,000 or 30,000 with an AGC target of 1e6 and MS2 resolution of 15,000 with an AGC target of 1e5. MS2 spectra were acquired based on an inclusion list ("do not pick others" option) containing the GIPC series 0–F (*m/z* values were calculated using enviPat Web 2.4 [29]). A more comprehensive picture of the GIPC composition of the analyzed plant material was obtained using several rounds of automatically generated exclusions lists for the sample pools [30].

4.5. Data Analysis

The GIPC assignment was performed using LDA (version 2.8.0) [17]; corresponding settings (Table A3), mass lists (Tables S3 and S4) and decision rule sets for series A (Folder S1) can be found in the Appendix A and Supplementary Materials. The correct GIPC annotation was ensured by a manual inspection of the results. MS1-based relative quantification of annotated GIPCs was performed with Skyline [25]. Total areas were divided by the corresponding calculated dry weights and areas of the IS, resulting in normalized ratios per g dry weight, of which the average was taken based on the number of replicates (3 for salad and spinach, 4 for strawberries and raspberries). More information can be found in Appendix B.

Supplementary Materials: The following are available online at http://www.mdpi.com/2218-1989/10/9/375/s1, Table S1. Application of the ECN model to ensure correct GIPC annotation, Table S2. List of annotated GIPCs including all detected MS2 fragments, Table S3. LDA mass list used for automated annotation of the series A GIPCs in positive mode, Table S4. LDA mass list used for automated annotation of the series A GIPCs in negative mode, Table S5. List of $[M + H]^+$ precursors comprising the GIPC series 0–F, Table S6. List of $[M - H]^-$ precursors comprising the GIPC series 0–F, Folder S1. Fragmentation rules for GIPC analysis by LDA.

Author Contributions: Conceptualization, E.R.; methodology, L.P. and E.R.; software, L.P., S.S., N.T. and J.H.; validation, L.P., E.R. and J.H.; formal analysis, L.P., S.S., N.T. and E.R.; investigation, L.P. and E.R.; resources, E.R. and G.K.; data curation, L.P. and E.R.; writing—original draft preparation, E.R. and L.P.; writing—review and editing, E.R., L.P., J.H. and G.K.; visualization, L.P.; supervision, E.R.; project administration, E.R. and G.K. All authors have read and agreed to the published version of the manuscript.

Funding: This research received no external funding.

Acknowledgments: This work was supported by the University of Vienna, the Faculty of Chemistry, the Vienna Metabolomics Center (VIME; http://metabolomics.univie.ac.at/), the research platform Chemistry Meets Microbiology and the Mass Spectrometry Centre of the University of Vienna. The authors thank (1) the Department of Food Chemistry and Toxicology (University of Vienna) for sharing their sample preparation equipment, (2) Sophia Mundigler for the preparation of a literature database for glycolipid fragmentation (3) Martin Schaier for graphical abstract support, as well as (4) all members of the Koellensperger lab (University of Vienna) for continuous support.

Conflicts of Interest: The authors declare no conflict of interest.

Appendix A

Table A1. Overview of GIPCs annotated in salad, spinach, strawberries and raspberries. The precursor ion *m/z* and retention times are listed as provided by the LDA display results function. In cases where only matching retention times but no *m/z* were explicitly shown by the LDA, because only one MS2 spectrum was annotated (level 3, level 3**), corresponding values (marked with an asterisk*) were manually assigned at the peak maximum using Thermo Scientific FreeStyle. In the Level column the levels of identification are listed. For all annotated GIPCs, accurate mass and retention times were observed. Their level of identification depends on the ddMS2 spectra. Level 2 annotation is based on the ddMS2 spectra in both ion modes. For level 3, ddMS2 spectra with characteristic fragments could only be detected in one ion mode. Putative hits, which cannot be annotated with such high confidence, because they were only observed with a sugar fragment in positive mode, but did not show the [IP]⁺/[IP+Na]⁺ fragment, are listed as level 3** at the end of the table and were not included in Figure 3.

Composition	Plant	m/z [M − H]⁻	m/z [M + H]⁺	Rt_neg [min]	Rt_pos [min]	Level	Normalized Ratio/g dw [g⁻¹]	CV [%]
A-NAc-t18:1/h16:0	Spinach	1189.6246	1191.6388	8.00	8.00	2	3.15	25
A-NAc-t18:1/h16:0	Salad	1189.6246	1191.6397	8.01	8.00	2	2.10	28
A-NAc-t18:1/h22:0	Spinach	1273.7179	1275.7359	14.25	14.25	2	27.43	19
A-NAc-t18:1/h22:0	Salad	1273.7193	1275.7344	14.19	14.26	2	3.57	44
A-NAc-t18:1/h22:0	Raspberry	1273.7172	1275.7341	14.21	14.19	2	0.94	6
A-NAc-t18:1/h23:0	Spinach	1287.7325	1289.7504	15.64	15.65	2	11.83	21
A-NAc-t18:0/h24:0	Raspberry	1303.7648	1305.7813	17.94	18.17	2	2.13	4
A-NAc-t18:1/h24:0	Spinach	1301.7502	1303.7657	16.98	16.98	2	70.89	21
A-NAc-t18:1/h24:0	Salad	1301.7498	1303.7637	16.98	16.94	2	17.23	32
A-NAc-t18:1/h24:0	Raspberry	1301.7491	1303.7649	16.96	16.94	2	3.40	5
A-NAc-t18:2/h24:0	Raspberry	1299.7322	1301.749	14.98	14.97	2	1.27	3
A-NAc-t18:1/h25:0	Spinach	1315.7652	1317.7811	18.41	18.41	2	11.08	20
A-NAc-t18:1/h26:0	Spinach	1329.7811	1331.797	19.75	19.75	2	20.54	16
A-NH2-t34:0	Raspberry	1133.6346	1135.6506	10.91	10.92	2	1.09	6
A-NH2-d34:1	Raspberry	1115.6248 *	1117.6415	10.89	10.90	3	1.85	7
A-NH2-t34:1	Raspberry	1131.6196 *	1133.6365	10.27	10.27	3	0.75	8
A-NH2-t18:1/h16:0	Salad	1147.6148	1149.6302	8.64	8.52	2	6.46	5
A-NH2-q40:0	Salad	1233.7225 *	1235.7372	16.82	16.76	3	3.87	27
A-NH2-t18:0/h22:0	Raspberry	1233.7242	1235.7389	16.64	16.55	2	3.33	7
A-NH2-t18:1/h22:0	Spinach	1231.7058 *	1233.7229	15.56	15.52	3	1.87	18
A-NH2-t18:1/h22:0	Salad	1231.7083	1233.7226	15.51	15.49	2	10.45	11
A-NH2-t18:1/h22:0	Raspberry	1231.7082	1233.7234	15.41	15.25	2	3.87	7
A-NH2-t18:2/h22:0	Raspberry	1229.6945 *	1231.7076	13.42	13.42	3	1.01	4
A-NH2-t18:1/h23:0	Salad	1245.7242 *	1247.7400	16.78	16.87	3	4.04	15
A-NH2-t18:1/h23:0	Raspberry	1245.7248	1247.7388	16.89	16.85	2	2.79	7
A-NH2-t18:0/h24:0	Raspberry	1261.7551	1263.7704	19.62	19.64	2	6.92	9
A-NH2-t18:0/h24:0	Salad	1261.7531*	1263.7701	19.82	19.53	3	3.80	27

Table A1. *Cont.*

Composition	Plant	*m/z* [M − H]⁻	*m/z* [M + H]⁺	Rt_neg [min]	Rt_pos [min]	Level	Normalized Ratio/g dw [g⁻¹]	CV [%]
A-NH2-t18:1/h24:0	Spinach	1259.7391	1261.7565	18.32	18.30	2	5.06	18
A-NH2-t18:1/h24:0	Salad	1259.7400	1261.7554	18.52	18.44	2	32.00	6
A-NH2-t18:1/h24:0	Raspberry	1259.7400	1261.7543	18.46	18.42	2	12.28	8
A-NH2-t18:2/h24:0	Raspberry	1257.7231	1259.7397	15.98	16.25	2	4.22	8
A-NH2-t18:1/h25:0	Raspberry	1273.7539 *	1275.7692	19.55	19.78	3	1.55	11
A-NH2-q43:2	Raspberry	1271.7419 *	1273.7571	17.71	17.73	3	0.87	9
A-OH-t34:0	Raspberry	1134.6192	1136.6369	10.21	10.21	2	0.58	8
A-OH-t18:0/h16:0	Salad	1150.6148	1152.6313	8.84	8.83	2	1.93	0
A-OH-t18:1/h16:0	Salad	1148.5984	1150.6138	8.14	8.13	2	10.27	16
A-OH-q34:1	Strawberry	1148.5982	1150.6144	8.11	8.13	2	1.51	6
A-OH-q40:0	Salad	1234.7095	1236.7239 *	15.66	15.60	3	3.00	2
A-OH-t18:0/h22:0	Raspberry	1234.7084	1236.7247	15.36	15.63	2	1.37	6
A-OH-t18:1/h22:0	Salad	1232.6921	1234.7071	14.47	14.44	2	15.16	19
A-OH-t18:1/h22:0	Strawberry	1232.6921	1234.7083	14.25	14.43	2	3.69	2
A-OH-t18:1/h22:0	Raspberry	1232.6924	1234.7055	14.43	14.43	2	1.44	6
A-OH-q41:0	Strawberry	1243.7232	1250.7395 *	17.01	16.98	3	0.83	2
A-OH-t18:1/h23:0	Raspberry	1245.7076	1248.78221	15.82	15.77	2	1.46	7
A-OH-t18:1/h23:0	Salad	1245.7079	1248.7209	15.85	15.84	2	4.45	18
A-OH-t18:1/h23:0	Strawberry	1245.7076	1248.7238	15.53	15.76	2	8.20	1
A-OH-t18:0/h24:0	Salad	1262.7397	1264.7564	18.55	18.50	2	12.30	4
A-OH-t18:0/h24:0	Raspberry	1262.7385	1264.754	18.51	18.53	2	2.58	6
A-OH-t18:0/h24:0	Strawberry	1262.7387	1264.7553	18.44	18.32	2	4.53	1
A-OH-t18:1/h24:0	Salad	1260.7237	1262.7389	17.30	17.26	2	50.77	14
A-OH-t18:1/h24:0	Strawberry	1260.7227	1262.7391	17.26	17.00	2	22.15	2
A-OH-t18:1/h24:0	Raspberry	1260.7240	1262.7388	17.25	17.23	2	3.89	6
A-OH-q42:2	Strawberry	1253.7073	1260.7238 *	15.21	15.22	3	1.22	2
A-OH-t18:1/h24:1	Salad	1253.7067	1260.7243	14.36	14.33	2	2.04	10
A-OH-t18:2/h24:0	Raspberry	1253.7082	1260.7234	15.23	15.23	2	1.47	5
A-OH-t18:0/h25:0	Strawberry	1275.7542	1278.7687	19.80	19.80	2	1.33	5
A-OH-t18:1/h25:0	Salad	1274.7387	1276.7533	18.76	18.75	2	3.72	11
A-OH-t18:1/h25:0	Strawberry	1274.7371	1276.7557	18.42	18.48	2	5.83	2
A-OH-t18:1/h25:0	Raspberry	1274.7381	1276.7548	18.71	18.73	2	0.84	9
A-OH-t18:1/h26:0	Salad	1288.7553	1290.7706	20.01	20.01	2	8.65	18
A-OH-t18:1/h26:0	Strawberry	1288.7540 *	1290.7697	19.76	19.94	3	5.49	2
A-NH2-d18:2/16:0	Raspberry	1113.6100 *	1115.6255	9.89	9.89	3 **	0.57	10
A-NH2-q41:0	Raspberry	1247.7383 *	1249.7544	18.16	18.17	3 **	2.03	7
A-NH2-q41:2	Raspberry	1243.7076 *	1245.7232	14.77	14.80	3 **	0.87	7

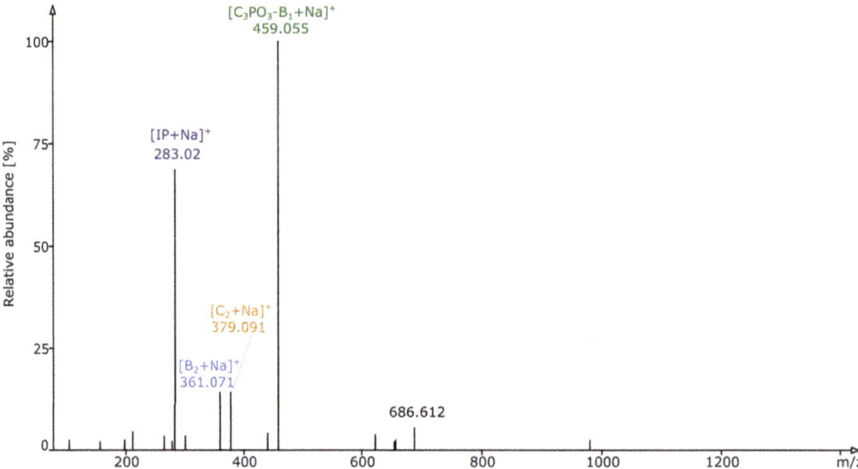

Figure A1. The ddMS2 spectrum of the [M + Na]$^+$ of GIPC A-OH-t18:1/h24:0 (*m/z* 1284.7193, Rt 17.30 min), measured in positive ion mode, showing the characteristic [IP + Na]$^+$ and additional sugar fragments.

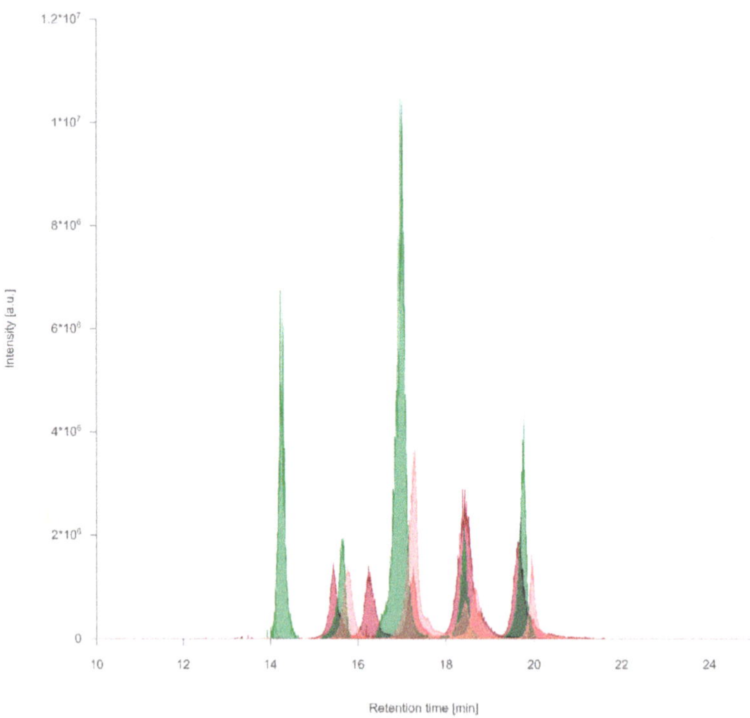

Figure A2. Comparison of the RP-HRMS/MS GIPC profiles in spinach (green), strawberry (rose) and raspberry (dark-red), showing the five most abundant GIPCs found in each plant sample measured in positive ion mode (detailed information can be found in Table A1).

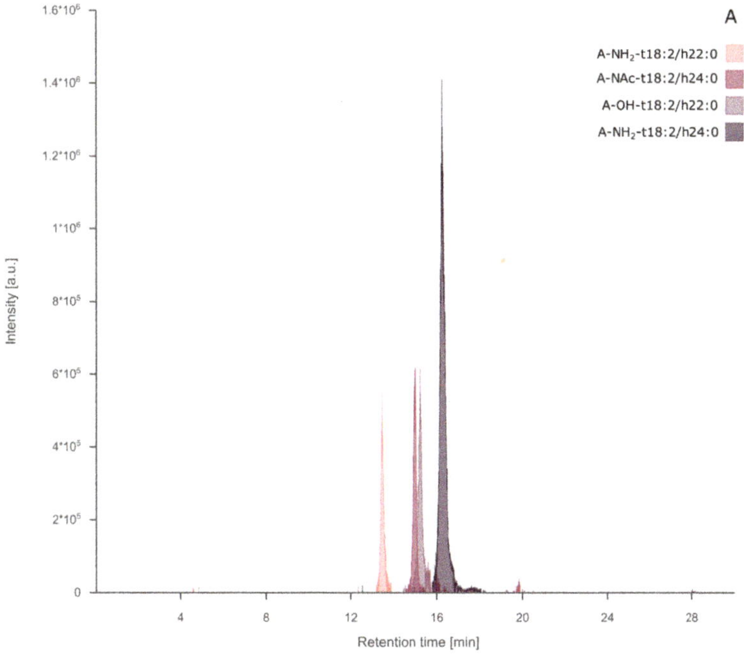

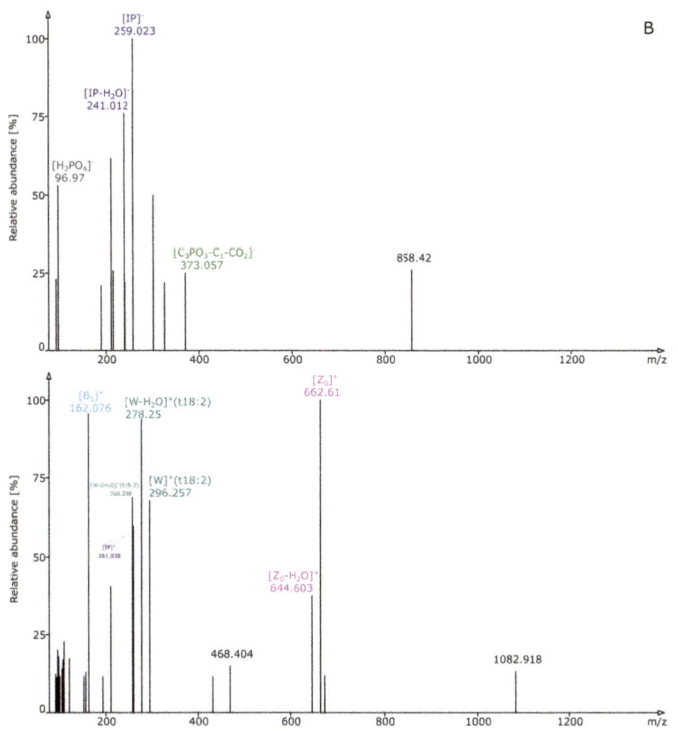

Figure A3. *Cont.*

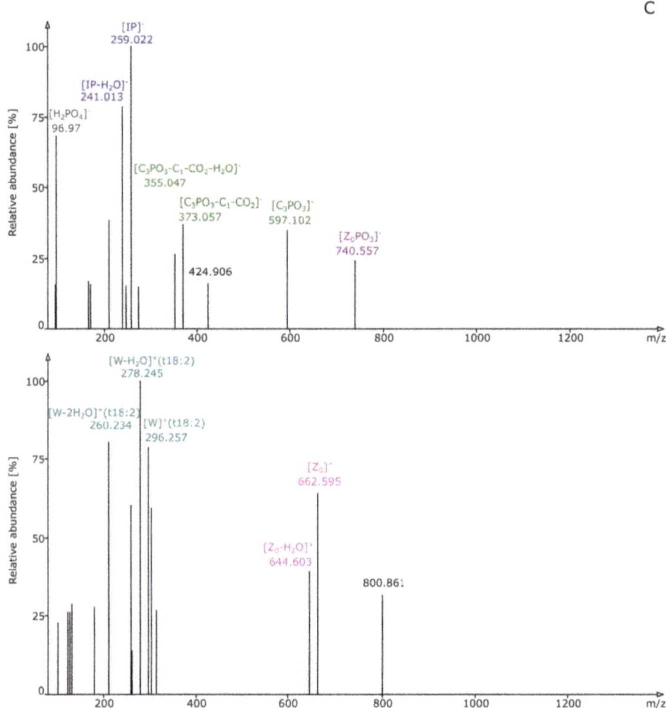

Figure A3. (**A**) Extracted ion chromatograms of GIPCs in positive ion mode, having a t18:2 LCB annotation on levels 2 and 3 in raspberries (A-NAc-t18:2/h24:0, A-NH₂-t18:2 h22:0 and h24:0, as well as A-OH-t18:2/h24:0). The ddMS2 spectra of GIPCs A-NH₂-t18:2/h24:0 at the retenion time of 16.1 min (**B**) A-OH-t18:2/h24:0 at the retention time of 15.2 min and (**C**) in negative and positive mode, showing characteristic fragments.

Table A2. The description of the plant samples, including plant species, origin, number of biological replicates, and the average fresh- and dry weights [g]. The extraction of GIPCs from strawberries and raspberries was performed one day after the collection was performed (28 June 2020).

Plant Species	Origin	Replicates	Fresh Weight [g]	Dry Weight [g]
Salad	Local supermarket	4	~1	~0.04
Spinach	Local supermarket	4	~1.2	~0.08
Strawberries	47°58′ N, 16°6′ O	5	~1.1	~0.10
Raspberries	47°58′ N, 16°6′ O	5	~1.1	~0.16

Table A3. The exemplary LDA parameters and settings used for automated GIPC annotation in negative ion mode.

Parameter	Setting
Time before tol.	1 min
Time after tol.	1 min
Rel. Base-peak cutoff	0.1‰
Rt-shift	0.0 min
Isotopic quantitation of _ isotopes where _ isotopic peak(s) have to match	2, 1
Find molecules where retention time is unknown	yes
LDA-version	2.8.0
machineName	OrbiTrap_exactive
neutronMass	1.005
coarseChromMzTolerance	0.015
MS2	true
basePeakCutoff	0.1
massShift	0.0
threeDViewerDefaultTimeResolution	2
threeDViewerDefaultMZResolution	0.005
ms2PrecursorTolerance	0.013
ms2MzTolerance	0.02
ms2MinIntsForNoiseRemoval	100
ms2IsobarSCExclusionRatio	0.01
ms2IsobarSCFarExclusionRatio	0.1
ms2IsobaricOtherRtDifference	2.0
chainCutoffValue	0.01
ms2ChromMultiplicationFactorForInt	10
threeDViewerMs2DefaultTimeResolution	1
threeDViewerMs2DefaultMZResolution	1
maxFileSizeForChromTranslationAtOnce	500
chromMultiplicationFactorForInt	1000
chromLowestResolution	1
chromSmoothRange	8.0
chromSmoothRepeats	4
use3D	true
isotopeCorrection	false
removeFromOtherIsotopes	true
respectIsotopicDistribution	true
checkChainLabelCombinationFromSpeciesName	false
useNoiseCutoff	true
noiseCutoffDeviationValue	2.0
scanStep	2
profileMzRangeExtraction	0.05
profileTimeTolerance	5.0
profileIntThreshold	5.0
broaderProfileTimeTolerance	3.0
profileSmoothRange	0.0025
profileSmoothRepeats	1
profileMeanSmoothRepeats	2
profileMzMinRange	0.002
profileSteepnessChange1	1.5
profileSteepnessChange2	1.8
profileIntensityCutoff1	0.15
profileIntensityCutoff2	0.2
profileGeneralIntCutoff	0.03
profilePeakAcceptanceRange	0.012
profileSmoothingCorrection	0.0
profileMaxRange	0.03
smallChromMzRange	0.004

Table A3. *Cont.*

Parameter	Setting
smallChromSmoothRepeats	3
smallChromMeanSmoothRepeats	0
smallChromSmoothRange	2.0
smallChromIntensityCutoff	0.03
broadChromSmoothRepeats	5
broadChromMeanSmoothRepeats	0
broadChromSmoothRange	2
broadChromIntensityCutoff	0.0
broadChromSteepnessChangeNoSmall	1.33
broadChromIntensityCutoffNoSmall	0.05
finalProbeTimeCompTolerance	0.1
finalProbeMzCompTolerance	5.0E-4
overlapDistanceDeviationFactor	1.5
overlapPossibleIntensityThreshold	0.15
overlapSureIntensityThreshold	0.7
overlapPeakDistanceDivisor	3.0
overlapFullDistanceDivisor	6.0
peakDiscardingAreaFactor	1000
isotopeInBetweenTime	30
isoInBetweenAreaFactor	3.0
isoNearNormalProbeTime	30
relativeAreaCutoff	0.05
relativeFarAreaCutoff	0.05
relativeFarAreaTimeSpace	30
relativeIsoInBetweenCutoff	0.5
isoInBetweenMaxTimeDistance	300
twinPeakMzTolerance	0.01
closePeakTimeTolerance	10
twinInBetweenCutoff	0.95
unionInBetweenCutoff	0.8
sparseData	false

Appendix B

Automated GIPC annotation was performed using LDA (version 2.8.0) [17] with the settings provided in Table A3. The mass-to-charge ratios included in the mass lists (see Tables S3 and S4) were calculated separately for negative and positive ion modes, with enviPat Web 2.4 [29] and decision rules (see Folder S1) were created based on fragments reported in the literature [12,13]. Please note that the raw data acquired in negative ion mode has to be analyzed using the mass list of Table S4 and the fragmentation rules ending with '-H.frag', while for positive mode the mass list of Table S3 and corresponding fragmentation rules ('H.frag' and 'Na.frag') should be used. Further information on working with the LDA can be found in [31].

References

1. Gronnier, J.; Germain, V.; Gouguet, P.; Cacas, J.-L.; Mongrand, S. GIPC: Glycosyl Inositol Phospho Ceramides, the major sphingolipids on earth. *Plant Signal. Behav.* **2016**, *11*, e1152438. [CrossRef] [PubMed]
2. Carter, H.E.; Gigg, R.H.; Laws, J.H. Structure of Phytoglyolipide. *J. Biol. Chem.* **1958**, *233*, 1309–1314. [PubMed]
3. Cacas, J.; Buré, C.; Grosjean, K.; Gerbeau-Pissot, P.; Lherminier, J.; Rombouts, Y.; Maes, E.; Bossard, C.; Gronnier, J.; Furt, F.; et al. Revisiting Plant Plasma Membrane Lipids in Tobacco: A Focus on Sphingolipids. *Plant Physiol.* **2016**, *170*, 367–384. [CrossRef] [PubMed]
4. Carter, H.E.; Strobach, D.R.; Hawthorne, J.N. Biochemistry of the Sphingolipids. XVIII. Complete Structure of Tetrasaccharide Phytoglycolipid. *Biochemistry* **1969**, *8*, 383–388. [CrossRef] [PubMed]

5. Buré, C.; Cacas, J.; Mongrand, S.; Schmitter, J.-M. Characterization of glycosyl inositol phosphoryl ceramides from plants and fungi by mass spectrometry. *Anal. Bioanal. Chem.* **2014**, *406*, 995–1010. [CrossRef]

6. LIPID MAPS® Lipidomics Gateway. Available online: https://lipidmaps.org/resources/lipidweb/index.php?page=lipids/sphingo/glyP_ino/index.htm (accessed on 4 August 2020).

7. Buré, C.; Cacas, J.-L.; Wang, F.; Gaudin, K.; Domergue, F.; Mongrand, S.; Schmitter, J.-M. Fast screening of highly glycosylated plant sphingolipids by tandem mass spectrometry. *Rapid Commun. Mass Spectrom.* **2011**, *25*, 3131–3145. [CrossRef]

8. Hastings, J.; Owen, G.; Dekker, A.; Ennis, M.; Kale, N.; Muthukrishnan, V.; Turner, S.; Swainston, N.; Mendes, P.; Steinbeck, C. ChEBI in 2016: Improved services and an expanding collection of metabolites. *Nucleic Acids Res.* **2016**, *44*, D1214–D1219. [CrossRef]

9. Sud, M.; Fahy, E.; Cotter, D.; Brown, A.; Dennis, E.A.; Glass, C.K.; Merrill, A.H.; Murphy, R.C.; Raetz, C.R.H.; Russell, D.W.; et al. LMSD: LIPID MAPS structure database. *Nucleic Acids Res.* **2007**, *35*, 527–532. [CrossRef]

10. Barrientos, R.C.; Zhang, Q. Recent advances in the mass spectrometric analysis of glycosphingolipidome—A review. *Anal. Chim. Acta* **2020**. [CrossRef]

11. Markham, J.E.; Jaworski, J.G. Rapid measurement of sphingolipids from Arabidopsis thaliana by reversed-phase high-performance liquid chromatography coupled to electrospray ionization tandem mass spectrometry. *Rapid Commun. Mass Spectrom.* **2007**, *21*, 1304–1314. [CrossRef]

12. Blaas, N.; Humpf, H.U. Structural profiling and quantitation of glycosyl inositol phosphoceramides in plants with fourier transform mass spectrometry. *J. Agric. Food Chem.* **2013**, *61*, 4257–4269. [CrossRef] [PubMed]

13. Cacas, J.; Buré, C.; Furt, F.; Maalouf, J.; Badoc, A.; Cluzet, S.; Schmitter, J.; Antajan, E.; Mongrand, S. Biochemical survey of the polar head of plant glycosylinositolphosphoceramides unravels broad diversity. *Phytochemistry* **2013**, *96*, 191–200. [CrossRef] [PubMed]

14. Domon, B.; Costello, C.E. Structure Elucidation of Glycosphingolipids and Gangliosides Using High-Performance Tandem Mass Spectrometry. *Biochemistry* **1988**, *27*, 1534–1543. [CrossRef] [PubMed]

15. Domon, B.; Costello, C.E. A systematic nomenclature for carbohydrate fragmentations in FAB-MS/MS spectra of glycoconjugates. *Glycoconj. J.* **1988**, *5*, 397–409. [CrossRef]

16. Hsu, F.; Turk, J. Glycosphingolipids as Lithiated Adducts by Dissociation on a Triple Stage Quadrupole Instrument. *J. Am. Soc. Mass Spectrom.* **2001**, *12*, 61–79. [CrossRef]

17. Hartler, J.; Triebl, A.; Ziegl, A.; Trötzmüller, M.; Rechberger, G.N.; Zeleznik, O.A.; Zierler, K.A.; Torta, F.; Cazenave-Gassiot, A.; Wenk, M.R.; et al. Deciphering lipid structures based on platform-independent decision rules. *Nat. Methods* **2017**, *14*, 1171–1174. [CrossRef]

18. Markham, J.E.; Li, J.; Cahoon, E.B.; Jaworski, J.G. Separation and Identification of Major Plant Sphingolipid Classes from Leaves. *J. Biol. Chem.* **2006**, *281*, 22684–22694. [CrossRef]

19. Dugo, P.; Cacciola, F.; Kumm, T.; Dugo, G.; Mondello, L. Comprehensive multidimensional liquid chromatography: Theory and applications. *J. Chromatogr. A* **2008**, *1184*, 353–368. [CrossRef]

20. Lísa, M.; Holčapek, M. Triacylglycerols profiling in plant oils important in food industry, dietetics and cosmetics using high-performance liquid chromatography–atmospheric pressure chemical ionization mass spectrometry. *J. Chromatogr. A* **2008**, *1198–1199*, 115–130. [CrossRef]

21. Ovčačíková, M.; Lísa, M.; Cífková, E.; Holčapek, M. Retention behavior of lipids in reversed-phase ultrahigh-performance liquid chromatography–electrospray ionization mass spectrometry. *J. Chromatogr. A* **2016**, *1450*, 76–85. [CrossRef]

22. Hartler, J.; Trötzmüller, M.; Chitraju, C.; Spener, F.; Köfeler, H.C.; Thallinger, G.G. Lipid Data Analyzer: Unattended identification and quantitation of lipids in LC-MS data. *Bioinformatics* **2011**, *27*, 572–577. [CrossRef] [PubMed]

23. LIPID MAPS®Lipidomics Gateway. Available online: https://www.lipidmaps.org/data/LMSDRecord.php?LMID=LMSP03030005 (accessed on 15 March 2020).

24. Salek, R.M.; Steinbeck, C.; Viant, M.R.; Goodacre, R.; Dunn, W.B. The role of reporting standards for metabolite annotation and identification in metabolomic studies. *Gigascience* **2013**, *2*, 2047-217X-2-13. [CrossRef] [PubMed]

25. Adams, K.J.; Pratt, B.; Bose, N.; Dubois, L.G.; St. John-Williams, L.; Perrott, K.M.; Ky, K.; Kapahi, P.; Sharma, V.; Maccoss, M.J.; et al. Skyline for Small Molecules: A Unifying Software Package for Quantitative Metabolomics. *J. Proteome Res.* **2020**, *19*, 1447–1458. [CrossRef] [PubMed]

26. Metabolomics Society. Available online: http://metabolomicssociety.org/ (accessed on 14 July 2020).

27. Shiva, S.; Enninful, R.; Roth, M.R.; Tamura, P.; Jagadish, K.; Welti, R. An efficient modified method for plant leaf lipid extraction results in improved recovery of phosphatidic acid. *Plant Methods* **2018**, *14*, 1–8. [CrossRef]

28. Peng, B.; Weintraub, S.T.; Coman, C.; Ponnaiyan, S.; Sharma, R.; Tews, B.; Winter, D.; Ahrends, R. A Comprehensive High-Resolution Targeted Workflow for the Deep Profiling of Sphingolipids. *Anal. Chem.* **2017**, *89*, 12480–12487. [CrossRef] [PubMed]

29. Loos, M.; Gerber, C.; Corona, F.; Hollender, J.; Singer, H. Accelerated Isotope Fine Structure Calculation Using Pruned Transition Trees. *Anal. Chem.* **2015**, *87*, 5738–5744. [CrossRef] [PubMed]

30. Koelmel, J.P.; Kroeger, N.M.; Gill, E.L.; Ulmer, C.Z.; Bowden, J.A.; Patterson, R.E.; Yost, R.A.; Garrett, T.J. Expanding Lipidome Coverage Using LC-MS/MS Data-Dependent Acquisition with Automated Exclusion List Generation. *J. Am. Soc. Mass Spectrom.* **2017**, *28*, 908–917. [CrossRef]

31. LDA User Manual. Available online: http://genome.tugraz.at/lda2/2.6/LDA_2.6.pdf (accessed on 17 September 2020).

 *metabolites*

Article

Plasma Lipid Profiling of Three Types of Drug-Induced Liver Injury in Japanese Patients: A Preliminary Study

Kosuke Saito [1], Tatehiro Kagawa [2], Keiji Tsuji [3], Yuji Kumagai [4], Ken Sato [5], Shotaro Sakisaka [6], Naoya Sakamoto [7], Mitsuhiko Aiso [8], Shunji Hirose [2], Nami Mori [3], Rieko Tanaka [4], Toshio Uraoka [5], Kazuhide Takata [6], Koji Ogawa [7], Kazuhiko Mori [9], Motonobu Sato [10], Takayoshi Nishiya [11], Kazuhiko Takamatsu [10], Noriaki Arakawa [1], Takashi Izumi [12], Yasuo Ohno [12], Yoshiro Saito [1,*] and Hajime Takikawa [8,13]

[1] Division of Medical Safety Science, National Institute of Health Sciences, Kanagawa 210-9501, Japan; saitok2@nihs.go.jp (K.S.); arakawa@nihs.go.jp (N.A.)
[2] Division of Gastroenterology and Hepatology, Department of Internal Medicine, Tokai University School of Medicine, Isehara 259-1193, Japan; kagawa@tokai.ac.jp (T.K.); s-hirose@is.icc.u-tokai.ac.jp (S.H.)
[3] Department of Gastroenterology, Hiroshima Red Cross Hospital and Atomic-Bomb Survivors Hospital, Hiroshima 730-8619, Japan; k-tsuji@hiroshima-med.jrc.or.jp (K.T.); morin@hiroshima-med.jrc.or.jp (N.M.)
[4] Kitasato University School of Medicine, 1-15-1 Kitasato, Minami-ku, Sagamihara, Kanagawa 252-0374, Japan; kuma-guy@za2.so-net.ne.jp (Y.K.); rieko_t@med.kitasato-u.ac.jp (R.T.)
[5] Department of Gastroenterology and Hepatology, Gunma University Graduate School of Medicine, Maebashi 371-8511, Japan; satoken@gunma-u.ac.jp (K.S.); uraoka@gunma-u.ac.jp (T.U.)
[6] Department of Gastroenterology, Fukuoka University Faculty of Medicine, Fukuoka 814-0180, Japan; sakisaka@fukuoka-u.ac.jp (S.S.); edihuzak_t@yahoo.co.jp (K.T.)
[7] Department of Gastroenterology and Hepatology, Hokkaido University Faculty of Medicine and Graduate School of Medicine, Sapporo 060-8648, Japan; sakamoto@med.hokudai.ac.jp (N.S.); k-ogawa@med.hokudai.ac.jp (K.O.)
[8] Department of Medicine, Teikyo University School of Medicine, Tokyo 173-8606, Japan; aiso@med.teikyo-u.ac.jp (M.A.); takikawa@med.teikyo-u.ac.jp (H.T.)
[9] Daiichi Sankyo Co., Ltd., Tokyo 134-8630, Japan; mori.kazuhiko.md@daiichisankyo.co.jp
[10] Astellas Pharma Inc., Tsukuba 305-8585, Japan; motonobu.sato@astellas.com (M.S.); kazuhiko.takamatsu@astellas.com (K.T.)
[11] Daiichi Sankyo RD Novare Co., Ltd., Tokyo 134-8630, Japan; nishiya.takayoshi.ec@rdn.daiichisankyo.co.jp
[12] Kihara Memorial Foundation, Yokohama 230-0045, Japan; t-izumi@kihara.or.jp (T.I.); yasuokunhe@ra3.so-net.ne.jp (Y.O.)
[13] Faculty of Medical Technology, Teikyo University, Tokyo 173-8606, Japan
* Correspondence: yoshiro@nihs.go.jp; Tel.: +81-44-270-6623; Fax: +81-44-270-6624

Received: 16 July 2020; Accepted: 28 August 2020; Published: 31 August 2020

Abstract: Drug-induced liver injury (DILI) is a major adverse event caused by drug treatment, which can be categorized into three types: hepatocellular, mixed, and cholestatic. Although nearly every class of drugs can cause DILI, an overall understanding of lipid profiles in DILI patients is lacking. We used lipidomics to analyze the plasma lipid profiles of patients to understand their hepatic pathophysiology and identify DILI biomarkers. We identified 463 lipids and compared their levels between the acute and recovery phases of the three types of DILI patients. Mixed and cholestatic types demonstrated specific plasma lipid alterations between the phases, but the hepatocellular type did not. Moreover, as specific indicators of mixed-type DILI, levels of several ceramides increased in the acute phase, while those of arachidonic acid-containing ether-linked phosphoglycerolipids decreased. In contrast, as specific indicators of cholestatic-type DILI, levels of palmitic acid-containing saturated or monounsaturated phosphatidylcholines increased in the acute phase, while those of arachidonic acid- or docosahexaenoic acid-containing ether-linked phosphoglycerolipids and phosphatidylinositols decreased. We also identified lipids with a relatively high capacity to discriminate the acute phase from the recovery phase and healthy subjects. These findings may help with understanding the pathophysiology of different DILI types and identify candidate biomarkers.

Keywords: lipidomics; drug-induced liver injury; biomarker; plasma lipid profiles

---

1. Introduction

Drug-induced liver injury (DILI) is a major adverse event caused by drug treatment and is the most frequent cause of acute liver failure in the U.S. [1,2]. Depending on the histological location of the tissue damage, DILI is categorized as hepatocellular, cholestatic, or mixed type, which is usually based on changes in blood levels of alanine transaminase (ALT) and alkaline phosphatase (ALP). The causal relationship between DILI and suspected drugs has been digitized by the CIOMS/RUCAM and DDW-J2004 scoring scales (in Japan), which are used in clinical practice [3–5]. The mechanisms of DILI are diverse and include direct toxicity by the administered drug or its metabolites and immune reactions against the drug or its metabolites [6,7]. The most studied drug causing DILI is acetaminophen, which is metabolized to a toxic and electrophilic intermediate by cytochrome P450 isoenzymes (such as CYP2E1 and CYP3A4); this intermediate interacts with intracellular proteins resulting in hepatocyte damage [8]. Although specific mechanisms of drugs with relatively high incidence of DILI have also been studied [6,7], nearly every class of drug can cause DILI. However, biomarkers and characteristics of DILI that are important to understand its pathophysiology are limited.

Lipids, such as phosphoglycerolipids, sphingolipids, and neutral lipids, are components of cellular membranes that also play important roles in multiple biological processes, including apoptosis, inflammation, proliferation, and differentiation [9–12]. The liver is a central organ in regulating lipid levels, and therefore, aberrations in lipid homeostasis are associated with hepatic injury and disease. In addition, a recent study demonstrated that the composition of plasma lipids correlates well with that of hepatic lipids [13]. Thus, plasma lipid profiles could be useful tools to understand the biological processes in the liver. To analyze plasma lipid profiles, lipidomics based on mass spectrometry has been established [14–17]. Plasma lipidomics has already been used to study hepatocellular carcinoma [18,19], liver phospholipidosis [20], nonalcoholic fatty liver disease [21], and other hepatic diseases and toxicities. For example, plasma lipidomics of hepatocellular carcinoma demonstrated decreased levels of lysophosphatidylcholine (LPC) in plasma, suggesting the hepatic activation of autotoxin and its involvement in hepatocarcinogenesis [22]. Moreover, plasma lipidomics of liver phospholipidosis demonstrated increased levels of d18:1/24:0 glucosylceramide (GluCer), which was proposed as a biomarker for the disease [20]. Therefore, the characterization of overall plasma lipid profiles could lead to a better understanding of hepatic pathophysiology and identify new DILI biomarkers.

In this study, we aimed to analyze the differences in lipid profiles among three DILI types (hepatocellular, mixed, and cholestatic) during acute and recovery phases in human patients. We present novel lipidomic data for the three injury types, which could be used to screen for DILI biomarkers and/or develop future novel therapies by understanding lipid homeostasis in DILI.

2. Results

2.1. DILI Patients Recruited in the Present Study

We recruited 54 DILI patients, comprising 33 hepatocellular, 13 mixed, and 8 cholestatic types (Table 1). Of these patients, 11 males and 22 females were diagnosed with hepatocellular type, 9 males and 4 females were diagnosed with mixed type, and 4 males and 4 females were diagnosed with cholestatic type. Their median ages were 56, 60, and 69 for the hepatocellular, mixed, and cholestatic types, respectively. The median CIOMS/RUCAM scores were eight, nine, and eight for the hepatocellular, mixed, and cholestatic types, respectively. In addition, the median DDW-J2004 scores were eight for each of the respective DILI patient types. The causal relationship between suspected drug and liver damage was definite in all patients using the CIOMS/RUCAM scale and all patients using the DDW-J 2004 score, except for one case.

Table 1. Clinical information of patients in this study.

DILI Type	Hepatocellular	Mixed	Cholestatic
no. of subjects	33	13	8
CIOMS/RUCAM scale; median (quartile)	8 (7–9)	9 (7–9)	8 (7.5–9)
DDW-J 2004 score; median (quartile)	8 (7–9)	8 (7–8)	8 (7–8.5)
Sex; male/female	11/22	9/4	4/4
Age; median (quartile)	56 (46–68)	60 (57–76)	69 (64.5–72.5)
BMI; median (quartile)	22.7 (19.8–24.1)	22.1 (21.3–23.3)	25.3 (22.8–26.1)
acute phase AST(U/L); median (quartile)	239 (102–526)	130 (96–191)	81 (55.75–200.5)
acute phase ALT(U/L); median (quartile)	336 (204–963)	196 (161–423)	97 (84.75–131)
acute phase ALP(U/L); median (quartile)	360 (277–410)	555 (448–914)	1465 (1129.5–1721.5)
acute phase T. Bil(mg/dl); median (quartile)	1 (0.7–2.2)	0.6 (0.5–1.3)	1.5 (0.8–3.65)
recovered phase AST(U/L); median (quartile)	23 (19–29)	31 (20–34)	28.5 (19.5–30.25)
recovered phase ALT(U/L); median (quartile)	25 (17–35)	32 (23–48)	22 (15–30.25)
recovered phase ALP(U/L); median (quartile)	249 (185.75–324.75)	316 (244–373)	266.5 (190.25–332)
recovered phase T. Bil(mg/dl); median (quartile)	0.7 (0.525–0.975)	0.65 (0.5–0.925)	0.7 (0.6–0.9)
Cause			
-Prescribed drugs	26	10	6
-Other	1	1	0
-Undefined	6	2	2
Suspected drugs; ad. in over 2 DILI patients			
-Acetaminophen	2	1	0
-Cefditoren	2	0	0
-Cyclophosphamide	2	0	0
-Febuxostat	0	1	1
-Gemcitabine	2	0	0
-Loxoprofen	2	1	1
-Nifedipine	1	0	1
ATC level 2 of suspected drugs; ad. in over 3 DILI patients (ATC code in parenthesis)			
-calcium channel blockers (C08)	2	0	1
-antibacterials for systemic use (J01)	6	2	0
-antineoplastic agents (L01)	5	1	0
-anti-inflammatory and antirheumatic products (M01)	3	2	1
-psycholeptics (N05)	4	1	1

AST; aspartate transaminase, ALT; alanine transaminase, ALP; alkaline phosphatase, T. Bil; total bilirubin, ad.; administrated. The reference ranges of the liver blood test were <30 for AST, <30 for ALT, 100–325 for ALP, and 0.2–1.2 for T. Bil. The threshold numbers of patients in "Suspected drugs" and "ATC level 2 of suspected drugs" were judged by the sum of all types of drug-induced liver injury (DILI).

The suspected drug with the highest frequency of culpability was loxoprofen, which was responsible for four cases out of the 54 patients (two, one, and one case in the hepatocellular-, mixed-, and cholestatic-type patients, respectively). In addition, when the prescribed drugs were categorized according to the World Health Organization (WHO) Anatomical Therapeutic Chemical (ATC) codes, the highest number of cases was found in antibacterial agents for systemic use (J01, eight cases), followed by antineoplastic agents (L01, six cases), anti-inflammatory and antirheumatic products (M01, six cases), and psycholeptics (N05, six cases). The causes of DILI were widely diverse among cases, which hindered the analysis of drug-specific or drug category-specific effects.

2.2. Global Plasma Lipid Profiling in the Three DILI Types

Global plasma lipidomic profiling using our lipidomics platform detected 463 lipids spanning 31 lipid classes (Table S1 and summarized in Table 2). Note that in our assay platform, unconjugated bile acids were detectable but not quantitative because their liquid chromatography (LC) retention time is close to the void fraction where ionization is unstable due to the presence of unretained salts. The exemplar LC/MS traces are shown in Figure S1. To distinguish stereoisomers, each quantified lipid was assigned a specific metabolite ID. The fatty acid side chains in the lipids were confirmed using mass spectrometry (MS), the confirmed fatty acid fragments were indicated after a semicolon in the name. The combination of fatty acid side chains was combined using a slash. If two different sets of fragments were confirmed, we provided both of them, separated by a comma. The identified lipids comprised 184 phospholipids, 80 sphingolipids, 180 neutral lipids, and 19 others, including coenzyme Q10 (CoQ10), free fatty acids (FAs), and acylcarnitines (Cars). The major phospholipid class, phosphatidylcholines (PCs), contained 56 lipids. In addition, the major sphingolipid class, sphingomyelins (SMs), comprised 37 lipids, and the major neutral lipid class, triacylglycerols (TGs), comprised 138 lipids. The identified lipid levels were compared between the acute and recovery phases in each DILI type. Lipids with both high effect sizes (g > 0.8) and statistically significant differences ($p < 0.05$) were defined as altered.

Table 2. Identified lipid classes and numbers of individual lipids.

Category	Class (Abbreviation)	Class	Number of Lipids
Phosphoglycerolipid	LPC	Lysophosphatidylcholine	12
Phosphoglycerolipid	LPCe	Ether-type lysophosphatidylcholine	2
Phosphoglycerolipid	LPE	Lysophosphatidylethanolamine	5
Phosphoglycerolipid	LPEe	Ether-type lysophosphatidylethanolamine	1
Phosphoglycerolipid	LPI	Lysophosphaidylinositol	2
Phosphoglycerolipid	PC	Phosphatidylcholine	56
Phosphoglycerolipid	PC+O	Oxidized phosphatidylcholine	2
Phosphoglycerolipid	ether-linked PC	Ether-type phosphatidylcholine	40
Phosphoglycerolipid	PE	Phosphatidylethanolamine	15
Phosphoglycerolipid	ether-linked PE	Ether-type phosphatidylethanolamine	29
Phosphoglycerolipid	PI	Phosphatidylinositol	18
Phosphoglycerolipid	PS	Phosphatidylserine	2
Sphingolipid	Cer	Ceramide	14
Sphingolipid	CerG1	Monoglycosylceramide	6
Sphingolipid	CerG1+O	Oxidized monoglycosylceramide	3
Sphingolipid	CerG2	Diglycosylceramide	4
Sphingolipid	CerG3	Triglycosylceramide	4
Sphingolipid	Gb4	Ganglioside Gb4	1
Sphingolipid	GM3	Ganglioside GM3	7
Sphingolipid	GM3+O	Oxidized ganglioside GM3	1
Sphingolipid	SM	Sphingomyelin	37
Sphingolipid	SM+O	Oxidized sphingomyelin	2
Sphingolipid	Su1G1	Sulfatide	1
Neutral lipid	ChE	Cholesterolester	19
Neutral lipid	DG	Diacylglycerol	22
Neutral lipid	TG	Triacylglycerol	138
Other lipid	Car	Acylcarnitine	6
Other lipid	CoQ	CoenzymeQ	1
Other lipid	FA	Fatty acid	5
Other lipid	FAA	Fatty amide	5
Other lipid	Other	Other	3

Although 112 lipids were significantly different between the phases, no lipid was defined as altered in the hepatocellular type (Figure 1a). In contrast, 9 and 20 lipids were defined as altered in the mixed and cholestatic types, respectively (Figure 1b,c). Thus, we focused on the mixed and cholestatic types for further analysis.

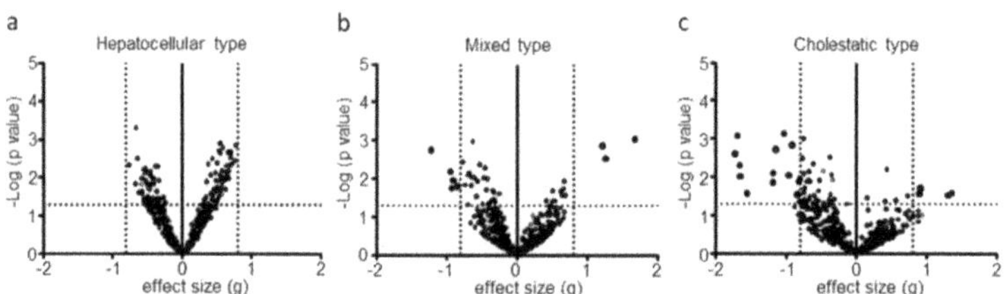

Figure 1. Volcano plot of lipid alterations in three types of DILI. Statistical probability (*p* value) and effect size (g) were determined by a comparison of lipid levels between acute phase and recovery phase of the DILI patients. Volcano plots show -log *p* value versus g value for (**a**) hepatocellular type, (**b**) mixed type, and (**c**) cholestatic type. Each dot represents an individual lipid.

2.3. Discrimination Ability for Mixed-Type DILI between Acute Phase and Recovery Phase or Healthy Volunteers

In the mixed-type DILI patients, three lipids, ceramide (Cer)(d34:1; d18:1/16:0), Cer(d36:1; d18:1/18:0), and oxidized ganglioside GM3 (GM3+O)(d34:1), were increased in the acute phase compared with the recovery phase, while six lipids, LPC(18:2), ether-linked LPC LPC(16:1e), ether-linked PC PC(38:6e; 18:2e/20:4, 16:1e/22:5), ether-linked phosphatidylethanolamine (PE) PE(36:4e; 16:0e/20:4), PE(38:4e; 18:0e/20:4), and PE(38:6e; 18:2e/20:4), were decreased in plasma (Table 3). The increased lipid of the highest effect size in the mixed-type patients was Cer(d34:1; d18:1/16:0), and the corresponding decreased lipid was PE(38:4e; 18:0e/20:4). All PCes and PEes contained the same FA (20:4: arachidonic acid).

Once we had characterized the specific lipids that were altered in mixed-type DILI, we next evaluated their discrimination ability between the acute phase and recovery phase by receiver operating characteristics (ROC) analysis. As shown in Table 3, four lipids, PE(38:4e; 18:0e/20:4), Cer(d34:1; d18:1/16:0), Cer(d36:1; d18:1/18:0), and GM3(d34:1)+O, had area under the curve (AUC) values over 0.8. The lipid with the highest AUC was Cer(d34:1; d18:1/16:0), with a value of 0.87.

We further compared the lipids levels of acute phase mixed-type DILI patients with the lipid levels of healthy subjects. Although the median ages of the three DILI patient types were approximately 60 years, we recruited the healthy subjects in four groups according to sex and age (HM1; middle-age male, HM2; old-age male, HF1; middle-age female, HF2; old-age female, where middle age was approximately 45 years and old age was approximately 60 years) (Table S2). The different lipids between mixed or cholestasis type DILI and all healthy subjects were listed in Table S3 (mixed) and Table S4 (cholestasis). As shown in Table 3, 6 lipids, LPC(18:2), LPC(16:1e), PE(38:6e; 18:2e/20:4), Cer(d34:1; d18:1/16:0), Cer(d36:1; d18:1/18:0), and GM3(d34:1)+O, were significantly different when comparing the acute phase DILI patients with all groups of healthy subjects. Cer(d34:1; d18:1/16:0), Cer(d36:1; d18:1/18:0), and GM3(d34:1)+O also had AUC values > 0.8 by ROC analysis versus all groups of healthy subjects. The representative individual plots of lipid levels discriminating the acute phase of mixed-type DILI from the recovery phase or the healthy volunteer groups are shown in Figure 2. Furthermore, we also calculated the ratio of altered specific lipids and evaluated their discriminating ability to acute phase mixed-type DILI patients from other groups, but no ratio of altered specific lipids further improved the discriminating ability. In addition, the absolute correlation coefficient of the altered specific lipids in mixed-type DILI with clinical parameters (AST, ALT, ALP, and total bilirubin) were all less than 0.6 (Table S5).

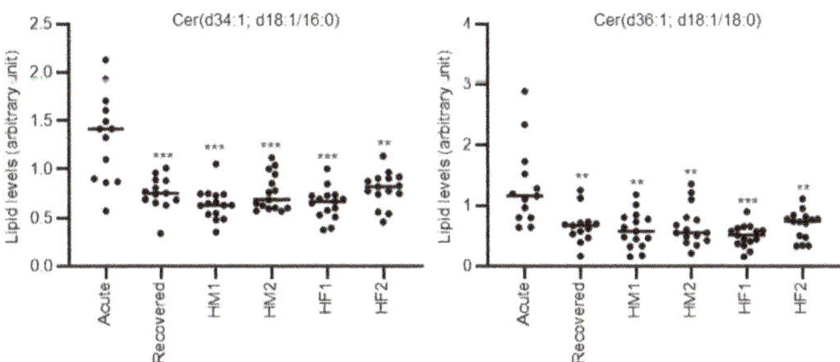

Figure 2. Representative individual plots of lipids levels discriminating the acute phase of mixed-type DILI from the recovery phase and the four healthy volunteer groups. Each dot represents an individual sample. Statistical significance is indicated as follows: ** $p < 0.01$, *** $p < 0.001$. Acute; acute phase DILI patients, Recovered; recovery phase DILI patients, HM1; healthy male subject group 1 (approximately 45 years old), HM2; healthy male subject group 2 (approximately 60 years old), HF1; healthy female subject group 1 (approximately 45 years old), HF2; healthy female subject group 2 (approximately 60 years old).

Table 3. Specific lipids altered in mixed-type DILI.

Metabolite ID	Name	vs. Recovered Phase			vs. Healthy M1			vs. Healthy M2			vs. Healthy F1			vs. Healthy F2		
		p Value	Effect Size	ROC–AUC	*p* Value	Effect Size	ROC–AUC	*p* Value	Effect Size	ROC–AUC	*p* Value	Effect Size	ROC–AUC	*p* Value	Effect Size	ROC–AUC
M008	LPC(18:2)	1.07×10^{-2}	**−0.91**	0.75	4.09×10^{-3}	**−1.17**	**0.81**	3.85×10^{-5}	**−1.85**	**0.91**	2.94×10^{-2}	**−0.89**	0.72	3.06×10^{-2}	**−0.85**	0.72
M014	LPC(16:1e)	6.48×10^{-3}	**−0.94**	0.78	3.40×10^{-4}	**−1.6**	**0.89**	3.16×10^{-5}	**−2**	**0.9**	1.24×10^{-2}	**−1**	0.79	8.68×10^{-4}	**−1.4**	**0.85**
M104	PC(38:6e)	1.74×10^{-2}	**−0.93**	0.78	8.23×10^{-2}	−0.67	0.71	3.46×10^{-2}	**−0.82**	0.72	2.84×10^{-2}	**−0.86**	0.75	3.72×10^{-3}	**−1.2**	0.78
M140	PE(36:4e)	1.52×10^{-2}	**−0.85**	0.76	2.70×10^{-1}	−0.43	0.69	3.60×10^{-1}	−0.36	0.66	5.05×10^{-3}	**−1.14**	0.84	2.13×10^{-1}	−0.49	0.74
M146	PE(38:4e)	1.74×10^{-3}	**−1.22**	0.82	2.74×10^{-1}	−0.4	0.56	1.00×10^{-1}	−0.62	0.67	2.60×10^{-3}	**−1.24**	0.82	3.24×10^{-2}	**−0.84**	0.73
M151	PE(38:6e)	1.61×10^{-2}	**−0.85**	0.75	2.21×10^{-2}	**−0.9**	0.73	6.88×10^{-3}	**−1.09**	0.76	5.98×10^{-3}	**−1.14**	0.79	1.60×10^{-2}	**−0.96**	0.75
M185	Cer(d34:1)	9.02×10^{-4}	**1.68**	0.87	1.21×10^{-4}	**2.08**	**0.93**	7.00×10^{-4}	**1.69**	**0.87**	1.33×10^{-4}	**2.06**	**0.93**	1.18×10^{-3}	**1.6**	**0.85**
M186	Cer(d36:1)	2.90×10^{-3}	**1.26**	0.86	2.62×10^{-3}	**1.42**	**0.88**	5.66×10^{-3}	**1.26**	**0.85**	8.77×10^{-4}	**1.71**	**0.96**	5.91×10^{-3}	**1.29**	**0.85**
M224	GM3(d34:1)+O	1.37×10^{-3}	**1.22**	0.83	9.78×10^{-5}	**2.23**	**0.94**	2.30×10^{-4}	**1.97**	**0.91**	6.28×10^{-4}	**1.75**	**0.89**	2.33×10^{-3}	**1.49**	**0.81**

The values fulfilled the threshold values ($p < 0.05$, effect size > 0.8, ROC–AUC > 0.8) are indicated by bold fonts. M1; middle-age male, M2; old-age male, F1; middle-age female, F2; old-age female, where middle age was approximately 45 years and old age was approximately 60 years, ROC: receiver operating characteristics, AUC: area under the curve.

2.4. Discrimination Ability for Cholestatic-Type DILI between Acute Phase and Recovery Phase or Healthy Volunteers

In the cholestatic-type DILI patients, 4 lipids, PC(30:0; 14:0/16:0), PC(31:0; 15:0/16:0), PC(32:1; 16:0/16:1), and PC (33:1; 15:0/18:1, 16:0/17:1), were increased in the acute phase then the recovery phase, while 16 lipids, PC(36:5e; 16:1e/20:4), PC(38:6e; 18:2e/20:4, 16:1e/22:5), PE(36:4e; 16:0e/20:4), PE(38:4e; 18:0e/20:4), PE(40:6e; 18:0e/22:6), PE(40:7e; 18:1e/22:6; M160), PE(40:7e; 18:1e/22:6; M161), phosphatidylinositol (PI)(38:3), PI(38:4; 18:0/20:4), PI(40:4), triglycosylceramide (CerG3)(d40:1), CerG3(d42:1), CerG3(d42:2), SM(d40:1; d18:1/22:0), TG(44:0; 14:0/14:0/16:0, 12:0/16:0/16:0), and CoQ10, were decreased (Table 4). The increased lipid of highest effect size in the cholestatic-type patients was PC(33:1; 15:0/18:1, 16:0/17:1) and the corresponding decreased lipid was PE(40:7e; 18:1e/22:6). Three FA(20:4)-containing ether-linked phospholipids, PC(38:6e; 18:2e/20:4, 16:1e/22:5), PE(36:4e; 16:0e/20:4), and PE(38:4e; 18:0e/20:4), were common with the mixed-type cases, but FA(22:6), corresponding to docosahexaenoic acid, was contained in three PEes, PE(40:6e; 18:0e/22:6), PE(40:7e; 18:1e/22:6; M160), and PE(40:7e; 18:1e/22:6; M161), which are specific for the cholestatic-type cases. In addition, all increased PCs in the acute phase of cholestatic-type patients contained the same FA(16:0: palmitic acid).

We also evaluated the discrimination ability of specific lipids that were altered in cholestatic-type DILI between the acute and recovery phases by ROC analysis. As shown in Table 4, 12 lipids, PC(31:0; 15:0/16:0), PC (33:1; 15:0/18:1, 16:0/17:1), PE(36:4e; 16:0e/20:4), PE(38:4e; 18:0e/20:4), PE(40:6e; 18:0e/22:6), PE(40:7e; 18:1e/22:6; M160), PI(38:3), PI(38:4; 18:0/20:4), PI(40:4), CerG3(d40:1), CerG3(d42:1), and CoQ10, had AUC values over 0.8. The lipid with the highest AUC was PE(40:7e; 18:1e/22:6; M160), with a value of 0.91.

We further compared the lipids levels of acute phase cholestatic-type DILI patients with the lipid levels of healthy subjects (grouped as indicated in Section 2.3). As shown in Table 4, eight lipids, PC(30:0; 14:0/16:0), PC(31:0; 15:0/16:0), PC(32:1; 16:0/16:1), PC(33:1; 15:0/18:1, 16:0/17:1), PC(36:5e; 16:1e/20:4), PI(38:3), PI(38:4; 18:0/20:4), and SM(d40:1; d18:1/22:0), were significantly different when comparing the acute phase DILI patients with all the compared groups of healthy subjects. PC(30:0; 14:0/16:0), PC(31:0; 15:0/16:0), PC(32:1; 16:0/16:1), PC (33:1; 15:0/18:1, 16:0/17:1), PI(38:3), PI(38:4; 18:0/20:4), and SM(d40:1; d18:1/22:0) also had AUC values >0.8 using ROC analysis versus all the groups of healthy subjects. The representative individual plots of lipid levels discriminating cholestatic-type DILI in the acute phase from the recovery phase or the healthy volunteer groups are shown in Figure 3. Furthermore, we also calculated the ratio of altered specific lipids and evaluated their discriminating ability to acute phase cholestatic-type DILI patients from other groups, but no ratio of altered specific lipids further improved the discriminating ability. In addition, the correlation coefficient of the altered specific lipids in mixed-type DILI with clinical parameters (AST, ALT, ALP, and total bilirubin) demonstrated over 0.6 (with $p < 0.05$) for three out of four palmitic acid-containing saturated or monounsaturated PCs, PC(31:0; 15:0/16:0), PC(32:1; 16:0/16:1), and PC (33:1; 15:0/18:1, 16:0/17:1) (Table S5). The absolute correlation coefficient of all other specific lipids was less than 0.6.

Table 4. Specific lipids altered in cholestatic-type DILI.

Metabolite ID	Name	vs. Recovered Phase			vs. Healthy M1			vs. Healthy M2			vs. Healthy F1			vs. Healthy F2		
		p Value	Effect Size	ROC–AUC	p value	Effect SIZE	ROC–AUC	p Value	Effect Size	ROC–AUC	p Value	Effect Size	ROC–AUC	p Value	Effect Size	ROC–AUC
M023	PC(30:0)	2.68×10^{-2}	**0.89**	0.72	3.05×10^{-3}	2.59	0.99	5.61×10^{-3}	2.23	0.96	1.54×10^{-2}	**1.66**	0.89	3.39×10^{-2}	**1.31**	0.82
M024	PC(31:0)	2.92×10^{-2}	1.31	0.84	5.89×10^{-4}	3.16	0.99	6.21×10^{-4}	3.06	0.98	3.25×10^{-3}	**2.16**	0.93	6.31×10^{-3}	**1.82**	0.88
M026	PC(32:1)	1.90×10^{-2}	0.91	0.72	7.82×10^{-3}	2.23	0.98	1.10×10^{-2}	2.05	0.94	1.86×10^{-2}	**1.74**	0.87	2.73×10^{-2}	**1.57**	0.83
M028	PC(33:1)	2.60×10^{-2}	1.36	0.88	1.29×10^{-3}	2.37	0.96	1.44×10^{-3}	2.88	0.99	4.99×10^{-3}	**2.05**	0.94	9.55×10^{-3}	**1.58**	0.88
M096	PC(36:5e)	4.36×10^{-2}	−0.8	0.69	1.27×10^{-2}	−1.45	0.83	8.67×10^{-3}	−1.14	0.8	6.55×10^{-3}	**−1.45**	0.86	2.30×10^{-2}	-1.24	0.77
M104	PC(38:6e)	1.45×10^{-3}	−0.92	0.73	5.38×10^{-2}	−1.06	0.78	2.23×10^{-2}	−1.09	0.81	2.41×10^{-2}	**−1.16**	0.78	1.46×10^{-2}	-1.49	0.83
M140	PE(36:4e)	9.57×10^{-3}	−1.65	0.86	7.10×10^{-2}	−0.78	0.73	1.03×10^{-1}	−0.7	0.74	1.18×10^{-3}	**−1.39**	0.88	5.27×10^{-2}	-0.85	0.76
M146	PE(38:4e)	2.68×10^{-2}	−1.55	0.89	9.48×10^{-2}	−0.62	0.68	2.93×10^{-2}	−0.85	0.78	8.03×10^{-4}	**−1.38**	0.91	9.08×10^{-3}	-1.2	0.84
M157	PE(40:6e)	4.92×10^{-3}	−1.66	0.89	6.09×10^{-2}	−0.69	0.72	8.24×10^{-3}	−0.99	0.88	6.23×10^{-5}	**−1.83**	0.95	2.43×10^{-4}	-1.61	0.98
M160	PE(40:7e)	2.45×10^{-3}	−1.73	0.91	4.44×10^{-2}	−0.76	0.74	2.30×10^{-4}	−1.63	0.96	8.09×10^{-5}	**−1.73**	0.93	2.99×10^{-4}	-1.65	0.94
M161	PE(40:7e)	1.11×10^{-2}	−0.8	0.75	3.58×10^{-1}	−0.42	0.61	1.12×10^{-2}	−1.24	0.81	4.52×10^{-2}	**−0.98**	0.73	6.81×10^{-3}	-1.44	0.84
M174	PI(38:3)	8.80×10^{-3}	−0.96	0.83	2.36×10^{-2}	−1.18	0.83	2.72×10^{-2}	−1.14	0.82	2.15×10^{-2}	**−1.17**	0.82	2.28×10^{-2}	-1.14	0.83
M175	PI(38:4)	7.65×10^{-3}	−1.18	0.83	1.24×10^{-3}	−1.85	0.9	2.65×10^{-4}	−2.2	0.95	2.56×10^{-3}	**−1.68**	0.88	3.88×10^{-3}	-1.54	0.86
M179	PI(40:4)	1.90×10^{-3}	−1.15	0.82	1.19×10^{-1}	−0.86	0.73	5.49×10^{-2}	−1.07	0.74	9.32×10^{-3}	**−1.41**	0.78	7.07×10^{-2}	-1.09	0.71
M213	CerG3(d40:1)	1.38×10^{-2}	−1.19	0.81	7.12×10^{-2}	−0.69	0.7	2.04×10^{-3}	−1.26	0.86	4.20×10^{-3}	**−1.19**	0.84	3.09×10^{-5}	-1.9	0.96
M214	CerG3(d42:1)	8.26×10^{-4}	−1.68	0.86	5.45×10^{-2}	−0.78	0.72	2.75×10^{-3}	−1.17	0.88	8.50×10^{-3}	**−1.06**	0.81	2.85×10^{-6}	-2.39	0.98
M215	CerG3(d42:2)	1.34×10^{-2}	−0.82	0.69	6.34×10^{-2}	0.88	0.78	9.13×10^{-1}	0.04	0.58	4.44×10^{-1}	0.34	0.62	2.92×10^{-1}	-0.53	0.61
M243	SM(d40:1)	2.86×10^{-2}	−0.84	0.77	6.17×10^{-3}	−1.82	0.89	8.18×10^{-4}	−2.33	0.97	8.43×10^{-3}	**−1.77**	0.89	7.56×10^{-4}	-1.95	0.97
M316	TG(44:0)	2.79×10^{-2}	−0.84	0.78	4.52E-01	0.38	0.67	3.30E-01	0.56	0.62	2.11×10^{-1}	0.79	0.68	8.15×10^{-1}	0.11	0.5
M450	CoQ10	7.22×10^{-4}	−1.04	0.8	5.17×10^{-3}	−1.55	0.85	1.79×10^{-3}	−1.96	0.92	5.59×10^{-2}	**−1.23**	0.82	1.05×10^{-1}	-0.81	0.78

The values fulfilled the threshold values ($p < 0.05$, effect size > 0.8, ROC–AUC > 0.8) are indicated by bold fonts. M1; middle-age male, M2; old-age male, F1; middle-age female, F2; old-age female, where middle age was approximately 45 years and old age was approximately 60 years, ROC: receiver operating characteristics, AUC: area under the curve.

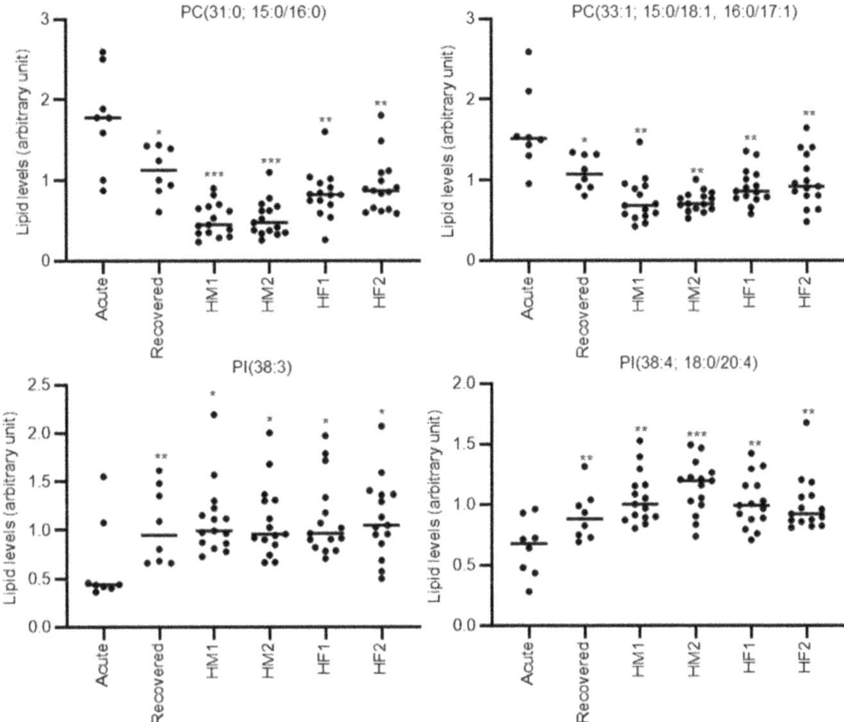

Figure 3. Representative individual plots of lipids levels discriminating the acute phase of cholestatic-type DILI from the recovery phase and the four healthy volunteer groups. Each dot represents an individual sample. Statistical significance is indicated as follows: * $p < 0.05$, ** $p < 0.01$, *** $p < 0.001$. Acute; acute phase DILI patients, Recovered; recovery phase DILI patients, HM1; healthy male subject group 1 (approximately 45 years old), HM2; healthy male subject group 2 (approximately 60 years old), HF1; healthy female subject group 1 (approximately 45 years old), HF2; healthy female subject group 2 (approximately 60 years old).

3. Discussion

In this study, we used plasma lipid profiling to characterize the pathophysiology of three different types of DILI in human patients and made five broad observations. First, the mixed and cholestatic types of DILI demonstrated specific plasma lipid alterations between acute and recovered phases, but the hepatocellular type did not. Second, as specific features of mixed-type DILI, when compared with levels in the recovery phase, several ceramides were increased in the acute phase, while arachidonic acid-containing ether-linked phosphoglycerolipids were decreased. Third, as specific features of cholestatic-type DILI, when compared with levels in the recovery phase, palmitic acid-containing saturated or monounsaturated PCs increased in the acute phase, while arachidonic acid- or docosahexaenoic acid-containing ether-linked phosphoglycerolipids and PIs decreased. Fourth, of the specific lipids altered in mixed-type DILI, the levels of Cer(d34:1; d18:1/16:0), Cer(d36:1; d18:1/18:0), and GM3(d34:1)+O demonstrated relatively high discrimination ability for the acute phase over the recovery phase in all groups of healthy subjects. Finally, of the specific lipids altered in cholestatic-type DILI, the levels of PC(31:0; 15:0/16:0), PC(33:1; 15:0/18:1, 16:0/17:1), PI(38:3), and PI(38:4; 18:0/20:4) demonstrated relatively high discrimination ability for the acute phase over the recovery phase in all groups of healthy subjects.

Although the number of subjects in the cholestatic-type DILI group was limited, the number of specific lipids altered was larger in this group than in the other two groups. This result suggests that the alteration in hepatic lipid homeostasis in cholestatic-type DILI shares a common mechanism among diverse suspected drugs. One representative plasma lipid that was increased in cholestatic-type DILI was palmitic acid (16:0)-containing saturated or monounsaturated PCs. Palmitic acid has been reported as the major fatty acid in biliary PCs [23]. In addition, the partner FAs of palmitic acid in the specifically altered PCs in cholestatic-type DILI, FA(16:1), and FA(17:1) are preferentially secreted into bile [24]. Thus, these increased levels of palmitic acid-containing saturated or monounsaturated PCs in the plasma were probably due to the reduced bile secretion of palmitic acid-containing saturated or monounsaturated PCs by cholestasis. This is also supported because the total bilirubin levels, which were also elevated by biliary structure, were highly correlated with the lipid levels in the DILI patients in this study.

Along with the palmitic acid-containing saturated or monounsaturated PCs, PIs, such as PI(38:3) and PI(38:4; 18:0/20:4), were also specifically increased in the cholestatic-type DILI patients. To date, the role of increased plasma PIs in cholestatic-type DILI remains unclear. However, supplementation with PIs decreases mRNA levels of the inflammatory cytokines/chemokines, tumor necrosis factor-alpha (TNF-α), and monocyte chemoattractant protein-1 (MCP-1), which are upregulated in steatosis [25]. In addition, blood and liver PIs were shown to increase with hepatic steatosis [26,27]. Therefore, one plausible reason for the increased PIs in plasma is to counteract the hepatic inflammation that can occur with lipid dysregulation.

Unlike other lipid classes, arachidonic acid-containing ether-linked phosphoglycerolipids were commonly altered in mixed and cholestatic-type DILI. Decreased levels of serum ether-linked phosphoglycerolipids have been reported in patients with nonalcoholic steatohepatitis and nonalcoholic fatty liver disease when compared to the levels in healthy controls [21]. In addition, plasma and liver ether-linked phosphoglycerolipid levels were decreased in a valproic acid-induced rat model of hepatic steatosis [28]. Thus, the decreased levels of ether-linked phosphoglycerolipids that we observed in the plasma of mixed and cholestatic-type DILI patients could be caused by mechanisms that are like those in steatosis and steatohepatitis, and they could reflect reduced levels in the liver. Arachidonic acid is well-known to be metabolized to inflammatory eicosanoids, such as prostaglandin E$_2$; thus, decreased levels of arachidonic acid-containing ether-linked phosphoglycerolipids in the plasma and the liver in the reference would implicate the inflammatory incidences in the liver of mixed and cholestatic-type DILI patients as well as patients with steatosis and steatohepatitis. Alternatively, ether-linked phosphoglycerolipids have been characterized as peroxisome-synthesized lipids and are a key component of peroxisome [29]. In fact, decreased levels of hepatic glyceronephosphate O-acyltransferase, which is a key peroxisomal enzyme for the synthesis of ether-linked phosphoglycerolipids, have been observed in a rat model of hepatic steatosis [28]. In addition, rescuing ether-linked phosphoglycerolipid levels by alkyl glycerol treatment could prevent impaired peroxisomal metabolism and hepatic steatosis [30,31]. Taken together, the decreased levels of plasma ether-linked phosphoglycerolipids that we observed may be caused by peroxisomal dysfunction in mixed and cholestatic types of DILI, and the rescue of ether-linked phosphoglycerolipid levels could be utilized for the therapeutic treatment of these DILI types.

Besides arachidonic acid-containing phosphoglycerolipids, increased plasma Cer was a characteristic feature of mixed-type DILI. To date, whether the increase in Cers plays a pivotal role in mixed-type DILI is unclear. However, Cers possess cell-signaling properties that are relevant to inflammation and apoptosis [32,33], and they may be involved in cystic fibrosis in the lung [34,35]. Thus, it is reasonable to speculate that increased Cer levels in mixed-type DILI patients contribute to hepatic inflammation and trigger subsequent pathological fibrosis.

In the present study, we also evaluated the differences in plasma lipids and their ability to discriminate between acute state DILI and healthy subjects divided into four age/sex groups. We identified 3 and 4 lipids in mixed and cholestatic types of DILI, respectively, as lipids with high discrimination ability. Although their scores did not exceed those of ALT and ALP (data not shown), these lipids could be utilized as biomarkers for DILI patients with ALT and ALP levels that are not diagnostic of liver disease. For example, ALT is elevated in patients with muscle injury and ALP is elevated in bone diseases. These lipids may also be helpful to discriminate DILI types and determine therapeutic approaches. Further analysis is needed to corroborate these speculations.

There are several limitations in the present study. First, it was performed with a few subjects of mixed and cholestatic types. Although we collected samples in both the acute and recovery phases from the same patients, the number of analyzed patients was limited, thus restricting the statistical power of our analysis. Second, due to the sparse number of events and limited ability to follow up patients, we recruited DILI patients from seven core hospitals. Although we used the same sampling protocol, hospital-to-hospital variation in sample preparation may have produced slightly different results in plasma lipid levels. Third, postprandial effect has been reported to have a global impact on lipidomics, although the impact is less than that of inter-individual variations [36,37]. Thus, this impact should be taken into consideration even though it is less than the impact caused by inter-individual variations. However, it is difficult to control the food intake of DILI patients, especially during the acute phase. Therefore, we believe that the state of fasting can be disregarded for this preliminary study. Fourth, although we recruited self-reported healthy subjects who had taken no medication for at least 1 week as controls, they may have been unaware of their disease status. Fifth, we did not control the alcohol and food intake of the patients, and the time of blood draw was not standardized, both of which might have affected plasma lipid levels. Sixth, since we used one internal standard (PC[1 2:0/12:0]) for all the classes of lipids, ionization efficiency should be different among the classes. Thus, the fold changes can effectively be calculated/estimated even using the same IS for all the lipids, but the comparison between lipid classes is not valid then. Last, it is difficult to consider the effects of other disease states and external factors, such as sexes and ages. In fact, several lipids, such as PEes and CerG3s, have high discriminant ability between acute phase and recovered phase DILI, while those lipids could not discriminate acute phase DILI and some groups of healthy subjects, which may be attributed to differences in sexes and ages. In addition, as was reported in the literature, many diseases, including liver-related diseases, which are possibly base diseases and complications, alter the plasma lipid levels [18–22]. Multivariate analysis including these potentially affecting factors should be performed using more patients' samples. Therefore, to address these limitations, a future, large-scale study with updated protocols should be performed.

In conclusion, we characterized the plasma lipid profiles of three types of DILI patients using a lipidomics approach. By comparing samples in acute and recovery phases, we revealed that mixed and cholestatic types of DILI produce specific alterations in plasma lipid profiles. In addition, by comparing these data to those of healthy subjects, we found several candidate markers of mixed and cholestatic DILI that discriminate the acute phase from the recovery phase and healthy state. Our study provides insights into the alterations in plasma lipidomic profiles, which reflect alterations in lipid homeostasis in the livers of DILI patients. These findings may help to understand the pathophysiology of different types of DILI.

4. Materials and Methods

4.1. Subjects and Sample Collection

DILI patients were recruited at the Teikyo University Hospital, Tokai University Hospital, Hiroshima Atomic-bomb Survivors Hospital, Kitasato University Hospital, Gunma University Hospital, Fukuoka University Hospital, and Hokkaido University Hospital. The inclusion criteria for DILI in the acute phase were ALT $\geq$150 U/L and/or ALP $\geq$ 2× upper limit of normal, as described previously [38,39]. In addition, each DILI patient was scored using the CIOMS/RUCAM [3] and DDW-J2004 [4,5] scales, and the highest probability cases in these scores were included in this study. The CIOMS/RUCAM scale involves a scoring system that categorizes the cases into "definite or highly probable" (score > 8), "probable" (score 6–8), "possible" (score 3–5), "unlikely" (score 1–2), and "excluded" (score ≤ 0). The DDW-J2004 scale involves a scoring system that categorizes the cases into "highly probable" (score > 5), "possible" (score 3–4), and "unlikely" (score ≤ 2). The DILI type and entry into the recovery phase were also diagnosed by DILI experts at each hospital. All healthy subjects were non-smoking, self-reported healthy volunteers who had taken no medications for at least 1 week before the study.

Blood samples were collected by venipuncture into 7 mL EDTA-2Na-containing vacuum blood collection tubes (VENOJECT II, TERUMO, Tokyo, Japan). The blood samples were immediately centrifuged (2500× g, 10 min, 4 °C); the resulting plasma was dispensed into screw-capped polypropylene tubes and stored in a deep freezer (−80 °C) before use. The plasma was typically frozen within 2 h from blood draw, although this occasionally extended to 4 h.

This study was conducted in accordance with the Declaration of Helsinki and approved by the Ethics Committee of the National Institute of Health Science (256, and 260 for Kihara Memorial Foundation), Teikyo University Hospital (15-127-2), Tokai University Hospital (15R-117), Hiroshima Atomic-bomb Survivors Hospital (H27-399-2), Kitasato University Hospital (B13-182), Gunma University Hospital (1487), Fukuoka University Hospital (18-8-04), Hokkaido University Hospital (016-0345), Daiichi Sankyo Co., Ltd. (15-0504-00), and Astellas Pharma Inc. (150028-01, 150047-01). Written informed consent was obtained from all participants.

4.2. Lipidomics

Lipid extraction was performed using the Microlab NIMBUS workstation (Hamilton, Binaduz, GR, Switzerland). The plasma samples were mixed with nine volumes of methanol/isopropanol (1/1) containing an internal standard (PC[12:0/12:0]), which is not detectable endogenously, at 2 μM. The mixed samples were filtered through a FastRemover Protein Removal Plate (GL Science, Tokyo, Japan) using an MPE2 automated liquid handling unit (Hamilton). The resulting lipid-containing filtrate was directly subjected to lipidomics. To obtain the lipidomics data, we performed reversed-phase LC (RPLC; Ultimate 3000, Thermo Fisher Scientific, Waltham, MA, USA) and MS (Orbitrap Fusion, Thermo Fisher Scientific), as described previously [40,41]. Compound Discoverer 2.1 (Thermo Fisher Scientific) was used with the raw data for peak extraction, annotation, identification, and lipid quantification, as described previously with a prior version of the software [40,41]. For isomers (same class, carbon length, and number of double bonds) showing different retention times in RPLC, each lipid was assigned a metabolite ID to distinguish it. Lipids with two different fatty acid combinations (e.g., 38:6e; 18:2e/20:4, 16:1e/22:5) indicate that the quantified lipid is a mixture of two different lipids that could not be separated. The quantified raw data were normalized to the internal standard. Since the lipidomics analysis was combined across two batches, the median value of each lipid in all samples was set to one in each batch to consolidate data from two batches after normalization. The processed data for the lipid levels are presented in Table S1.

4.3. Statistical Analysis

Significant differences in lipid levels were assessed by paired t-tests and Welch's t-test, and the effect size, which is calculated by Hedge's g, was considered. In this study, due to the limitation of sample size, a lipid level was considered specifically altered if its p value was <0.05 and its absolute effect size was >0.8. The discrimination ability was assessed by AUC score in ROC analysis using GraphPad Prism 6 (GraphPad Software, San Diego, CA, USA). The correlation coefficient was calculated as Pearson's correlation coefficient.

Supplementary Materials: The following are available online at http://www.mdpi.com/2218-1989/10/9/355/s1, Table S1: Lipidomics data set used in the present study. Table S2: Details of healthy subjects in the present study. Table S3: Lipids different between mixed-type DILI and healthy subjects. Table S4: Lipids different between cholestatic-type DILI and healthy subjects. Table S5: Correlation coefficient of altered specific lipids to clinical parameters in DILI patients. Figure S1: The exemplar LC/MS traces of plasma lipid profiles. Text S1: The details on the lipidomic analyses and associated data processing.

Metabolites **2020**, *10*, 355

Author Contributions: Conceptualization, Y.O., Y.S., and H.T.; subject recruitment, T.K., K.T. (Keiji Tsuji), Y.K., K.S. (Ken Sato), S.S., N.S., M.A., S.H., N.M., R.T., T.U., K.T. (Kazuhide Takata), K.O., and H.T.; blood sample preparation, T.K., K.T. (Keiji Tsuji), Y.K., K.S. (Ken Sato), S.S., N.S., M.A., S.H., N.M., R.T., T.U., K.T. (Kazuhide Takata), K.O., N.A., and H.T.; lipidomics analysis K.S. (Kosuke Saito); writing—original draft preparation, K.S. (Kosuke Saito); writing—review and editing, T.K., K.T. (Keiji Tsuji), Y.K., K.S. (Ken Sato), S.S., N.S., M.A., S.H., N.M., R.T., T.U., K.T. (Kazuhide Takata), K.O., K.M., M.S., T.N., K.T. (Kazuhiko Takamatsu), T.I., Y.O., Y.S., and H.T.; funding acquisition, K.S. (Kosuke Saito), T.K., K.T. (Keiji Tsuji), Y.K., K.S. (Ken Sato), S.S., N.S., N.A., Y.O., Y.S., and H.T. All authors have read and agreed to the published version of the manuscript.

Funding: This research was funded by the Japan Agency for Medical Research and Development, grant number (19mk0101045h0005, 19mk0101045s0105, 19mk0101045s0205, 19mk0101045s0305, 19mk0101045s1205, 19mk0101045s1405, 19mk0101045s1505, 19mk0101045s1605, 19mk0101045j0105, 19mk0101045j0305, 19mk0101045j0405).

Acknowledgments: We thank C. Sudo (National Institute of Health Sciences) for administrative assistance; M. Kojima, R. Iiji, R. Kaneko (National Institute of Health Sciences) for analytical assistance; K. Kubota, T. Hirata (Daiichi Sankyo RD Novare Co., Ltd.), K. Hashimoto (Daiichi Sankyo Co., Ltd.), Y. Hashimoto (Astellas Pharma Inc.), for technical advice.

Conflicts of Interest: Tatehiro Kagawa has received research funding from AbbVie GK, Sumitomo Dainippon Pharma, Eisai, and Daiichi Sankyo and honoraria from AbbVie GK, Sumitomo Dainippon Pharma, Bayer Yakuhin, Eisai, Daiichi Sankyo, and Gilead Sciences. The funders had no role in the design of the study; in the collection, analyses, or interpretation of data; in the writing of the manuscript, or in the decision to publish the results. All the other authors declare no conflict of interest.

References

1. Smith, D.A.; Schmid, E.F. Drug withdrawals and the lessons within. *Curr. Opin. Drug Discov. Dev.* **2006**, *9*, 38–46.

2. Ostapowicz, G.; Fontana, R.J.; Schiødt, F.V.; Larson, A.; Davern, T.J.; Han, S.H.; McCashland, T.M.; Shakil, A.O.; Hay, J.E.; Hynan, L.; et al. Results of a prospective study of acute liver failure at 17 tertiary care centers in the United States. *Ann. Intern. Med.* **2002**, *137*, 947–954. [CrossRef] [PubMed]

3. Danan, G.; Benichou, C. Causality assessment of adverse reactions to drugs-I. A novel method based on the conclusions of international consensus meetings: Application to drug-induced liver injuries. *J. Clin. Epidemiol.* **1993**, *46*, 1323–1330. [CrossRef]

4. Takikawa, H.; Onji, M.; Takamori, Y.; Murata, Y.; Taniguchi, H.; Ito, T.; Watanabe, M.; Ayada, M.; Maeda, N.; Nomoto, M.; et al. Proposal of diagnostic criteria for drug-induced liver injury revised by the DDW-J 2004 Workshop. *Kanzo* **2005**, *46*, 85–90. (In Japanese) [CrossRef]

5. Takikawa, H.; Onji, M. A proposal of the diagnostic scale of drug-induced liver injury. *Hepatol. Res.* **2005**, *2*, 250–251. [CrossRef] [PubMed]

6. David, S.; Hamilton, J.P. Drug-induced Liver Injury. *US Gastroenterol. Hepatol. Rev.* **2010**, *6*, 73–80. [PubMed]

7. Katarey, D.; Verma, S. Drug-induced liver injury. *Clin. Med.* **2016**, *16* (Suppl. 6), s104–s109. [CrossRef]

8. Park, B.K.; Kitteringham, N.R.; Maggs, J.L.; Pirmohamed, M.; Williams, D.P. The role of metabolic activation in drug-induced hepatotoxicity. *Annu. Rev. Pharmacol. Toxicol.* **2005**, *45*, 177–202. [CrossRef]

9. Mené, P.; Simonson, M.S.; Dunn, M.J. Phospholipids in signal transduction of mesangial cells. *Am. J. Physiol.* **1989**, *256*, F375–F386. [CrossRef]

10. Hannun, Y.A.; Linardic, C.M. Sphingolipid breakdown products: Anti-proliferative and tumor-suppressor lipids. *Biochim. Biophys. Acta* **1993**, *1154*, 223–236. [CrossRef]

11. Kolesnick, R.N.; Haimovitz-Friedman, A.; Fuks, Z. The sphingomyelin signal transduction pathway mediates apoptosis for tumor necrosis factor, Fas, and ionizing radiation. *Biochem. Cell Biol.* **1994**, *72*, 471–474. [CrossRef] [PubMed]

12. Ridgway, N.D. The role of phosphatidylcholine and choline metabolites to cell proliferation and survival. *Crit. Rev. Biochem. Mol. Biol.* **2013**, *48*, 20–38. [CrossRef] [PubMed]

13. Kotronen, A.; Seppänen-Laakso, T.; Westerbacka, J.; Kiviluoto, T.; Arola, J.; Ruskeepää, A.L.; Yki-Järvinen, H.; Oresic, M. Comparison of lipid and fatty acid composition of the liver, subcutaneous and intra-abdominal adipose tissue, and serum. *Obesity* **2010**, *18*, 937–944. [CrossRef] [PubMed]

14. Houjou, T.; Yamatani, K.; Imagawa, M.; Shimizu, T.; Taguchi, R. A shotgun tandem mass spectrometric analysis of phospholipids with normal-phase and/or reverse-phase liquid chromatography/electrospray ionization mass spectrometry. *Rapid Commun. Mass Spectrom.* **2005**, *19*, 654–666. [CrossRef]

15. Han, X.; Gross, R.W. Shotgun lipidomics: Electrospray ionization mass spectrometric analysis and quantitation of cellular lipidomes directly from crude extracts of biological samples. *Mass Spectrom. Rev.* **2005**, *24*, 367–412. [CrossRef]

16. Meikle, P.J.; Christopher, M.J. Lipidomics is providing new insight into the metabolic syndrome and its sequelae. *Curr. Opin. Lipidol.* **2011**, *22*, 210–215. [CrossRef]

17. Quehenberger, O.; Armando, A.M.; Brown, A.H.; Milne, S.B.; Myers, D.S.; Merrill, A.H.; Bandyopadhyay, S.; Jones, K.N.; Kelly, S.; Shaner, R.L.; et al. Lipidomics reveals a remarkable diversity of lipids in human plasma. *J. Lipid Res.* **2010**, *51*, 3299–3305. [CrossRef]

18. Patterson, A.D.; Maurhofer, O.; Beyoglu, D.; Lanz, C.; Krausz, K.W.; Pabst, T.; Gonzalez, F.J.; Dufour, J.F.; Idle, J.R. Aberrant lipid metabolism in hepatocellular carcinoma revealed by plasma metabolomics and lipid profiling. *Cancer Res.* **2011**, *71*, 6590–6600. [CrossRef]

19. Chen, S.; Yin, P.; Zhao, X.; Xing, W.; Hu, C.; Zhou, L.; Xu, G. Serum lipid profiling of patients with chronic hepatitis B, cirrhosis, and hepatocellular carcinoma by ultra-fast LC/IT-TOF MS. *Electrophoresis* **2013**, *34*, 2848–2856. [CrossRef]

20. Saito, K.; Maekawa, K.; Ishikawa, M.; Senoo, Y.; Urata, M.; Murayama, M.; Nakatsu, N.; Yamada, H.; Saito, Y. Glucosylceramide and lysophosphatidylcholines as potential blood biomarkers for drug-induced hepatic phospholipidosis. *Toxicol. Sci.* **2014**, *141*, 377–386. [CrossRef]

21. Puri, P.; Wiest, M.M.; Cheung, O.; Mirshahi, F.; Sargeant, C.; Min, H.K.; Contos, M.J.; Sterling, R.K.; Fuchs, M.; Zhou, H.; et al. The plasma lipidomic signature of nonalcoholic steatohepatitis. *Hepatology* **2009**, *50*, 1827–1838. [CrossRef]

22. Wu, J.M.; Xu, Y.; Skill, N.J.; Sheng, H.; Zhao, Z.; Yu, M.; Saxena, R.; Maluccio, M.A. Autotaxin expression and its connection with the TNF-alpha-NF-kappaB axis in human hepatocellular carcinoma. *Mol. Cancer* **2010**, *9*, 71. [CrossRef] [PubMed]

23. Agellon, L.B.; Walkey, C.J.; Vance, D.E.; Kuipers, F.; Verkade, H.J. The unique acyl chain specificity of biliary phosphatidylcholines in mice is independent of their biosynthetic origin in the liver. *Hepatology* **1999**, *30*, 725–729. [CrossRef]

24. Robins, S.J.; Fasulo, J.M.; Robins, V.F.; Patton, G.M. Utilization of different fatty acids for hepatic and biliary phosphatidylcholine formation and the effect of changes in phosphatidylcholine molecular species on biliary lipid secretion. *J. Lipid Res.* **1991**, *32*, 985–992. [PubMed]

25. Shirouchi, B.; Nagao, K.; Inoue, N.; Furuya, K.; Koga, S.; Matsumoto, H.; Yanagita, T. Dietary phosphatidylinositol prevents the development of nonalcoholic fatty liver disease in Zucker (fa/fa) rats. *J. Agric. Food Chem.* **2008**, *56*, 2375–2379. [CrossRef] [PubMed]

26. Ma, D.W.; Arendt, B.M.; Hillyer, L.M.; Fung, S.K.; McGilvray, I.; Guindi, M.; Allard, J.P. Plasma phospholipids and fatty acid composition differ between liver biopsy-proven nonalcoholic fatty liver disease and healthy subjects. *Nutr. Diabetes* **2016**, *6*, e220. [CrossRef] [PubMed]

27. Saito, K.; Uebanso, T.; Maekawa, K.; Ishikawa, M.; Taguchi, R.; Nammo, T.; Nishimaki-Mogami, T.; Udagawa, H.; Fujii, M.; Shibazaki, Y.; et al. Characterization of hepatic lipid profiles in a mouse model with nonalcoholic steatohepatitis and subsequent fibrosis. *Sci. Rep.* **2015**, *5*, 12466. [CrossRef]

28. Goda, K.; Saito, K.; Muta, K.; Kobayashi, A.; Saito, Y.; Sugai, S. Ether-phosphatidylcholine characterized by consolidated plasma and liver lipidomics is a predictive biomarker for valproic acid-induced hepatic steatosis. *J. Toxicol. Sci.* **2018**, *43*, 395–405. [CrossRef]

29. da Silva, T.F.; Sousa, V.F.; Malheiro, A.R.; Brites, P. The importance of ether-phospholipids: A view from the perspective of mouse models. *Biochim. Biophys. Acta* **2012**, *1822*, 1501–1508. [CrossRef]

30. Brites, P.; Ferreira, A.S.; Ferreira da Silva, T.; Sousa, V.F.; Malheiro, A.R.; Duran, M.; Waterham, H.R.; Baes, M.; Wanders, R.J.A. 2 Alkyl-Glycerol Rescues Plasmalogen Levels and Pathology of Ether-Phospholipid Deficient Mice. *PLoS ONE*, **2011**; *6*, e28539.

31. Jang, J.E.; Park, H.S.; Yoo, H.J.; Baek, I.J.; Yoon, J.E.; Ko, M.S.; Kim, A.R.; Kim, H.S.; Park, H.S.; Lee, S.E.; et al. Protective role of endogenous plasmalogens against hepatic steatosis and steatohepatitis in mice. *Hepatology* **2017**, *66*, 416–431. [CrossRef]

32. Teichgräber, V.; Ulrich, M.; Endlich, N.; Riethmüller, J.; Wilker, B.; Conceição De Oliveira-Munding, C.; van Heeckeren, A.M.; Barr, M.L.; von Kürthy, G.; Schmid, K.W.; et al. Ceramide accumulation mediates inflammation, cell death and infection susceptibility in cystic fibrosis. *Nat. Med.* **2008**, *14*, 382–391. [CrossRef] [PubMed]

33. Colombini, M. Ceramide channels and their role in mitochondria-mediated apoptosis. *Biochim. Biophys. Acta* **2010**, *1797*, 1239–1244. [CrossRef] [PubMed]

34. Shea, B.S.; Tager, A.M. Sphingolipid regulation of tissue fibrosis. *Open Rheumatol. J.* **2012**, *6*, 123–129. [CrossRef]

35. Ziobro, R.; Henry, B.; Edwards, M.J.; Lentsch, A.B.; Gulbins, E. Ceramide mediates lung fibrosis in cystic fibrosis. *Biochem. Biophys. Res. Commun.* **2013**, *434*, 705–709. [CrossRef] [PubMed]

36. Moriya, T.; Satomi, Y.; Kobayashi, H. Metabolomics of postprandial plasma alterations: A comprehensive Japanese study. *J. Biochem.* **2018**, *163*, 113–121. [CrossRef]

37. Saito, K.; Hattori, K.; Andou, T.; Satomi, Y.; Gotou, M.; Kobayashi, H.; Hidese, S.; Kunugi, H. Characterization of postprandial effects on CSF metabolomics: A pilot study with parallel comparison to plasma. *Metabolites* **2020**, *10*, 185. [CrossRef]

38. Takikawa, H.; Murata, Y.; Horiike, N.; Fukui, H.; Onji, M. Drug-induced liver injury in Japan: An analysis of 1676 cases between 1997 and 2006. *Hepatol. Res.* **2009**, *39*, 427–431. [CrossRef]

39. Aiso, M.; Takikawa, H.; Tsuji, K.; Kagawa, T.; Watanabe, M.; Tanaka, A.; Sato, K.; Sakisaka, S.; Hiasa, Y.; Takei, Y.; et al. Analysis of 307 cases with drug-induced liver injury between 2010 and 2018 in Japan. *Hepatol. Res.* **2019**, *49*, 105–110. [CrossRef]

40. Saito, K.; Ikeda, M.; Kojima, Y.; Hosoi, H.; Saito, Y.; Kondo, S. Lipid profiling of pre-treatment plasma reveals biomarker candidates associated with response rates and hand-foot skin reactions in sorafenib-treated patients. *Cancer Chemother. Pharmacol.* **2018**, *82*, 677–684. [CrossRef]

41. Saito, K.; Ohno, Y.; Saito, Y. Enrichment of resolving power improves ion-peak quantification on a lipidomics platform. *J. Chromatogr. B Analyt. Technol. Biomed. Life Sci.* **2017**, *1055*, 20–28. [CrossRef]

 *metabolites*

Article

Lipidomic Profiling of the Epidermis in a Mouse Model of Dermatitis Reveals Sexual Dimorphism and Changes in Lipid Composition before the Onset of Clinical Disease

Jackeline Franco [1], Bartek Rajwa [2,*], Christina R. Ferreira [3], John P. Sundberg [4] and Harm HogenEsch [1,5,*]

[1] Department of Comparative Pathobiology, Purdue University, West Lafayette, IN 47907, USA; francoj@purdue.edu
[2] Bindley Bioscience Center, Purdue University, West Lafayette, IN 47907, USA
[3] Metabolite Profiling Facility, Bindley Bioscience Center, Purdue University, West Lafayette, IN 47907, USA; cferrei@purdue.edu
[4] The Jackson Laboratory, Bar Harbor, ME 04609, USA; john.sundberg@jax.org
[5] Purdue Institute of Inflammation, Immunology and Infectious Diseases, Purdue University, West Lafayette, IN 47907, USA
* Correspondence: brajwa@purdue.edu (B.R.); hogenesc@purdue.edu (H.H.)

Received: 16 June 2020; Accepted: 18 July 2020; Published: 21 July 2020

Abstract: Atopic dermatitis (AD) is a multifactorial disease associated with alterations in lipid composition and organization in the epidermis. Multiple variants of AD exist with different outcomes in response to therapies. The evaluation of disease progression and response to treatment are observational assessments with poor inter-observer agreement highlighting the need for molecular markers. SHARPIN-deficient mice (*Sharpin*^{cpdm}) spontaneously develop chronic proliferative dermatitis with features similar to AD in humans. To study the changes in the epidermal lipid-content during disease progression, we tested 72 epidermis samples from three groups (5-, 7-, and 10-weeks old) of *cpdm* mice and their WT littermates. An agnostic mass-spectrometry strategy for biomarker discovery termed multiple-reaction monitoring (MRM)-profiling was used to detect and monitor 1,030 lipid ions present in the epidermis samples. In order to select the most relevant ions, we utilized a two-tiered filter/wrapper feature-selection strategy. Lipid categories were compressed, and an elastic-net classifier was used to rank and identify the most predictive lipid categories for sex, phenotype, and disease stages of *cpdm* mice. The model accurately classified the samples based on phospholipids, cholesteryl esters, acylcarnitines, and sphingolipids, demonstrating that disease progression cannot be defined by one single lipid or lipid category.

Keywords: lipidomics; atopic dermatitis; SHARPIN-deficient mice; flow-injection mass-spectrometry; predictive elastic net

---

1. Introduction

Atopic dermatitis (AD) is a multifactorial inflammatory skin disease that affects people and domestic animals worldwide [1]. Multiple variants (endotypes) of AD occur based on differences in the genetic background of patients, environment, immune activation pathways, and epidermal barrier status [1–3]. The classical AD presentation includes increased IgE serum levels, increased concentration of type 2 cytokines [4,5], and filaggrin (*FLG*) mutations that underlie skin barrier dysfunction [6–8]. However, variants of AD with normal levels of serum IgE and an increase of Th22 and Th17 cytokines instead of type 2 cytokines also exist [7,9]. In addition, *FLG* mutations occur in only 10% to 30% of

AD patients [10,11]. The less common variants of AD may require different therapeutic approaches as standard forms of therapy could result in unsatisfactory outcomes. Currently, clinical assessment of disease severity and diagnosis of AD relies on subjective observation of clinical signs, which change with the chronicity of the disease phase [6,7]. Several assessment indices are used to diagnose and score the disease, but these have poor inter-observer agreement highlighting the need for molecular disease biomarkers [12–15].

Alterations in the skin lipid composition have been reported in AD patients regardless of the genetic background, immune response, and clinical presentation [16,17]. Investigation of the lipid composition of the stratum corneum of the skin across different analytical platforms revealed changes in ceramide (CER) structure and presence of shorter and more unsaturated free fatty acids (FFA) in AD patients compared to healthy subjects [18–21]. Others reported changes in the amounts of phospholipids (PL), cholesteryl esters (CE), and triacylglycerides (TAG) in atopic skin, sweat, and sebum compared with healthy controls [22–24]. Alterations in the lipid composition lead to a disorganized stratum corneum lipid matrix and impaired barrier function of the skin [18], which permits increased allergen penetration that induces or aggravates the inflammatory reaction [25,26]. The cause of these lipid changes is not well understood, and it remains uncertain whether they result from a primary defect or downregulation of lipid processing enzymes by type 2 cytokines released in the course of dermatitis [27,28].

Sharpin^{cpdm} mice (hereafter referred to as *cpdm* mice), which have a mutation that causes absence of the SHARPIN protein, develop a chronic proliferative dermatitis that is very similar to human AD. The condition is characterized by pruritus, alopecia, and thickening of the skin, as well as accumulation of eosinophils, mast cells, M2 macrophages, and increased expression of type 2 cytokines [29,30]. In a previous study, we identified specific changes in ceramides and fatty acids in the epidermis of female SHARPIN-deficient mice with chronic proliferative dermatitis using a novel accelerated mass spectrometry strategy, multiple reaction monitoring (MRM)-profiling [31]. As the severity of the dermatitis rapidly increases with age, *cpdm* mice present a suitable model to identify lipid changes in the skin before the onset of clinical signs of inflammation and during progression of the dermatitis.

Lipidomics allows the detection and identification of a large number of molecules in a high-throughput manner aimed at the identification of new biomarkers for diagnosis and disease progression as well as novel targets for treatment [32]. These systems biology approaches yield complicated, high-dimensional data that should not be analyzed using naive univariate statistical methods as they may produce a high false-positive rate when predicting and classifying phenotypes. Consequently, this data requires multivariate approaches [33,34].

Although predicting phenotype from lipidomic data can be performed using various machine learning approaches, the critical question asked by biologists searching for a mechanistic model is the meaning of the statistical prediction. The black box predictors may be entirely accurate, but they do not allow easy formation of post-classification hypotheses regarding the causal relationship between the employed features, and the produced prediction. On the other hand, ante-hoc explainable models such as regression-based approaches can be used not only for supervised classification but also for the identification of critically important covariates, which can be further studied in pursuit of a mechanistic model [35]. Therefore, feature selection and reduction employing methods such as elastic-net (ENET) regularized regression are beneficial for finding key predictive features in the rich biological data and for identifying potential biomarkers amid the vast number of responses produced by systems biology methodologies [36–39]. Here we report the postulated biomarkers of AD, delivered via a multi-tiered feature selection strategy that processed the data generated by MRM-profiling in order to characterize lipid changes in the skin before the onset of clinical signs, both at the level of lipid categories and individual lipids ions. The method was used to investigate the association of the identified features with disease progression in male and female *cpdm* mice and their age and sex-matched wild type (WT) littermates. The study identified alterations in lipid composition preceding the onset of clinical dermatitis and a subset of lipid ions predictive of the disease stage of each sample. Additionally, the data demonstrated that the epidermis of female and male mice had distinct lipid profiles and differed in the lipid changes associated with disease progression.

2. Results

2.1. Association of Sex and Genotype to the Lipid Composition of the Mice Skin

Epidermal samples ($n = 72$, 36 *cpdm*, and 36 WT) were monitored for the presence of 1030 lipid ions belonging to multiple lipid categories. First, the collected data were pre-processed as described in the Methods section by executing log-ratio transformations, followed by single decomposition value (SVD)-driven principal component analysis. The result was visualized in the compositional principal component (CPC) space.

The CPC projection clearly differentiated samples by sex with the first component explaining 30.2% of the data variance, whereas the second component accounting for 22.4% of the variance was mostly associated with the genotype (Figure 1). The list of transitions driving the separation of samples in the CPC score plot is provided in Table S1.

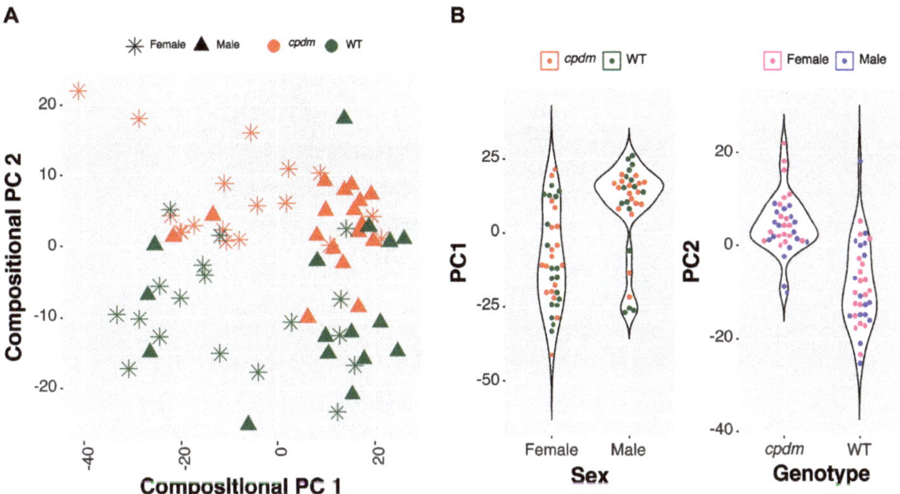

Figure 1. Monitored lipid ions in male and female *cpdm* and wild type (WT) epidermis by multiple reaction monitoring (MRM) scans in positive ion mode. Discrimination of the sex, as well as the genotypes of WT and *cpdm* mice (including non-lesional samples), was observed by compositional principal components (CPC) projection. (**A**) score plot of CPC analysis. (**B**) violin plots representing the separation of samples by sex and genotype. CPC 1 explained 30.2% of the variability of the data separating the samples by sex. CPC 2 explained 22.4% of the variance and was aligned with the genotype.

The visualization demonstrates that the information regarding the sex and genotype is encoded in the lipidomic profile of the sample. However, the CPC projection does not provide an actionable input from the perspective of feature selection or causal explanation. The 296 lipids in the top 25-percentile of accounted variance in CPC 1 contribute only between 0.167 and 0.31 percent to the representation. Similarly, for CPC 2, the individual contribution of each lipid in the top 25-percentile group ranges from 0.137 to 0.37. Therefore, a feature selection strategy is necessary.

The feature selection involved a two-tier selection, including a univariate step followed by the creation of a feature-ranking ENET regression model able to separate the samples into classes based on sex and genotype. The analysis was performed assuming a binary case (for sex and genotype data).

These tasks were approached using two different methods: Analysis of CPC-compressed compositional features representing lipid categories, and analysis of individual ions regardless of the category.

2.1.1. Selection of Predictive Lipid Categories for Sex and Genotype

The compressed features identified glycerolipids CPC 1, phospholipids CPC 1, and sphingolipids CPC 4 as most capable of separating samples by sex in the first selection step. For these features, the effect size expressed as η^2 ranged from 0.12 to 0.22 (Figure 2A). The η^2 of 0.22 is equivalent to Cohen's $f = 0.53$, which in a univariate model with two groups is equal to Cohen's $d = 1.06$, signifying a very substantial effect size. The subsequent feature-ranking ENET selected the sphingolipids CPC 4 and 5, phospholipids CPC 1, and glycerolipids CPC 1 as the most critical features to classify the samples by sex (Figure 2B). The classifier built using the 20-top composite CPC features had an overall accuracy of 0.76, $CI_{0.95} = (0.74, 0.77)$.

The univariate selection of compressed features for the binary genotype classification (WT vs. *cpdm*) identified phospholipids CPC 3, glycerolipids CPC 5, cholesteryl esters CPC 2, acylcarnitine CPC 2, and sphingolipids CPC 3 and 1, as the most predictive. The observed η^2 ranged from 0.26 to 0.73 for the top features (Figure 2A). The subsequently trained ENET identified phospholipids CPC 3, glycerolipids CPC 5, sphingolipids CPC 3 and 1, and cholesteryl esters CPC 2 as the top features in terms of importance (Figure 2B) and the model approached 100% accuracy ($CI_{0.95}$ from 1 to 0.95).

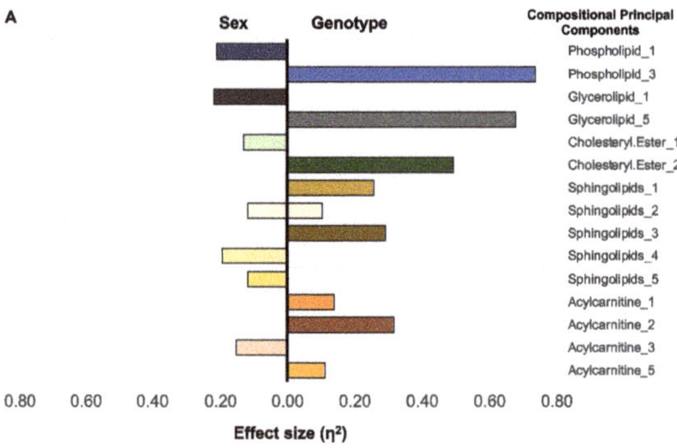

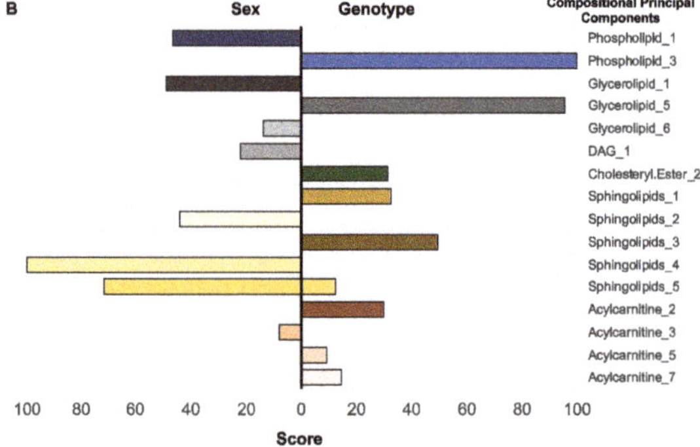

Figure 2. Importance of compressed categories for sex and genotype classification. (**A**) univariate linear model—selected lipid categories compositional principal component (CPC) based on their effect size (η^2). (**B**) top 10 CPC of lipid categories ranked by the multivariate elastic net model for sex and genotype, which generate a prediction accuracy of 76% for the sex and 100% for the genotype.

These results demonstrate that the composition and abundance of the same lipid categories carry information regarding the sex and genotype status of the tested animals. Similar to the CPC visualization incorporating all the lipid ions simultaneously, we observed that the information regarding sex and genotype is present in multiple compositional principal components.

2.1.2. Individual Lipid Ions Feature Selection for Genotype

Following the analysis of lipid categories, we performed a feature importance analysis for individual lipid ions. The study was conducted for genotypes (*cpdm* vs. WT) in a binary setting, filtering out the influence of sex. It is important to emphasize that even though this model used the individual lipid ion features, the tentative chemical attribution of the measured ions was not independently confirmed to eliminate the likelihood of isotopic interferences.

In the univariate step, we applied a linear model-based filter retaining only the features associated with genotype class (adjusted $p < 0.01$) and not strongly associated with sex (adjusted $p > 0.05$). The 100 lipid ions with the highest partial η^2 (ranging from 0.26 to 0.76) were selected for further analysis. In the top-100 group, acylcarnitines and phospholipids were by far the most prevalent. However, among the top ten features, there were nine phospholipids and one ion associated with sphingolipids. Table 1 shows the highest-scoring lipid ions as well as their weak effect size associated with sex.

Table 1. List of top 10 lipid ions ranked by the effect size of the univariate models linking genotype with the lipidomic profile.

Category	Tentative Attributions	MRM	Genotype (η^2)	Sex (η^2)
Phospholipid	PC(37:2), PC(O-38:2), PC(P-38:1)	800.6→184.1	0.76	0.001
Phospholipid	SM(d41:1) *	801.6→184.1	0.72	0.002
Sphingolipid	Cer[AS](d18:1/24:0)2OH	666.4→264.3	0.63	0.025
Phospholipid	PC(38:2), PC(P-39:1)	814.6→184.1	0.63	0.018
Phospholipid	SM(d36:0) *	733.6→184.1	0.61	0.010
Phospholipid	SM(d42:1) *	815.6→184.1	0.60	0.001
Phospholipid	PC(38:1), PC(P-38:1), PC(O-38:2) *	816.6→184.1	0.60	0.022
Phospholipid	PC (32:1), PC(O-33:1), PC(P-33:0)	732.1→184.1	0.59	0.008
Phospholipid	PC(40:8), PCo(40:1)	830.1→184.1	0.57	0.014

* Subject of possible isotopic interferences.

As in the previous analysis task, the second filtering step included an ENET regression used to filter and rank the lipids pre-selected by the univariate step. The trained ENET achieved an overall accuracy of 0.99, $CI_{95\%} = (0.924, 1)$. The most predictive lipid ions are summarized in Table 2. Among the selected lipids, five were phospholipids, two glycerolipids, and one was identified as a sphingolipid.

Table 2. List of top lipid ions ranked by importance score for prediction of genotype using the elastic net model.

Category	Tentative Attributions	MRM	Importance Score
Phospholipid	PC(37:2), PC(O-38:2), PC(P-38:1)	800.6→184.1	100.00
Glycerolipid	Glycerolipid containing 22:5 residue	627.1→280	50.58
Sphingolipid	Cer[AS](d18:1/24:0)2OH	666.4→264.3	48.40
Phospholipid	PC(38:1), PC(P-38:1), PC(O-38:2)	816.6→184.1	32.26
Phospholipid	SM(d41:1) *	801.6→184.1	29.57
Phospholipid	PC(38:2), PC(P-39:1)	814.6→184.1	19.86
Glycerolipid	Glycerolipid containing 18:2 residue	895.1→598	3.35
Phospholipid	SM(d37:0)	745.6→184.1	0.61

* Subject of possible isotopic interferences.

2.2. Selection of Features Associated with Disease Progression

2.2.1. Compositional Principal Component Analysis and Data Visualization

To study epidermal lipid changes associated with disease progression, a multiclass case was considered instead of a binary case. The *cpdm* mice were further divided into subclasses defined by the disease stage as non-lesional, established, and advanced. For general visualization of the data, we first computed CPC values using as input only the lipid data pre-selected in the previous binary step with the ENET filtering. The plot was prepared using the disease stage markings in a CPC space demonstrated that such a simple model was able to partially delineate the controls (independently of their age) and the levels of the *cpdm* genotype (Figure 3).

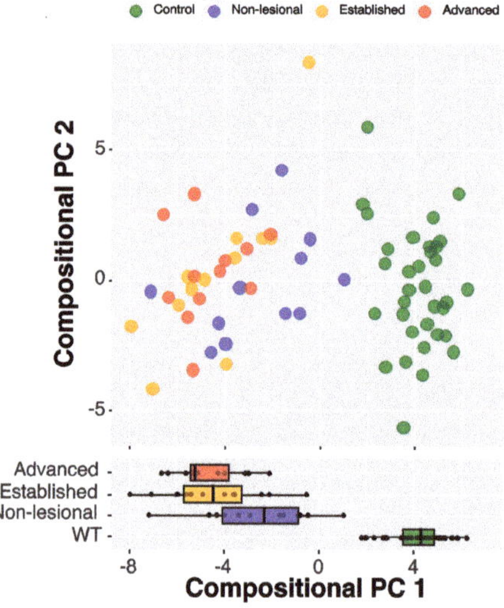

Figure 3. Lipid ions delineate disease stage groups. CPC analysis of all lipid ions plotted vs. disease progression. The model was able to delineate the controls and the three experimental groups of *cpdm* mice ($\eta^2 = 0.86$, *p*-value < 0.001).

2.2.2. Compressed-Feature Selection for Disease Progression

The disease-progression analysis, performed in a univariate setting, pointed to phospholipids CPC 3, glycerolipids CPC 5, and cholesteryl esters CPC 2 as the most informative compressed features (Figure 4A). The top features associated with disease progression produced η^2 ranging from 0.63 to 0.8. It is important to note that the features predicting disease progression were the same as those that separated WT from the broad *cpdm* group containing animals in all disease stages. The feature selection and ranking task performed by the ENET again identified phospholipid CPC 3, cholesteryl esters CPC 2, and glycerolipids CPC 5, as the top features in terms of importance. Interestingly, the highly ranked features were not equally important for all the disease stages (Figure 4B).

The disease progression prediction with an ENET classifier using a multinomial model achieved an overall accuracy of over 0.81, $CI_{95\%} = (0.71, 0.9)$. The substantial part of the observed inaccuracy was caused by the high similarity between samples from the adjacent "established" and "advanced" stages of the disease. This effect is also demonstrated by the difference between the unweighted and weighted Cohen's κ values (0.725 and 0.841, respectively).

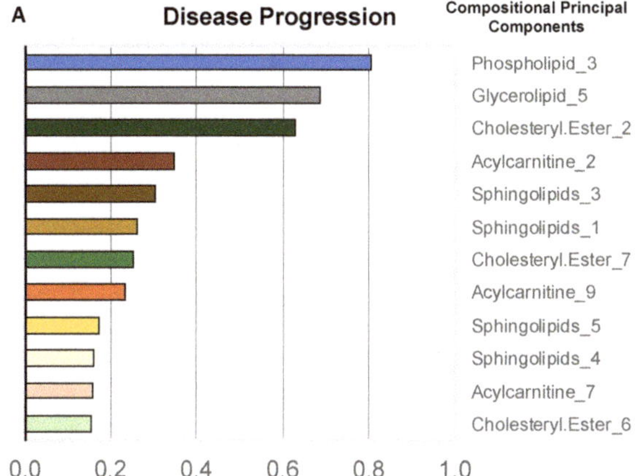

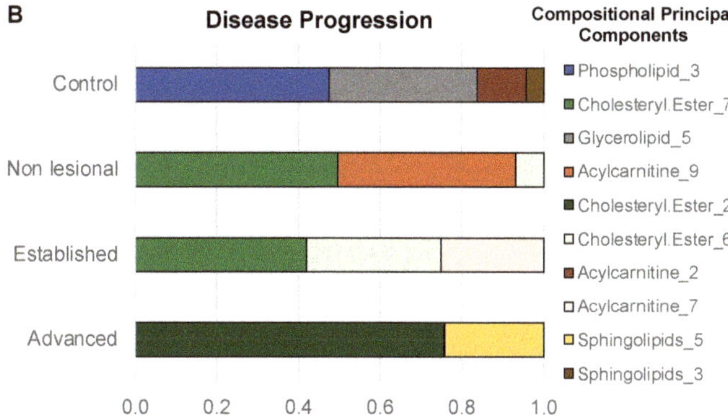

Figure 4. Importance of compressed categories for disease progression classification. (**A**) univariate linear model—selected lipid categories CPC based on their effect size (n^2). (**B**) contribution of the top 10 CPC to the classification of disease progression categories by the multivariate elastic net model with an accuracy of 81%.

2.2.3. Individual Lipid Ions Feature Selection for Disease Progression

The univariate feature selection step for disease progression selected ions with η^2 ranging from 0.22 to 0.73 and phospholipids dominated the very top of the list. The following multivariate analysis, performed by training an ENET, found a more diverse set of ions, some of them distinctly associated with a particular disease stage, but less useful for predicting others. It is an expected characteristic of a multivariate model, which combines all the features and their predictions to form a functional classifier. The ions contributing highly to the prediction of progression are listed in Table 3, and the results are illustrated in Figure 5.

Table 3. List of top 10 lipid ions ranked by importance score for prediction of disease progression in the elastic net model.

Category	Tentative Attributions	MRM	Importance Scores				
			Control	Non-Lesional	Established	Advanced	Overall
Phospholipid	PC(37:2), PC(O-38:2), PC(P-38:1)	800.6→184.1	100.00	0.00	0.00	0.00	100.00
Acylcarnitine	O-behenoylcarnitine	484.4→85.1	0.00	0.00	77.26	0.00	77.26
Sphingolipid	Sphingosine	300.2→282.2	0.00	71.50	0.00	0.00	71.50
Phospholipid	PC(31:0), PC(O-31:1), PC(P-31:0)	720.4→184.1	0.00	0.00	0.00	66.55	66.55
Cholesteryl Ester	22:1 Cholesteryl ester	725.4→369.1	0.00	0.00	0.00	60.59	60.59
Sphingolipid	Cer(d27:2)	438.2→266.2	0.00	0.00	44.09	0.00	44.09
Glycerolipid	Glycerolipids containing 30:0 residue	624.1→155.1	0.00	0.00	42.86	0.00	42.86
Phospholipid	SM(d42:3)	811.6→184.1	0.00	30.66	0.00	7.66	38.32
Acylcarnitine	Non attributed	837→85.1	0.00	36.96	0.00	0.00	36.96
Phospholipid	PC (36:0), PCp(38:6)	790.4→184.1	0.00	26.66	0.00	0.00	26.66

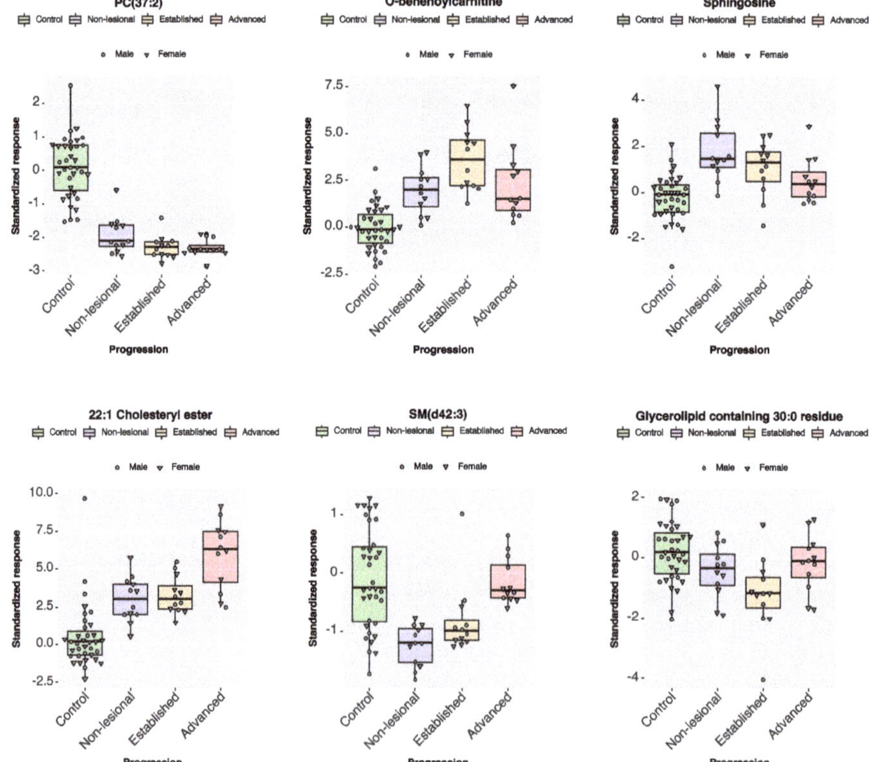

Figure 5. Epidermal lipid ions predictive of disease progression in mice. Representation of six lipids from the epidermis of WT and *cpdm* mice identified as predictive of disease stage in a sex-independent manner. Lipid features emphasize differences between controls and the various stages of the disease.

The ENET classifier trained on the disease progression data was able to classify the 36 *cpdm* and 36 WT samples into groups, including the control and the three disease stages with an overall accuracy of 0.79, $CI_{95\%}$ = (0.67, 0.87) when classified using weighted classes and 0.95, $CI_{95\%}$ = (0.88, 1) if the synthetic minority sampling technique (SMOTE) algorithm was used for correcting the class imbalance (Figure 6). The training used the variations transformed features corresponding to the presence of phosphatidylcholines, cholesteryl esters, acylcarnitines, and a glycerolipid-containing triacontanoic acid fatty acyl residue.

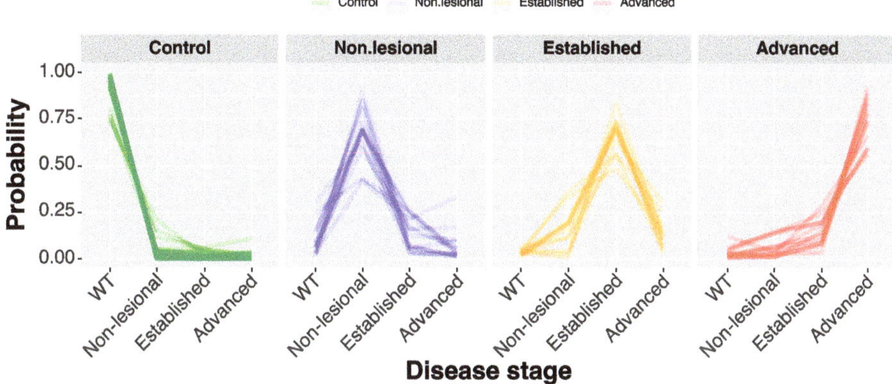

Figure 6. Classification of samples into disease progression groups. Parallel plot illustrating the classification of individual samples by elastic net regression. Each line represents a sample, and the highest point in the line corresponds to the group where the sample would be classified with higher probability.

3. Discussion

Lipids comprise a highly diverse group of molecules that play an essential role in the biology of the skin, and the relative proportions of different lipids are associated with the normal physiological functions of this organ [40–42]. In this study, we analyzed the relation between the lipids detected by MRM-profiling (lipidomic profile) and the observed genotypes using two machine learning feature-selection approaches. First, we computed a set of compressed features using compositional principal components to represent each of the lipid categories analyzed. These features easily separated male and female samples indicating a strong influence of sex on the epidermal lipid composition in mice. The first CPC summarizing variance in all the lipids was associated with clustering by sex, rather than by genotype. This result shows that the biochemical variability related to sex was dispersed among many lipids creating an effect more substantial than the one associated with the genotype. This result is in agreement with studies of skin-surface lipid clusters in humans where samples from males and females were distinguishable, but no significant difference between atopic or healthy subjects was observed [23,43]. However, it is necessary to note that the large variance visualized by CPC 1 (Figure 1) does not unequivocally demonstrate the importance of the differences between lipid composition in males and females, as it may also emerge from the fact that the sexual dichotomy has a high signal-to-noise ratio in the lipidomics data.

The multivariate analysis performed by the ENET demonstrated that classification to the male and female group was influenced mostly by sphingolipids (specifically, the compressed feature sets CPC 4 and CPC 5). The biological function of sphingolipids is determined by their composition, particularly the type of sphingoid base and the number of carbons and hydroxyl groups on the acyl chains, and their synthesis is affected by gonadal hormones in mice [44]. Several studies have shown alterations in ceramides, a sphingolipid, in the epidermis of AD patients [17,45–47]. However, conflicting results have been reported for changes in ceramides in non-lesional skin, probably because not all ceramide species are altered at the same stage of the disease and/or by the same mechanisms in males and females [23,48]. Our results show that sexual dimorphism is related strongly to the relative amounts of epidermal lipids in mice, and suggest that sex-related differences in the lipid biology of AD should be further investigated as they may partially explain the contradictory results regarding changes in ceramides in AD patients' skin [47].

A comparison of lipid categories in the *cpdm* and WT phenotypes by either univariate method or ENET showed differences driven by phospholipids and glycerolipids. The alteration in the lipid composition of the epidermis is a hallmark of AD associated with impaired barrier function of the skin [41,49], but whether these changes are primary or caused by the inflammatory process remains elusive [50,51]. Lipidomic and transcriptomic analysis of atopic patients have shown a global alteration of fatty acids caused by the interrelationship of type 2 cytokines and lipid elongase enzymes [52]. Our analysis demonstrates that the presence of phospholipids and glycerolipids in the epidermis was altered before any clinical signs of disease in the skin of the *cpdm* mice. Phospholipids and glycerolipids are essential for cellular and subcellular membrane dynamics and share common metabolic intermediates [53]. Phospholipids can also be secreted in the lamellar bodies of the epidermis along with the enzymes that use them as a substrate for ceramide synthesis [54]. Alterations in their concentrations may affect skin barrier function, cell metabolism, and inflammatory cell signaling, as they can carry esterified fatty acids that generate lipid mediators of inflammation by undergoing fatty acyl remodeling [55]. Likewise, lipids resulting from an aberrant lipid metabolism may be incorporated into membranes as phospholipids are in constant flux [55,56]. In agreement with our results, changes in phospholipids were reported to be present in the serum of atopic patients compared with controls [57]. In another mouse model of AD, NC/Nga mice, phospholipids were decreased in plasma, and oral supplementation of plasmalogens increased the phospholipid concentration in the skin of the mice and improved the skin condition [58]. In human skin, one study reported an increase of phospholipid content in AD patients compared to healthy subjects [24], while others showed a global change in phospholipids with an increase in the presence of shorter acylated fatty acids [52]. In addition to species differences, the disparities in sample preparation approach (epidermis vs. total skin), variations in analytical methods, and identification of the individual lipid species may account for these inconsistent results.

Changes in different categories of lipids were associated with stages of the disease as the severity of dermatitis increased. The control epidermis could be discriminated from the *cpdm* samples mostly by phospholipids and glycerolipids. However, other lipid categories were required to separate the disease stages of the *cpdm* mice samples. Classification of stages was influenced by acylcarnitines, cholesteryl esters, and sphingolipids. Acylcarnitines are fundamental for β-oxidation by delivering fatty acids to mitochondria and peroxisomes as an energy source; therefore, their dysregulation could potentially redirect fatty acids towards increased biosynthesis of phospholipids and other lipids, rather than being used as a source of energy [59,60]. Acylcarnitines were found dysregulated along with the phospholipid content in the serum of atopic patients [57]. Acylcarnitines also play a role in inflammatory processes, and their accumulation has been linked to lipotoxicity that results in apoptosis [61]. Cholesteryl esters, like acylcarnitines, are carriers of fatty acids and serve to store cholesterol in lipid droplets for its transport [62,63]. Free cholesterol is an essential constituent of the epidermis [19]; however, cholesteryl esters are less polar than free cholesterol conferring the skin with enhanced hydrophobicity for the barrier function [64]. Increased levels of free cholesterol in the skin of atopic patients [51], along with reduced levels of cholesteryl esters associated with high-density lipoproteins [65] suggest alterations in systemic cholesterol homeostasis, as reported in cardiovascular diseases [66]. The data showing an influence of sphingolipids on the classification model reproduces our previous results in which ceramides were identified as necessary for the classification of samples from *cpdm* mice with advanced dermatitis [31]. Changes in ceramide content are correlated with impaired barrier function and increased transepidermal water loss [51,67]. They are also associated with the disease severity, resulting in considerable changes in lesional compared to non-lesional skin [68]. Interestingly, sphingosine was an exception as its concentration was increased more in non-lesional than in lesional skin.

This study demonstrated an association between the presence of particular lipids in the epidermis and the occurrence of chronic proliferative dermatitis in mice. The link was displayed by a data-driven selection of MRM-extracted lipid features, training of a classifier, and identifying the key features contributing to the high classification accuracy. Although the presented procedure relies entirely on a statistical model, we hypothesize that the selected features will contribute to mechanistic insights if studied further. This argument is built upon the notion that the lipids providing high classification accuracy must be involved in the processes causing the phenotypic changes. Given that the lipid composition fingerprint differed not only between visibly lesional *cpdm* skin specimens and controls but was also altered in asymptomatic samples, the identified lipids could be considered candidates for predictive disease biomarkers. The current study involved an exploratory screening design that compares the lipid profiles of the groups using similar amounts of samples. The data analysis was performed employing compositional (relative) representation. Our study's scope did not include the validation of informative lipids by LC-MS/MS with the addition of internal standards; therefore, the isotopic and isobaric overlap may have occurred, and other lipids than the ones labeled with tentative attributions may have contributed to the reported predictive features. However, our previous work utilizing the relative amounts of ceramides demonstrated a concurrence between the relative values and the results obtained using quantitative LC-MS/MS [31]. Our analysis showed that not a single lipid (or lipid category) is modified sufficiently to be the sole differentiating factor. The accurate separation between healthy and diseased animals required an entire vector of lipid features, including phospholipids, acylcarnitines, cholesteryl esters, and sphingolipids. This result points to a multifaceted and multivariate nature of AD-associated lipid alterations in the skin.

The reliance on a classification tool to extract the most predictive lipids also defines the limitation of the presented approach. The task becomes particularly challenging when facing a severe class imbalance, as in the case of disease progression [69]. It is evident that sample availability determines the training performance, which in turn affects the robustness of the feature selection. We are aware of this limitation and hope that continuing research will allow for larger sample sizes, and correspondingly more confident analysis.

4. Materials and Methods

4.1. Animals

72 male and female C57BL/KaLawRij-*Sharpin*^{cpdm}/*Sharpin*^{cpdm} RijSunJ (*cpdm*) mice and WT littermates were obtained from the Jackson Laboratory and housed at 2 to 4 animals per box with food (Envigo) and water ad libitum. Room temperature was maintained at 20 ± 2 °C and relative humidity at $50 \pm 15\%$ with a 12/12-h light/dark cycle. Then, 18 WT males and 18 WT females and their *cpdm* littermates were divided into three groups of six males and six females with different ages and disease stages. The disease progression corresponded to non-lesional (5 weeks of age), established (7 weeks), and advanced (10 weeks) stages. Mice from the non-lesional group had no clinical signs of dermatitis on the dorsal or abdominal skin. The mice in the established group displayed erythema, moderate scaling, and mild alopecia of the dorsal and ventral skin. At 10 weeks, dermatitis covered most of the body with significant hair loss, erythema, thickening, and scaling. Mice were euthanized at 5, 7, or 10 weeks of age by CO_2 asphyxiation and cervical dislocation. The animal experiments and procedures were conducted in accordance with the Guide for the Care and Use of Laboratory Animals of the National Institutes of Health. The protocol was approved by the Purdue University Animal Care and Use Committee (PACUC protocol 111001019).

4.2. Epidermis Isolation and Lipid Extraction

Sample collection and lipid extraction were performed as previously described [31]. Briefly, a 1 by 2 cm slice of dorsal skin was collected, and after incubation with Thermolysin (from *Geobacillus stearothermophilus*, Sigma-Aldrich, St. Louis, MO, USA) dissolved in HEPES buffer, the epidermis was peeled off and stored at −80 °C until extraction. Tissue was weighed and homogenized in 250 µL of ultra-pure water using Precellys24 tissue homogenizer (Bertin Technologies, Rockville, MD, USA). The homogenate was submitted to a Bligh and Dyer [70] liquid–liquid extraction, and the organic phase was collected and dried in a concentrator. Samples were resuspended in 40 µL of 3:1 (*v/v*) acetonitrile (ACN)/chloroform, then diluted 50× with ACN/methanol/ammonium acetate 300 mM at 3:6.65:0.35 volume ratio for mass spectrometry analysis.

4.3. MRM-Profiling Method Development and Sample Screening

A composite sample of each group from the testing set was created by pooling aliquots of 5 µL from each specimen in the group. The composite samples were analyzed using a previously described methodology of MRM-profiling discovery experiments [31]. Briefly, neutral loss (NL) and precursor ion (Prec) scans were used to profile phospholipids, acylcarnitines (AC), sulfatides, cholesteryl esters, ceramides, glycerolipids with diverse fatty acid acyl residues, triacylglycerides, and free fatty acids in positive and negative ion modes [31,71–75]. Using a micro-autosampler (G1367A), 8 µL of the sample was directly delivered into a QQQ6410 triple quadrupole mass spectrometer (Agilent Technologies, San Jose, CA, USA) equipped with an ESI ion source. A cap pump (G1376A) was used to flow acetonitrile plus 0.1% formic acid at a rate of 5 µL/min. The source capillary and multiplier voltages were 3500 V and 300 V, respectively. The collision energy voltage was 2 V for the negative ion mode methods. In positive ion mode, the collision energies varied according to the lipid classes. For ceramides, phosphatidylethanolamines (PE), and lipids with arachidonate acyl residue and oleate acyl residue, the collision energy was set at 22 V, for phosphatidylcholines and sphingomyelins (SM) at 20 V, for phosphatidylserines (PS) and phosphatidylinositols (PI) at 16 V, for CE at 17 V and for acylcarnitines the collision energy was set at 30 V. The fragmentation voltage of all the methods was 100 V. In total, 80 different discovery scans were performed, producing 1030 informative lipid ions. The parent and the fragment were collected and organized as transitions in 6 different methods of 2 min each (Table S2). The individual samples were flow-injected 6 times to cover all the monitored lipid ions. The raw data files are deposited in the public proteomics repository MassIVE (http://massive.ucsd.edu) using the identifier: MSV000083884. The tentative identification of lipid ions was performed through MS/MS experiments and by using reference databases, such as the Lipid Maps database (http://www.lipidmaps.org/) and METLIN (https://metlin.scripps.edu). Validation of the method by liquid chromatography-mass spectrometry has been previously reported along with the linearity and dynamic range of over four orders of magnitude from 1 to 10,000 ppm [31].

4.4. Data Analysis

Using MSConvert (http://proteowizard.sourceforge.net), the files were converted into the mzML open-source format, and an in-house script was used to obtain the ion intensities of each *m/z* monitored. The relative amounts of each *m/z* were used for data analysis.

The visualization and subsequent selection of lipid categories /groups and individual lipid ions associated with the *cpdm* genotype were performed following normalization of the signals to 1 using the total ion count and subsequent transformation using the isometric log-ratio function or centered log-ratio function. For visualization of all the overall characteristics of the data, the pre-processed input was compressed using singular value decomposition. The resultant first two compositional principal components were employed to illustrate the tendency of the samples to separate themselves in the reduced dimensionality space into clusters according to the sex, the genotype, and the disease severity [76].

4.4.1. Selection of Predictive Lipid Categories

To identify the categories of lipids associated with the sex or genotype, the pre-processing and compression were followed by a two-tier selection including a univariate step, and a multivariate step driven by ENET regression.

In the beginning, the measured lipid ions were annotated and assigned to one of the following categories: (1) acylcarnitine, (2) acylcarnitine or glycerolipids, (3) cholesteryl esters, (4) DAG, (5) glycerolipids, (6) phospholipids, (7) phospholipids or cholesteryl esters, (8) phospholipids or glycerolipids, (9) sphingolipids, and (10) sphingolipids or glycerolipids. The overlap in categories reflects the uncertainty of the attribution due to the use of only one MRM related to a lipid candidate. Each of the sub-dataset was compressed using SVD to create compressed features sets. The number of retained columns (and by extension, the number of principal components used to represent the categories) was selected to retain 95% of the variance in each category. Therefore, the number of reduced composite features per class varied from 4 to 21. The resultant composite features for every lipid class were named "CPC", followed by the component number. For instance, "sphingolipids CPC 4" denotes the fourth principal component of the log-ratio transformed sphingolipids-class data.

It is important to emphasize that the SVD of the compositional data was not utilized here to enable a PCA-driven feature selection, but rather to produce a highly compressed input for the separate feature selection step. In other words, we are not claiming a direct association between the lipids that happen to display the most variation and the lipids (or lipid classes) that are most likely to be predictive and biologically significant.

The described data reduction process resulted in the creation of 57 compressed lipid-class features describing each of the samples. Subsequently, 57 linear models linking the computed features with sex, and another 57 models linking the features with the genotype (*cpdm* vs. control) were created. Finally, the third set of 57 linear models was computed to relate the features with the disease progression of *cpdm* mice (using the class assignment of control < non-lesional < established < advanced disease status). Benjamini–Hochberg p-value adjustment [77] was used to correct for false discovery. The features associated with genotype models having p-value < 0.05 (and η^2 effect sizes ranging from 0.73 to 0.1) were picked for the further feature selection step. For the sex-dependent changes in lipids, we also picked features with p-value < 0.05 (and η^2 effect sizes from 0.22 to 0.12).

4.4.2. Feature Selection of Predictive Individual Lipid Ions

In a similar procedure, in order to recover the most predictive individual lipid ions (rather than lipid categories), we first created 1,030 linear models linking the log-ratio transformed relative amounts of every transition to genotype and sex. We pre-selected the features that might be associated with disease progression (either in sex-dependent or sex-independent manner) by selecting p value < 0.01 for the criteria that were included and p-value > 0.05 for the factors that were ruled out. The lipids present in linear models connecting significantly with disease progression after Benjamini–Hochberg p-value adjustment ($p < 0.01$), but not being significant for sex ($p > 0.05$) were selected as sex-independent predictive lipids. About 50 ions were selected as possibly predictive and represented η^2 effect sizes ranging from 0.724 to 0.29.

4.4.3. Predictive Elastic Net Regression

The final but critical step of these feature selection procedures involved the use of ENET regression [78]. ENET was employed either as a binary (for sex and *cpdm* vs. WT separation) or a multiclass classifier. This regression approach includes LASSO L_1 and ridge L_2 penalty terms leading to a predictive model operating in a reduced dimensionality of the data produced by the MRM-profiling:

$$\hat{\beta} = \underset{\beta}{\operatorname{argmin}}\left(\|\mathbf{y} - \mathbf{X}\beta\|^2 + \lambda((1-\alpha)\|\beta\|^2/2 + \alpha\|\beta\|_1)\right) \tag{1}$$

In the ENET formula above, the input matrix **X** consists of all the pre-selected measured lipid ions (or pre-selected composite lipid categories), the output vector **y** describes the stages of the disease, and $\alpha, \lambda \geq 0$ are tuning parameters. The penalties included in the mathematical model are in the $\|\beta\|_1$ term which generates a sparse model by shrinking some regression coefficients to zero and, in the $\|\beta\|^2$ term which removes the limitation on the number of selected variables but encourages grouping effect, allowing similar features to be selected together. The individual lipid ions or the composite lipid categories with the larger absolute value of β are considered to be more predictive. The ENET simplifies to ridge regression when $\alpha = 1$ and to the LASSO regression when $\alpha = 0$.

The ENET regression was trained using the leave-one-out approach. Due to significant data imbalance, we used class weights (imposing a lesser penalty for errors in the majority class) or the SMOTE approach during training [79]. The resultant classifier allowed us to rank the lipid ions in terms of importance (ability to influence ENET prediction) using the absolute value of the non-zero coefficients.

To visualize the changes in the selected features, a central log-ratio transformation followed by standardization to the female WT subgroup was performed. Therefore, the y-axis in the figures shows the difference in relative lipid abundance as the number of standard deviations away from the WT-female group. The multiclass ENET prediction was illustrated using a parallel plot.

The statistical analyses were performed using R-language for statistical computing.

5. Conclusions

In this study, we paired an exploratory high-throughput lipidomics technique with rigorous machine learning analysis to rapidly screen for potential biomarkers in a mouse model of dermatitis. The measurements were performed using flow injection to the ion source of a triple quadrupole mass spectrometer, providing highly sensitive, but low-resolution mass data. The exploratory approach relied on product ions and neutral losses expected to be specific to the lipid classes, but not individual lipids; therefore, the detected lipids are assigned only tentative attributions. The approach revealed sexual dimorphism in the epidermal lipid profile, which was distributed throughout the lipid categories and identified sphingolipids as the best predictors for sex classification. Furthermore, epidermal lipid analysis allowed accurate classification of samples, not only by the genotype of the mice, cpdm vs. WT, but by the stages of disease progression. A panel of lipids comprised of phospholipids, acylcarnitines, sphingolipids, and cholesteryl esters was necessary to achieve successful classification into the different disease stages, showing that a single lipid or lipid category was not altered sufficiently to be the sole classifier. These results highlight the need to consider sex-related differences in the pathobiology of AD and the importance of building lipid panels that include lipids from different categories when investigating predictive biomarkers for AD.

Supplementary Materials: The following are available online at http://www.mdpi.com/2218-1989/10/7/299/s1, Table S1: Transitions for compositional principal component (CPC) score plot, Table S2: Compiled method for individual analysis of samples.

Author Contributions: Conceptualization, J.F., B.R. and H.H.; data curation, J.F. and B.R.; formal analysis, J.F. and B.R.; funding acquisition, J.F., J.P.S., and H.H.; investigation, J.F. and C.R.F.; methodology, J.F., B.R., C.R.F.; validation, J.F. and B.R.; project administration, H.H.; resources, J.P.S.; supervision, B.R. and H.H.; visualization, J.F.; writing (original draft), J.F., B.R. and H.H.; writing (review and editing), J.F., B.R., C.R.F., J.P.S. and H.H. All authors have read and agreed to the published version of the manuscript.

Funding: This research was supported in part by grants from the National Institutes of Health (AR049288) and the Purdue Institute of Inflammation, Immunology, and Infectious Disease (PI4D). JF was supported in part by a Colciencias fellowship.

Conflicts of Interest: The authors declare no conflict of interest. The funders had no role in the design of the study; in the collection, analyses, or interpretation of data; in the writing of the manuscript, or in the decision to publish the results.

References

1. Thijs, J.L.; Strickland, I.; Bruijnzeel-Koomen, C.A.; Nierkens, S.; Giovannone, B.; Csomor, E.; Sellman, B.R.; Mustelin, T.; Sleeman, M.A.; De Bruin-Weller, M.; et al. Moving toward endotypes in atopic dermatitis: Identification of patient clusters based on serum biomarker analysis. *J. Allergy Clin. Immunol.* **2017**, *140*, 730–737. [CrossRef] [PubMed]

2. Brunner, P.M.; Guttman-Yassky, E. Racial differences in atopic dermatitis. *Ann. Allergy Asthma Immunol.* **2019**, *122*, 449–455. [CrossRef] [PubMed]

3. Ardern-Jones, M.; Bieber, T. Biomarkers in atopic dermatitis: It is time to stratify. *Br. J. Dermatol.* **2014**, *171*, 207–208. [CrossRef] [PubMed]

4. Elias, P.M.; Gruber, R.; Crumrine, D.; Menon, G.; Williams, M.L.; Wakefield, J.S.; Holleran, W.M.; Uchida, Y. Formation and functions of the corneocyte lipid envelope (CLE). *Biochim. Biophys. Acta (BBA) Bioenerg.* **2013**, *1841*, 314–318. [CrossRef] [PubMed]

5. Elias, P.M. Epidermal Lipids, Barrier Function, and Desquamation. *J. Investig. Dermatol.* **1983**, *80*, S44–S49. [CrossRef]

6. Malajian, D.; Guttman-Yassky, E. New pathogenic and therapeutic paradigms in atopic dermatitis. *Cytokine* **2015**, *73*, 311–318. [CrossRef] [PubMed]

7. Czarnowicki, T.; He, H.; Krueger, J.G.; Guttman-Yassky, E. Atopic dermatitis endotypes and implications for targeted therapeutics. *J. Allergy Clin. Immunol.* **2019**, *143*, 1–11. [CrossRef] [PubMed]

8. Brunner, P.M.; Guttman-Yassky, E.; Leung, D.Y. The immunology of atopic dermatitis and its reversibility with broad-spectrum and targeted therapies. *J. Allergy Clin. Immunol.* **2017**, *139*, S65–S76. [CrossRef]

9. Kaufman, B.P.; Guttman-Yassky, E.; Alexis, A.F. Atopic dermatitis in diverse racial and ethnic groups-Variations in epidemiology, genetics, clinical presentation and treatment. *Exp. Dermatol.* **2018**, *27*, 340–357. [CrossRef]

10. Ong, P.Y.; Leung, D.Y.M. The Infectious Aspects of Atopic Dermatitis. *Immunol. Allergy Clin. N. Am.* **2010**, *30*, 309–321. [CrossRef]

11. O'Regan, G.M.; Sandilands, A.; McLean, W.I.; Irvine, A.D. Filaggrin in atopic dermatitis. *J. Allergy Clin. Immunol.* **2008**, *122*, 689–693. [CrossRef] [PubMed]

12. Schmitt, J.; Langan, S.; Williams, H.C. What are the best outcome measurements for atopic eczema? A systematic review. *J. Allergy Clin. Immunol.* **2007**, *120*, 1389–1398. [CrossRef] [PubMed]

13. Chopra, R.; Vakharia, P.P.; Sacotte, R.; Patel, N.; Immaneni, S.; White, T.; Kantor, R.; Hsu, D.Y.; Silverberg, J.I. Severity strata for Eczema Area and Severity Index (EASI), modified EASI, Scoring Atopic Dermatitis (SCORAD), objective SCORAD, Atopic Dermatitis Severity Index and body surface area in adolescents and adults with atopic dermatitis. *Br. J. Dermatol.* **2017**, *177*, 1316–1321. [CrossRef] [PubMed]

14. Silverberg, J.I.; Gelfand, J.M.; Margolis, D.J.; Fonacier, L.; Boguniewicz, M.; Schwartz, L.B.; Simpson, E.; Grayson, M.H.; Ong, P.Y.; Fuxench, Z.C.C.; et al. Severity strata for POEM, PO-SCORAD, and DLQI in US adults with atopic dermatitis. *Ann. Allergy, Asthma Immunol.* **2018**, *121*, 464.e3–468.e3. [CrossRef] [PubMed]

15. Futamura, M.; Leshem, Y.A.; Thomas, K.S.; Nankervis, H.; Williams, H.C.; Simpson, E.; Information, P.E.K.F.C. A systematic review of Investigator Global Assessment (IGA) in atopic dermatitis (AD) trials: Many options, no standards. *J. Am. Acad. Dermatol.* **2016**, *74*, 288–294. [CrossRef] [PubMed]

16. Elias, P.M.; Hatano, Y.; Williams, M.L. Basis for the barrier abnormality in atopic dermatitis: Outside-inside-outside pathogenic mechanisms. *J. Allergy Clin. Immunol.* **2008**, *121*, 1337–1343. [CrossRef]

17. Van Smeden, J.; Janssens, M.; Gooris, G.; Bouwstra, J. The important role of stratum corneum lipids for the cutaneous barrier function. *Biochim. Biophys. Acta (BBA) Mol. Cell Boil. Lipids* **2014**, *1841*, 295–313. [CrossRef]

18. Janssens, M.; Van Smeden, J.; Gooris, G.S.; Bras, W.; Portale, G.; Caspers, P.J.; Vreeken, R.J.; Hankemeier, T.; Kezic, S.; Wolterbeek, R.; et al. Increase in short-chain ceramides correlates with an altered lipid organization and decreased barrier function in atopic eczema patients[S]. *J. Lipid Res.* **2012**, *53*, 2755–2766. [CrossRef]

19. Yamamoto, A.; Serizawa, S.; Ito, M.; Sato, Y. Stratum corneum lipid abnormalities in atopic dermatitis. *Arch. Dermatol. Res.* **1991**, *283*, 219–223. [CrossRef]

20. Imokawa, G.; Abe, A.; Jin, K.; Higaki, Y.; Kawashima, M.; Hidano, A. Decreased Level of Ceramides in Stratum Corneum of Atopic Dermatitis: An Etiologic Factor in Atopic Dry Skin? *J. Investig. Dermatol.* **1991**, *96*, 523–526. [CrossRef]

21. Van Smeden, J.; Janssens, M.; Kaye, E.C.J.; Caspers, P.J.; Lavrijsen, A.P.; Vreeken, R.J.; Bouwstra, J. The importance of free fatty acid chain length for the skin barrier function in atopic eczema patients. *Exp. Dermatol.* **2014**, *23*, 45–52. [CrossRef]

22. Agrawal, K.; Hassoun, L.A.; Foolad, N.; Pedersen, T.L.; Sivamani, R.K.; Newman, J.W. Sweat lipid mediator profiling: A noninvasive approach for cutaneous research. *J. Lipid Res.* **2016**, *58*, 188–195. [CrossRef]

23. Agrawal, K.; Hassoun, L.A.; Foolad, N.; Borkowski, K.; Pedersen, T.L.; Sivamani, R.; Newman, J.W. Effects of atopic dermatitis and gender on sebum lipid mediator and fatty acid profiles. *Prostaglandins Leukot. Essent. Fat. Acids* **2018**, *134*, 7–16. [CrossRef]

24. Schäfer, L.; Kragballe, K. Abnormalities in Epidermal Lipid Metabolism in Patients with Atopic Dermatitis. *J. Investig. Dermatol.* **1991**, *96*, 10–15. [CrossRef]

25. Newell, L.; Polak, M.E.; Perera, J.; Owen, C.; Boyd, P.; Pickard, C.; Howarth, P.H.; Healy, E.; Holloway, J.W.; Friedmann, P.S.; et al. Sensitization via Healthy Skin Programs Th2 Responses in Individuals with Atopic Dermatitis. *J. Investig. Dermatol.* **2013**, *133*, 2372–2380. [CrossRef]

26. Elias, P.M.; Schmuth, M. Abnormal skin barrier in the etiopathogenesis of atopic dermatitis. *Curr. Opin. Allergy Clin. Immunol.* **2009**, *9*, 437–446. [CrossRef]

27. Danso, M.O.; Van Drongelen, V.; Mulder, A.A.; Van Esch, J.; Scott, H.; Van Smeden, J.; El Ghalbzouri, A.; Bouwstra, J.A. TNF-α and Th2 Cytokines Induce Atopic Dermatitis–Like Features on Epidermal Differentiation Proteins and Stratum Corneum Lipids in Human Skin Equivalents. *J. Investig. Dermatol.* **2014**, *134*, 1941–1950. [CrossRef] [PubMed]

28. Danso, M.; Boiten, W.; Van Drongelen, V.; Meijling, K.G.; Gooris, G.; El Ghalbzouri, A.; Absalah, S.; Vreeken, R.J.; Kezic, S.; Van Smeden, J.; et al. Altered expression of epidermal lipid bio-synthesis enzymes in atopic dermatitis skin is accompanied by changes in stratum corneum lipid composition. *J. Dermatol. Sci.* **2017**, *88*, 57–66. [CrossRef] [PubMed]

29. HogenEsch, H.; Dunham, A.; Seymour, R.; Renninger, M.; Sundberg, J.P. Expression of chitinase-like proteins in the skin of chronic proliferative dermatitis (cpdm/cpdm) mice. *Exp. Dermatol.* **2006**, *15*, 808–814. [CrossRef]

30. HogenEsch, H.; Torregrosa, S.E.; Boggess, D.; Sundberg, B.A.; Carroll, J.; Sundberg, J.P. Increased expression of type 2 cytokines in chronic proliferative dermatitis (cpdm) mutant mice and resolution of inflammation following treatment with IL-12. *Eur. J. Immunol.* **2001**, *31*, 734–742. [CrossRef]

31. Franco, J.; Ferreira, C.; Sobreira, T.J.P.; Sundberg, J.P.; HogenEsch, H. Profiling of epidermal lipids in a mouse model of dermatitis: Identification of potential biomarkers. *PLoS ONE* **2018**, *13*, e0196595. [CrossRef] [PubMed]

32. Stephenson, D.J.; Hoeferlin, L.A.; Chalfant, C.E. Lipidomics in translational research and the clinical significance of lipid-based biomarkers. *Transl. Res.* **2017**, *189*, 13–29. [CrossRef] [PubMed]

33. Majd, T.M.; Kalantari, S.; Shahraki, H.R.; Nafar, M.; Almasi, A.; Samavat, S.; Parvin, M.; Hashemian, A. Application of Sparse Linear Discriminant Analysis and Elastic Net for Diagnosis of IgA Nephropathy: Statistical and Biological Viewpoints. *Iran. Biomed. J.* **2018**, *22*, 374–384. [CrossRef]

34. Kirpich, A.; Ainsworth, E.A.; Wedow, J.M.; Newman, J.; Michailidis, G.; McIntyre, L.M. Variable selection in omics data: A practical evaluation of small sample sizes. *PLoS ONE* **2018**, *13*, e0197910. [CrossRef] [PubMed]

35. Holzinger, A.; Langs, G.; Denk, H.; Zatloukal, K.; Müller, H. Causability and explainability of artificial intelligence in medicine. *Wiley Interdiscip. Rev. Data Min. Knowl. Discov.* **2019**, *9*, e1312. [CrossRef]

36. Basu, A.; Mitra, R.; Liu, H.; Schreiber, S.L.; Clemons, P.A. RWEN: Response-weighted elastic net for prediction of chemosensitivity of cancer cell lines. *Bioinformatics* **2018**, *34*, 3332–3339. [CrossRef] [PubMed]

37. Gonzales, G.B.; De Saeger, S. Elastic net regularized regression for time-series analysis of plasma metabolome stability under sub-optimal freezing condition. *Sci. Rep.* **2018**, *8*, 1–10. [CrossRef] [PubMed]

38. Bujak, R.; Daghir-Wojtkowiak, E.; Kaliszan, R.; Markuszewski, M.J. PLS-Based and Regularization-Based Methods for the Selection of Relevant Variables in Non-targeted Metabolomics Data. *Front. Mol. Biosci.* **2016**, *3*, 1–10. [CrossRef] [PubMed]

39. Furman, D.; Davis, M.M. New approaches to understanding the immune response to vaccination and infection. *Vaccine* **2015**, *33*, 5271–5281. [CrossRef]

40. Joo, K.-M.; Hwang, J.-H.; Bae, S.; Nahm, D.-H.; Park, H.; Ye, Y.-M.; Lim, K.-M. Relationship of ceramide–, and free fatty acid–cholesterol ratios in the stratum corneum with skin barrier function of normal, atopic dermatitis lesional and non-lesional skins. *J. Dermatol. Sci.* **2015**, *77*, 71–74. [CrossRef]

41. Agrawal, R.; Woodfolk, J.A. Skin barrier defects in atopic dermatitis. *Curr. Allergy Asthma Rep.* **2014**, *14*, 433. [CrossRef]

42. Loiseau, N.; Obata, Y.; Moradian, S.; Sano, H.; Yoshino, S.; Aburai, K.; Takayama, K.; Sakamoto, K.; Holleran, W.M.; Elias, P.M.; et al. Altered sphingoid base profiles predict compromised membrane structure and permeability in atopic dermatitis. *J. Dermatol. Sci.* **2013**, *72*, 296–303. [CrossRef]

43. Cotterill, J.A.; Cunliffe, W.J.; Williamson, B.; Bulusu, L. AGE AND SEX VARIATION IN SKIN SURFACE LIPID COMPOSITION AND SEBUM EXCRETION RATE. *Br. J. Dermatol.* **1972**, *87*, 333–340. [CrossRef]

44. Norheim, F.; Bjellaas, T.; Hui, S.T.; Krishnan, K.C.; Lee, J.; Gupta, S.; Pan, C.; Hasin-Brumshtein, Y.; Parks, B.W.; Li, D.; et al. Genetic, dietary, and sex-specific regulation of hepatic ceramides and the relationship between hepatic ceramides and IR. *J. Lipid Res.* **2018**, *59*, 1164–1174. [CrossRef]

45. Blaess, M.; Deigner, H.-P. Derailed Ceramide Metabolism in Atopic Dermatitis (AD): A Causal Starting Point for a Personalized (Basic) Therapy. *Int. J. Mol. Sci.* **2019**, *20*, 3967. [CrossRef]

46. Rabionet, M.; Gorgas, K.; Sandhoff, R. Ceramide synthesis in the epidermis. *Biochim. Biophys. Acta (BBA) Mol. Cell Boil. Lipids* **2014**, *1841*, 422–434. [CrossRef]

47. Borodzicz, S.; Rudnicka, L.; Mirowska-Guzel, D.; Cudnoch-Jędrzejewska, A. The role of epidermal sphingolipids in dermatologic diseases. *Lipids Heal. Dis.* **2016**, *15*, 13. [CrossRef]

48. Sacotte, R.; Silverberg, J.I. Epidemiology of adult atopic dermatitis. *Clin. Dermatol.* **2018**, *36*, 595–605. [CrossRef]

49. Elias, P.M. Lipid abnormalities and lipid-based repair strategies in atopic dermatitis. *Biochim. Biophys. Acta.* **2013**, *1841*, 323–330. [CrossRef]

50. Elias, P.M.; Wakefield, J.S.; Man, M.-Q. Moisturizers versus Current and Next-Generation Barrier Repair Therapy for the Management of Atopic Dermatitis. *Ski. Pharmacol. Physiol.* **2018**, *32*, 1–7. [CrossRef]

51. Di Nardo, A. Ceramide and cholesterol composition of the skin of patients with atopic dermatitis. *Acta Derm. Venereol.* **1998**, *78*, 27–30. [CrossRef]

52. Berdyshev, E.; Goleva, E.; Bronova, I.A.; Dyjack, N.; Rios, C.; Jung, J.; Taylor, P.; Jeong, M.; Hall, C.F.; Richers, B.N.; et al. Lipid abnormalities in atopic skin are driven by type 2 cytokines. *JCI Insight* **2018**, *3*, 1–15. [CrossRef]

53. Fantini, J.; Yahi, N. Chemical Basis of Lipid Biochemistry. In *Brain Lipids in Synaptic Function and Neurological Disease*; Elsevier: Amsterdam, The Netherlands, 2015; pp. 1–28. ISBN 978-0-12-800111-0.

54. Radner, F.; Fischer, J. The important role of epidermal triacylglycerol metabolism for maintenance of the skin permeability barrier function. *Biochim. Biophys. Acta (BBA) Mol. Cell Boil. Lipids* **2014**, *1841*, 409–415. [CrossRef]

55. Han, X. Lipidomics for studying metabolism. *Nat. Rev. Endocrinol.* **2016**, *12*, 668–679. [CrossRef] [PubMed]

56. Piotto, S.; Trapani, A.; Bianchino, E.; Ibarguren, M.; Lopez, D.J.; Busquets, X.; Concilio, S. The effect of hydroxylated fatty acid-containing phospholipids in the remodeling of lipid membranes. *Biochim. Biophys. Acta (BBA) Biomembr.* **2014**, *1838*, 1509–1517. [CrossRef]

57. Ottas, A.; Fishman, D.; Okas, T.-L.; Püssa, T.; Toomik, P.; Märtson, A.; Kingo, K.; Soomets, U. Blood serum metabolome of atopic dermatitis: Altered energy cycle and the markers of systemic inflammation. *PLoS ONE* **2017**, *12*, e0188580. [CrossRef]

58. Watanabe, N.; Suzuki, T.; Yamazaki, Y.; Sugiyama, K.; Koike, S.; Nishimukai, M. Supplemental feeding of phospholipid-enriched alkyl phospholipid from krill relieves spontaneous atopic dermatitis and strengthens skin intercellular lipid barriers in NC/Nga mice. *Biosci. Biotechnol. Biochem.* **2019**, *83*, 717–727. [CrossRef] [PubMed]

59. Mårtensson, C.U.; Doan, K.N.; Becker, T. Effects of lipids on mitochondrial functions. *Biochim. Biophys. Acta (BBA) Mol. Cell Boil. Lipids* **2017**, *1862*, 102–113. [CrossRef]

60. Chegary, M.; Brinke, H.T.; Ruiter, J.P.; Wijburg, F.A.; Stoll, M.S.; Minkler, P.E.; Van Weeghel, M.; Schulz, H.; Hoppel, C.L.; Wanders, R.J.; et al. Mitochondrial long chain fatty acid β-oxidation in man and mouse. *Biochim. Biophys. Acta (BBA) Mol. Cell Boil. Lipids* **2009**, *1791*, 806–815. [CrossRef] [PubMed]

61. Rutkowsky, J.M.; Knotts, T.A.; Ono-Moore, K.D.; McCoin, C.S.; Huang, S.; Schneider, D.; Singh, S.; Adams, S.H.; Hwang, D.H. Acylcarnitines activate proinflammatory signaling pathways. *Am. J. Physiol. Metab.* **2014**, *306*, E1378–E1387. [CrossRef]

62. Choi, S.-H.; Sviridov, D.; Miller, Y.I. Oxidized cholesteryl esters and inflammation. *Biochim. Biophys. Acta (BBA) Mol. Cell Boil. Lipids* **2017**, *1862*, 393–397. [CrossRef]

63. Korber, M.; Klein, I.; Daum, G. Steryl ester synthesis, storage and hydrolysis: A contribution to sterol homeostasis. *Biochim. Biophys. Acta (BBA) Mol. Cell Boil. Lipids* **2017**, *1862*, 1534–1545. [CrossRef] [PubMed]

64. Tachi, M.; Iwamori, M. Mass spectrometric characterization of cholesterol esters and wax esters in epidermis of fetal, adult and keloidal human skin. *Exp. Dermatol.* **2008**, *17*, 318–323. [CrossRef]

65. Trieb, M.; Wolf, P.; Knuplez, E.; Weger, W.; Schuster, C.; Peinhaupt, M.; Holzer, M.; Trakaki, A.; Eichmann, T.; Lass, A.; et al. Abnormal composition and function of high-density lipoproteins in atopic dermatitis patients. *Allergy* **2019**, *74*, 398–402. [CrossRef] [PubMed]

66. Gerl, M.J.; Vaz, W.L.C.; Domingues, N.; Klose, C.; Surma, M.A.; Sampaio, J.L.; Almeida, M.S.; Rodrigues, G.; Araújo-Gonçalves, P.; Ferreira, J.; et al. Cholesterol is Inefficiently Converted to Cholesteryl Esters in the Blood of Cardiovascular Disease Patients. *Sci. Rep.* **2018**, *8*, 14764. [CrossRef] [PubMed]

67. Marsella, R.; Olivry, T.; Carlotti, D.-N. For the International Task Force on Canine Atopic Dermatitis Current evidence of skin barrier dysfunction in human and canine atopic dermatitis. *Veter Dermatol.* **2011**, *22*, 239–248. [CrossRef] [PubMed]

68. Tončić, R.J.; Jakasa, I.; Hadžavdić, S.L.; Goorden, S.M.; Der Vlugt, K.J.G.-V.; Stet, F.S.; Balic, A.; Petkovic, M.; Pavicic, B.; Žužul, K.; et al. Altered Levels of Sphingosine, Sphinganine and Their Ceramides in Atopic Dermatitis Are Related to Skin Barrier Function, Disease Severity and Local Cytokine Milieu. *Int. J. Mol. Sci.* **2020**, *21*, 1958. [CrossRef]

69. Blagus, R.; Lusa, L. Class prediction for high-dimensional class-imbalanced data. *BMC Bioinform.* **2010**, *11*, 523. [CrossRef] [PubMed]

70. Bligh, E.G.; Dyer, W.J. A rapid method of total lipid extraction and purification. *Can. J. Biochem. Physiol.* **1959**, *37*, 911–917. [CrossRef] [PubMed]

71. Ferreira, C.R.; Yannell, K.E.; Mollenhauer, B.; Espy, R.D.; Cordeiro, F.B.; Ouyang, Z.; Cooks, R.G. Chemical profiling of cerebrospinal fluid by multiple reaction monitoring mass spectrometry. *Anal.* **2016**, *141*, 5252–5255. [CrossRef] [PubMed]

72. Cordeiro, F.; Ferreira, C.R.; Sobreira, T.; Yannell, K.E.; Jarmusch, A.K.; Cedenho, A.P.; Turco, E.G.L.; Cooks, R.G. Multiple reaction monitoring (MRM)-profiling for biomarker discovery applied to human polycystic ovarian syndrome. *Rapid Commun. Mass Spectrom.* **2017**, *31*, 1462–1470. [CrossRef]

73. Xie, Z.; Gonzalez, L.E.; Ferreira, C.R.; Vorsilak, A.; Frabutt, D.; Sobreira, T.J.P.; Pugia, M.; Cooks, R.G. Multiple Reaction Monitoring Profiling (MRM-Profiling) of Lipids To Distinguish Strain-Level Differences in Microbial Resistance in Escherichia coli. *Anal. Chem.* **2019**, *91*, 11349–11354. [CrossRef]

74. Dipali, S.S.; Ferreira, C.R.; Zhou, L.T.; Pritchard, M.T.; Duncan, F.E. Histologic analysis and lipid profiling reveal reproductive age-associated changes in peri-ovarian adipose tissue. *Reprod. Boil. Endocrinol.* **2019**, *17*, 46. [CrossRef]

75. Yannell, K.E.; Ferreira, C.R.; Tichy, S.E.; Cooks, R.G. Multiple reaction monitoring (MRM)-profiling with biomarker identification by LC-QTOF to characterize coronary artery disease. *Analyst* **2018**, *143*, 5014–5022. [CrossRef]

76. Filzmoser, P.; Hron, K.; Reimann, C. Principal component analysis for compositional data with outliers. *Environmetrics* **2009**, *20*, 621–632. [CrossRef]

77. Benjamini, Y.; Hochberg, Y. Controlling the False Discovery Rate: A Practical and Powerful Approach to Multiple Testing. *J. R. Stat. Soc. Ser. B Stat. Methodol.* **1995**, *57*, 289–300. [CrossRef]

78. Zou, H.; Hastie, T. Regularization and variable selection via the elastic net. *J. R. Stat. Soc. Ser. B Stat. Methodol.* **2005**, *67*, 301–320. [CrossRef]

79. Chawla, N.; Bowyer, K.W.; Hall, L.O.; Kegelmeyer, W.P. SMOTE: Synthetic Minority Over-sampling Technique. *J. Artif. Intell. Res.* **2002**, *16*, 321–357. [CrossRef]

 *metabolites*

Article

LC–MS Lipidomics: Exploiting a Simple High-Throughput Method for the Comprehensive Extraction of Lipids in a Ruminant Fat Dose-Response Study

Benjamin Jenkins [1], Martin Ronis [2] and Albert Koulman [1,*]

[1] NIHR BRC Core Metabolomics and Lipidomics Laboratory, University of Cambridge, Pathology Building Level 4, Addenbrooke's Hospital, Cambridge CB2 0QQ, UK; bjj25@medschl.cam.ac.uk

[2] College of Medicine, Department of Pharmacology & Experimental Therapeutics, Louisiana State University Health Sciences Centre, 1901 Perdido Str., New Orleans, LA 70112, USA; mronis@lsuhsc.edu

*   Correspondence: AK675@medschl.cam.ac.uk

Received: 22 June 2020; Accepted: 16 July 2020; Published: 17 July 2020

Abstract: Typical lipidomics methods incorporate a liquid–liquid extraction with LC–MS quantitation; however, the classic sample extraction methods are not high-throughput and do not perform well at extracting the full range of lipids especially, the relatively polar species (e.g., acyl-carnitines and glycosphingolipids). In this manuscript, we present a novel sample extraction protocol, which produces a single phase supernatant suitable for high-throughput applications that offers greater performance in extracting lipids across the full spectrum of species. We applied this lipidomics pipeline to a ruminant fat dose–response study to initially compare and validate the different extraction protocols but also to investigate complex lipid biomarkers of ruminant fat intake (adjoining onto simple odd chain fatty acid correlations). We have found 100 lipids species with a strong correlation with ruminant fat intake. This novel sample extraction along with the LC–MS pipeline have shown to be sensitive, robust and hugely informative (>450 lipids species semi-quantified): with a sample preparation throughput of over 100 tissue samples per day and an estimated ~1000 biological fluid samples per day. Thus, this work facilitating both the epidemiological involvement of ruminant fat, research into odd chain lipids and also streamlining the field of lipidomics (both by sample preparation methods and data presentation).

Keywords: odd chain lipids; lipid profiling; Folch; protein precipitation; sample preparation; relative lipid composition (Mol%)

---

1. Introduction

Lipids are generally understood as a class of molecules that have a high solubility in organic solvents and typically contain or originate from fatty acids. Although, lipids may be commonly derived, research has shown that there is a huge variety both structurally and functionally (potentially >40,000 [1]); where they play a vital role in energy production and storage [2,3], regulation and signalling [4,5], provide structure and support and membrane formation [6]. Lipids are now emerging as biomarkers of dietary/nutritional intakes [7] as well as indicators of pathophysiological status [8–11]. As a reaction of lipid-pathophysiological involvements, the field of lipidomics has emerged as a discipline that examines and quantifies a large proportion of the lipids present in a given sample set.

Lipidomics requires an effective isolation protocol that comprehensively extracts lipids from the sample as well as an analytical method that allows their identification and quantitation. The typical analyte isolation protocols (with/without minor adaptations) that are often used in the literature include three different liquid–liquid extractions: Folch and colleagues [12] (cited >65,000 times), Bligh and

Dyer [13] (cited >52,000 times) or Matyash and colleagues [14] (cited >1000 times). Although these extraction protocols are heavily cited and do result in adequate results, there are several caveats with their use. Firstly, there is the need to perform duplicate extraction in non-fluid samples to ensure optimal recovery of the lipid analytes. This is extremely time consuming, especially for the Folch and the Bligh and Dyer methods. Secondly, there are reasonable concerns that using a biphasic extraction (producing immiscible aqueous and organic phases) may result in a loss of relatively polar lipids (e.g., acyl-carnitines and gangliosides) into the disposed aqueous fractions (consisting of mostly methanol and water in these extraction protocols: Folch and the Bligh and Dyer). There are publications that use a single phase extraction protocol but they do not appear to solve the problem of extracting the relatively more polar lipids since a mixture of methanol, chloroform and tert-butyl methyl ether were used [15,16].

The technique overwhelmingly used tor analysing the lipidome is mass spectrometry hyphenated with chromatography (LC–MS) due to its sensitivity and selectivity; furthermore by using a high-resolution accurate mass instrument (e.g., Orbitrap or Time-of-Flight instruments), a huge number of analytes can be analysed simultaneously. Reversed phase chromatography is the predominant chromatographic technique employed to separate the analytes before entering the mass spectrometer to determine their structure and concentration. Variants of a liquid chromatography method utilising a C18-column with a water and acetonitrile mix for the weak eluting mobile phase and acetonitrile and propan-2-ol for the strong eluting mobile phase are the most commonly used [17–22]. These reversed phase C18-column methods both separate lipid based on their lipid class assignment (i.e., either phosphatidylcholines or phosphatidyethanolamines head group) and their fatty-acyl composition (i.e., chain length and degree of unsaturation) with some degree of isomeric separation.

In this study, we present a lipidomics pipeline that including a novel analyte isolation protocol utilising a single phase, which results in a comprehensive lipid extraction suitable for a full range of lipid polarities (from polar to non-polar lipids species). This lipidomics method was tested, validated and then applied in a rat model investigating ruminant fat biomarkers via a beef tallow dose response dietary investigation.

2. Results

This lipidomics LC–MS method incorporating both of the described sample preparation protocols: protein precipitation (chloroform: methanol: acetone, ~7:3:4) and Folch liquid–liquid (chloroform: methanol: water, ~7:3:4), were utilised for the quantitation of lipids in liver samples from Sprague–Dawley rats who received one of four experimental diets overfed at 17% above matched growth.

A comparison of the two sample preparation methods on the extraction of the stable isotope-labelled internal standards are shown in the figure below (see Figure 1). A comparison on the samples' endogenous individual lipid classes are shown in the Supplementary Materials (see Supplementary Figure S1 and Table S1).

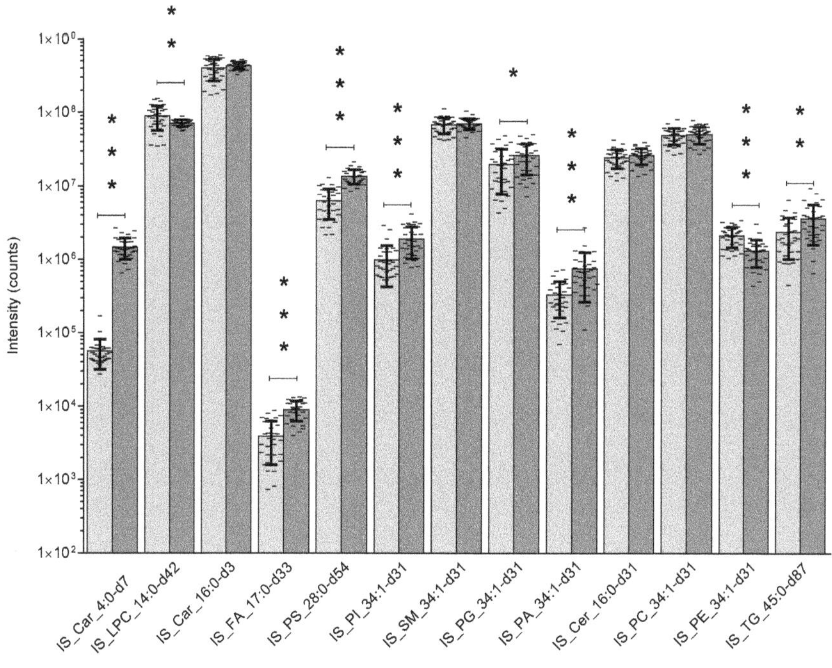

Figure 1. This figure shows the comparison between the two lipid extraction techniques regarding their extraction efficiency of the stable isotope internal standards from the rat liver samples (Folch liquid–liquid extraction: chloroform: methanol: water, ~7:3:4 ▢ , and Protein precipitation liquid extraction: chloroform: methanol: acetone, ~7:3:4 ▨). n = 34 rat liver samples per extraction method. The intensity of the internal standards were measured by liquid chromatography with mass spectrometry. The significance of the difference between the two extraction protocols are shown by the p-value star system; where $p \leq 0.05$ was considered statistically significant (* $p < 0.05$, ** $p < 0.01$, *** $p < 0.001$). Error bars represent ± standard deviation. Lipid internal standard include: Butyryl-d7-L-carnitine (abbreviated to IS_Car_4:0-d7), N-tetradecylphosphocholine-d42 (abbreviated to IS_LPC_14:0-d42), hexadecanoyl-L-carnitine-d3 (abbreviated to IS_Car_16:0-d3), heptadecanoic-d33 acid (abbreviated to IS_FA_17:0-d33), 1,2-dimyristoyl-d54-sn-glycero-3-[phospho-L-serine] (abbreviated to IS_PS_28:0-d54), 1-palmitoyl-d31-2-oleoyl-sn-glycero-3-phosphoinositol (abbreviated to IS_PI_34:1-d31), N-palmitoyl-d31-D-erythro-sphingosylphosphorylcholine (abbreviated to IS_SM_34:1-d31), 1-palmitoyl-d31-2-oleoyl-sn-glycero-3-[phospho-rac-(1-glycerol)] (abbreviated to IS_PG_34:1-d31), 1-palmitoyl-d31-2-oleoyl-sn-glycero-3-phosphate (abbreviated to IS_PA_34:1-d31), N-palmitoyl-d31-D-erythro-sphingosine (abbreviated to IS_Cer_16:0-d31), 1-palmitoyl-d31-2-oleoyl-sn-glycero-3-phosphocholine (abbreviated to IS_PC_34:1-d31), 1-palmitoyl-d31-2-oleoyl-sn-glycero-3-phosphoethanolamine (abbreviated to IS_PE_34:1-d31), glyceryl tri(pentadecanoate-d29) (abbreviated to IS_TG_45:0-d87).

As shown the intensity of seven of the internal standards are statistically significantly higher in the protein precipitation protocol when compared to the Folch liquid–liquid protocol (between ~30% to ~2500% higher), whereas, only two of the internal standards were higher in the Folch liquid–liquid protocol (between ~20% and ~37% higher). Additionally, there is far less variation in the protein precipitation liquid extraction protocol: 10 out of 13 internal standards had ~2% to ~72% less variation in their coefficient of variation (CV). The protein precipitation liquid extraction protocol has also been shown to produce a significantly higher detection of the samples' endogenous lipids, both producing a higher total number of lipid detected (Folch-LLE: 455 lipid species, PPLE: 472 lipid species) and a statistically significantly higher total intensity for twelve of the sixteen lipid classes detected (see Supplementary Figure S1 and Table S1). Taken as a whole, the protein precipitation protocol (chloroform: methanol: acetone, ~7:3:4) showed a greater extraction capability across the full lipid hydro-philicity/phobicity range and across the internal standards. Additionally, the high throughput of the protein precipitation protocol allows for over 100 tissue samples to be extracted per day (including dissection, weighing and tissue extraction ready for LC–MS analysis), whereas, the Folch liquid–liquid protocol could take up to three- to four-times longer due to the necessity of duplicate extractions and the delicacy of liquid–liquid phase separation. The throughput of the protein precipitation protocol (chloroform: methanol: acetone, ~7:3:4) on fluid samples allows an estimated ~1000 biological fluid sample extractions per day (including aliquoting, sample extraction ready for LC–MS analysis) when utilising basic laboratory fluid handling equipment/robots (throughput data not shown here).

The liver lipid concentration (nM/mg) for each experimental diet group of rats are shown in the table below (see Table 1), along with the correlation (trendline equation, slope significance, R^2 and successive change across the groups) of the measured lipid concentration with the percentage composition of ruminant fat (beef tallow) in each experimental diet. An R^2 threshold of 0.75, slope significance p-value < 0.05 and successive increase/decrease were set to establish if there was a strong correlation between the lipid concentration and the ruminant fat composition.

Table 1. This table shows the liver lipid concentrations from the Sprague–Dawley rats who received one of four experimental diets (n = 8–9 per group). Diet 1: 50% corn oil, 16.4% MCT oil and 3.6% beef tallow; Diet 2: 35% corn oil, 28.7% MCT oil and 6.3% beef tallow; Diet 3: 20% corn oil, 41.0% MCT oil and 9.0% beef tallow; Diet 4: 5% corn oil, 53.3% MCT oil and 11.7% beef tallow. MCT: medium chain triglyceride oil. Lipid are shown in their shorthand notations with the number of carbons and unsaturated bonds in the fatty acid moiety separated by a colon; acyl-carnitines (Carn), ceramides (Cer), cardiolipins (CL), diacylglycerols (DG), gangliosides (GM1), hexosylceramides (Hex-Cer), lyso-phosphatidylcholines (LPC), lyso-phosphatidyethanolamines (LPE), lyso-phosphatidylinositols (LPI), lyso-cardiolipins (Lyso_CL), phosphatidic acids (PA), phosphatidylcholines (PC), phosphatidylethanolamines (PE), phosphatidylglycerol (PG), phosphatidylinositols (PI), phosphatidylserines (PS), sulfatides (S), sphingomyelins (SM), triacylglycerides (TG). Lipid concentrations (nM/mg) are shown as mean ± standard deviation and were extracted via the protein precipitation liquid extraction protocol (chloroform: methanol: acetone, ~7:3:4). Correlation between the ruminant fat composition of the diet and the lipid concentration are depicted by the trendline equation (* denotes statistical significance of the slope: p-value < 0.05) and coefficient of determination (R^2). Lipid concentrations continually increasing/decreasing across the groups as the ruminant fat composition of the diet increase are emphasised in the 'Successive change across group' column. Lipids with a significant slope (p-value < 0.05) with an R^2 greater than 0.75 and successively increasing/decreasing are in bold and highlighted.

	Diet 1	Diet 2	Diet 3	Diet 4	Trendline Equation	R^2	Successive Change Across Groups
Carn_(C00:0)	99,100 ± 19,200	90,200 ± 38,800	126,000 ± 29,000	92,400 ± 49,600	y = 581x + 97,500	0.02	
Carn_(C02:0)	10,400 ± 4150	10,700 ± 3280	12,900 ± 6740	11,700 ± 5300	y = 226x + 9700	0.49	
Carn_(C03:0)	**3430 ± 1380**	**3590 ± 1820**	**3760 ± 2170**	**3830 ± 2240**	**y = 50.7x + 3260 ***	**0.97**	**Increasing**
Carn_(C03:0-2COOH)	713 ± 326	436 ± 299	943 ± 438	700 ± 604	y = 17.3x + 565	0.08	
Carn_(C03:0-OH)	ND ± ND	1.46 ± 4.39	2.1 ± 5.94	6.16 ± 13	y = 0.87x − 4.59	0.85	
Carn_(C03:1)	108 ± 64.2	105 ± 58.2	101 ± 75.6	75.5 ± 35.3	y = −3.76x + 126	0.78	Decreasing
Carn_(C04:0)	1040 ± 847	675 ± 342	1560 ± 659	793 ± 309	y = 5.33x + 976	0.00	
Carn_(C04:0-OH)	952 ± 429	702 ± 503	1730 ± 944	1230 ± 619	y = 69x + 626	0.30	
Carn_(C04:1)	134 ± 56	102 ± 94.4	89.6 ± 63.3	104 ± 78.4	y = −3.79x + 136	0.49	
Carn_(C05:0)	490 ± 534	359 ± 173	857 ± 577	553 ± 437	y = 25.4x + 370	0.18	
Carn_(C05:1)	319 ± 125	255 ± 87.3	317 ± 205	227 ± 99.5	y = −7.93x + 340	0.36	
Carn_(C06:0)	324 ± 287	183 ± 61.5	635 ± 296	234 ± 123	y = 6.74x + 292	0.01	
Carn_(C06:0-2COOH)	511 ± 145	362 ± 124	742 ± 246	477 ± 213	y = 10.3x + 444	0.05	
Carn_(C08:0)	168 ± 174	107 ± 66.6	412 ± 185	186 ± 215	y = 13.3x + 117	0.12	
Carn_(C08:1)	81.7 ± 50.9	102 ± 56.5	124 ± 54.8	88.3 ± 62.7	y = 1.55x + 87.2	0.08	
Carn_(C10:0-2COOH)	353 ± 178	361 ± 124	378 ± 146	296 ± 177	y = −5.7x + 391	0.31	
Carn_(C12:0)	0.265 ± 0.339	0.147 ± 0.353	0.495 ± 0.291	0.412 ± 0.402	y = 0.0292x + 0.106	0.43	
Carn_(C14:0)	11.1 ± 6.52	9.39 ± 8.48	14.8 ± 5.36	13 ± 5.69	y = 0.411x + 8.92	0.38	
Carn_(C15:0)	4.33 ± 2.64	3.76 ± 2.87	4.88 ± 1.52	3.65 ± 0.927	y = −0.0341x + 4.42	0.04	
Carn_(C16:0)	618 ± 339	457 ± 315	739 ± 411	462 ± 226	y = −6.89x + 622	0.03	
Carn_(C16:0-OH)	1.2 ± 1.56	1.52 ± 1.98	1.61 ± 1.35	2.46 ± 2.18	y = 0.143x + 0.601	0.86	Increasing
Carn_(C16:2)	9.89 ± 7.09	9.69 ± 4.49	7.61 ± 4.01	10.9 ± 4.92	y = 0.0352x + 9.25	0.01	
Carn_(C17:0)	19.2 ± 8.93	13.9 ± 6.7	22.3 ± 9.04	18.4 ± 7.41	y = 0.222x + 16.8	0.05	
Carn_(C18:0)	627 ± 220	541 ± 170	656 ± 154	620 ± 221	y = 3.48x + 584	0.06	
Carn_(C18:0-OH)	5.47 ± 3.31	6.8 ± 3.5	5.41 ± 4.18	8.14 ± 4.52	y = 0.245x + 4.58	0.44	

Table 1. *Cont.*

	Diet 1	Diet 2	Diet 3	Diet 4	Trendline Equation	R^2	Successive Change Across Groups
Carn_(C18:1)	1530 ± 1090	1230 ± 931	1900 ± 958	1130 ± 449	$y = -19.6x + 1600$	0.04	
Carn_(C18:2)	824 ± 754	631 ± 525	374 ± 157	505 ± 259	$y = -45x + 927$	0.67	
Carn_(C18:3)	16.8 ± 16.4	14.7 ± 13.3	11.5 ± 6.38	14.8 ± 5.16	$y = -0.341x + 17.1$	0.29	
Carn_(C20:0)	43.1 ± 38.9	41.9 ± 28.2	32.5 ± 20.4	45.9 ± 22.9	$y = -0.037x + 41.1$	0.00	
Carn_(C22:5)	**2.6 ± 3.13**	**2.34 ± 2.81**	**1.7 ± 1.25**	**1.66 ± 1.06**	**$y = -0.128x + 3.06$ ***	**0.91**	**Decreasing**
Cer_(32:1)	5.17 ± 1.26	4.31 ± 0.783	5.34 ± 1.78	5.4 ± 1.75	$y = 0.0637x + 4.57$	0.19	
Cer_(33:1)	**2.29 ± 1.92**	**1.68 ± 1.26**	**1.48 ± 0.544**	**1.27 ± 0.925**	**$y = -0.121x + 2.6$ ***	**0.92**	**Decreasing**
Cer_(34:0)	0.918 ± 0.57	1.19 ± 1.13	2.53 ± 2.19	1.93 ± 1.51	$y = 0.162x + 0.402$	0.60	
Cer_(34:1)	1.49 ± 0.581	1.23 ± 0.881	1.79 ± 1.01	2.56 ± 0.989	$y = 0.14x + 0.699$	0.71	
Cer_(35:0)	ND ± ND	ND ± ND	0.0674 ± 0.191	ND ± ND			
Cer_(35:1)	**1.03 ± 1.06**	**1.61 ± 1.06**	**1.63 ± 1.08**	**2.16 ± 1.45**	**$y = 0.126x + 0.641$ ***	**0.91**	**Increasing**
Cer_(36:0)	**7.26 ± 2.79**	**9.98 ± 5.3**	**10.5 ± 4.07**	**13.1 ± 2.89**	**$y = 0.668x + 5.1$ ***	**0.95**	**Increasing**
Cer_(36:1)	54.5 ± 26.2	55.4 ± 10.4	59.7 ± 16.3	71.7 ± 24.5	$y = 2.07x + 44.5$	0.83	Increasing
Cer_(36:2)	6.59 ± 3.55	5.23 ± 1.18	5.94 ± 2.23	6.81 ± 3.13	$y = 0.0507x + 5.75$	0.06	
Cer_(37:1)	2.23 ± 0.767	1.9 ± 0.723	2.31 ± 0.772	4.23 ± 1.09	$y = 0.237x + 0.851$	0.61	
Cer_(37:2)	ND ± ND	ND ± ND	0.066 ± 0.187	ND ± ND			
Cer_(38:0)	2.14 ± 1.19	3.82 ± 3.78	2.1 ± 0.715	2.9 ± 0.626	$y = 0.0207x + 2.58$	0.01	
Cer_(38:1)	47.9 ± 18.8	34.1 ± 7.21	39.5 ± 10.3	56 ± 13	$y = 1.1x + 36$	0.16	
Cer_(38:2)	6.76 ± 3.87	4.76 ± 1.57	4.7 ± 0.967	8.02 ± 1.98	$y = 0.138x + 5.01$	0.09	
Cer_(39:0)	0.188 ± 0.0814	0.221 ± 0.149	0.243 ± 0.153	0.724 ± 0.454	$y = 0.0604x - 0.118$	0.68	Increasing
Cer_(39:1)	10.5 ± 3.44	8.36 ± 1.6	10.6 ± 1.62	20.3 ± 3.17	$y = 1.17x + 3.48$	0.58	
Cer_(39:2)	0.436 ± 0.447	0.332 ± 0.333	0.416 ± 0.342	1.65 ± 0.755	$y = 0.138x - 0.347$	0.58	
Cer_(40:0)	**3.71 ± 0.85**	**4.12 ± 0.933**	**4.74 ± 1.44**	**6.17 ± 0.98**	**$y = 0.296x + 2.42$ ***	**0.92**	**Increasing**
Cer_(40:1)	126 ± 37.9	116 ± 18.7	146 ± 51	227 ± 39.1	$y = 12.3x + 59.4$	0.73	
Cer_(40:2)	65.5 ± 24.7	50.8 ± 8.2	63 ± 13.4	105 ± 12.7	$y = 4.84x + 34$	0.52	
Cer_(41:0)	**2.95 ± 0.547**	**3.91 ± 1.03**	**4.94 ± 1.2**	**7.37 ± 1.36**	**$y = 0.529x + 0.744$ ***	**0.94**	**Increasing**
Cer_(41:1)	**72.1 ± 18.6**	**89.7 ± 20.4**	**130 ± 88.3**	**210 ± 48.4**	**$y = 16.8x - 3.18$ ***	**0.91**	**Increasing**
Cer_(41:2)	51.4 ± 19.1	49.6 ± 11.4	63.3 ± 6.75	144 ± 22.5	$y = 10.8x - 5.52$	0.70	
Cer_(42:0)	**6.21 ± 2.01**	**8.11 ± 2.54**	**11.5 ± 5.17**	**12.2 ± 3.76**	**$y = 0.791x + 3.45$ ***	**0.95**	**Increasing**
Cer_(42:1)	**310 ± 63.3**	**444 ± 130**	**624 ± 539**	**773 ± 305**	**$y = 58.1x + 93.2$ ***	**1.00**	**Increasing**
Cer_(42:2)	575 ± 161	599 ± 87.4	721 ± 120	1180 ± 187	$y = 71.7x + 220$	0.79	Increasing
Cer_(42:3)	105 ± 40.2	92 ± 19.5	103 ± 26.1	151 ± 16.5	$y = 5.52x + 70.5$	0.54	
Cer_(43:0)	**1.31 ± 0.425**	**2.05 ± 0.699**	**2.43 ± 0.806**	**3.07 ± 0.982**	**$y = 0.21x + 0.611$ ***	**0.99**	**Increasing**
Cer_(43:1)	**131 ± 31.2**	**218 ± 62.9**	**253 ± 87.4**	**396 ± 111**	**$y = 30.7x + 14.3$ ***	**0.94**	**Increasing**
Cer_(43:2)	42.9 ± 7.62	58.9 ± 11.9	64.2 ± 12.8	115 ± 17.2	$y = 8.21x + 7.46$	0.84	Increasing
Cer_(44:1)	**16.7 ± 4.31**	**28.5 ± 8.34**	**38.9 ± 26.1**	**43 ± 14.8**	**$y = 3.31x + 6.47$ ***	**0.96**	**Increasing**
Cer_(44:2)	**0.908 ± 0.271**	**1.19 ± 0.288**	**1.72 ± 1.08**	**2.08 ± 0.644**	**$y = 0.15x + 0.328$ ***	**0.99**	**Increasing**
Cer_(45:1)	0.46 ± 0.215	1.19 ± 0.531	1.32 ± 0.618	1.58 ± 0.746	$y = 0.129x + 0.149$	0.88	Increasing

Table 1. *Cont.*

	Diet 1	Diet 2	Diet 3	Diet 4	Trendline Equation	R^2	Successive Change Across Groups
Cer_(45:2)	ND ± ND	0.00348 ± 0.0104	0.00503 ± 0.0142	ND ± ND	$y = 0.000574x - 0.000137$	1.00	
Cer_(46:1)	**0.0023 ± 0.0069**	**0.0192 ± 0.0266**	**0.0409 ± 0.0568**	**0.0568 ± 0.0824**	$\mathbf{y = 0.00686x - 0.0227}$ *	**1.00**	**Increasing**
Cer_(46:2)	0.83 ± 0.187	1.53 ± 0.439	1.57 ± 0.516	1.79 ± 0.679	$y = 0.108x + 0.603$	0.82	Increasing
CL_(66:02)	**8.65 ± 6.02**	**22.3 ± 17.1**	**32.3 ± 17.1**	**37.7 ± 31.3**	$\mathbf{y = 3.6x - 2.29}$ *	**0.97**	**Increasing**
CL_(66:03)	15.8 ± 10.1	23.6 ± 16.4	24.2 ± 13.2	17.3 ± 8.8	$y = 0.189x + 18.8$	0.02	
CL_(66:04)	15.5 ± 10.7	21.8 ± 10.4	29.7 ± 16.7	27 ± 15.3	$y = 1.57x + 11.5$	0.76	
CL_(66:05)	**40.5 ± 20**	**39.1 ± 10.7**	**18.5 ± 10.3**	**1 ± 0.911**	$\mathbf{y = -5.15x + 64.2}$ *	**0.92**	**Decreasing**
CL_(66:06)	66.1 ± 37	67.9 ± 26.4	65.1 ± 38.1	22.6 ± 13.2	$y = -4.94x + 93.2$	0.62	
CL_(67:02)	4.6 ± 5.88	15.8 ± 13.3	11.3 ± 6.02	10.1 ± 10	$y = 0.444x + 7.05$	0.11	
CL_(67:03)	13.4 ± 9.85	22.4 ± 14	14.1 ± 6.36	5.97 ± 4.69	$y = -1.13x + 22.6$	0.35	
CL_(67:05)	**0.848 ± 1.74**	**1.4 ± 2.57**	**2.58 ± 2.02**	**3.16 ± 1.62**	$\mathbf{y = 0.301x - 0.303}$ *	**0.98**	**Increasing**
CL_(68:00)	0.484 ± 0.502	0.898 ± 1.1	0.998 ± 1.12	0.519 ± 1.01	$y = 0.00759x + 0.667$	0.01	
CL_(68:01)	105 ± 91.1	260 ± 222	190 ± 107	142 ± 163	$y = 1.52x + 163$	0.01	
CL_(68:02)	469 ± 385	1010 ± 799	664 ± 355	434 ± 480	$y = -16.7x + 772$	0.05	
CL_(68:03)	631 ± 421	1030 ± 670	560 ± 261	237 ± 199	$y = -61.2x + 1080$	0.43	
CL_(68:04)	367 ± 179	439 ± 197	294 ± 113	184 ± 112	$y = -25.7x + 518$	0.68	
CL_(69:04)	40.4 ± 31.9	66.9 ± 41.7	56.3 ± 27.3	28.8 ± 18.6	$y = -1.68x + 61$	0.12	
CL_(69:05)	64.8 ± 37.5	67.7 ± 29.7	51.2 ± 20.4	31.1 ± 15.2	$y = -4.36x + 87$	0.83	
CL_(69:06)	49.7 ± 32.3	49.7 ± 20.5	29.3 ± 11	25.1 ± 10.8	$y = -3.49x + 65.1$	0.86	
CL_(69:07)	0.245 ± 0.736	1.06 ± 0.942	1.33 ± 1.17	0.685 ± 0.79	$y = 0.0589x + 0.38$	0.19	
CL_(70:01)	1.45 ± 1.87	6.38 ± 7.09	4.32 ± 2.94	3.91 ± 5.53	$y = 0.197x + 2.51$	0.12	
CL_(70:02)	61.2 ± 59.7	102 ± 90.1	54.6 ± 31.2	33.4 ± 40.7	$y = -4.84x + 99.9$	0.35	
CL_(70:03)	470 ± 379	796 ± 600	559 ± 271	340 ± 303	$y = -23.2x + 719$	0.18	
CL_(70:04)	1270 ± 814	2010 ± 1300	1510 ± 656	950 ± 761	$y = -54.1x + 1850$	0.18	
CL_(70:05)	1770 ± 889	2210 ± 1060	1670 ± 641	1200 ± 780	$y = -83.3x + 2350$	0.49	
CL_(70:06)	1280 ± 629	1570 ± 595	1560 ± 688	1440 ± 897	$y = 17.4x + 1330$	0.20	
CL_(70:07)	882 ± 480	1110 ± 399	1210 ± 657	971 ± 551	$y = 13.6x + 939$	0.11	
CL_(70:08)	**256 ± 199**	**245 ± 108**	**147 ± 56.7**	**89.7 ± 52.6**	$\mathbf{y = -22.1x + 354}$ *	**0.93**	**Decreasing**
CL_(70:09)	25.8 ± 18.3	27.1 ± 13.6	17.7 ± 8.62	14.1 ± 7.15	$y = -1.65x + 33.8$	0.83	
CL_(71:02)	0.744 ± 1.36	2.59 ± 3	1.41 ± 1.38	0.451 ± 0.714	$y = -0.0763x + 1.88$	0.08	
CL_(71:03)	7.35 ± 7.45	20.2 ± 18.2	18.2 ± 11	9.9 ± 8.38	$y = 0.209x + 12.3$	0.01	
CL_(71:04)	19.5 ± 12.7	51.1 ± 39.5	65.7 ± 41.9	43 ± 33	$y = 3.15x + 20.7$	0.32	
CL_(71:05)	56.9 ± 35	115 ± 76.6	130 ± 56.9	107 ± 69.8	$y = 6.09x + 55.9$	0.44	
CL_(71:06)	106 ± 65.4	164 ± 78.1	192 ± 74.8	164 ± 88.4	$y = 7.48x + 99.3$	0.52	
CL_(71:07)	76.3 ± 55.8	104 ± 37.6	120 ± 42.8	101 ± 54.8	$y = 3.34x + 74.8$	0.41	
CL_(71:08)	**23.9 ± 25.3**	**17.1 ± 6.59**	**5.98 ± 4.09**	**0.44 ± 0.928**	$\mathbf{y = -3.02x + 34.9}$ *	**0.98**	**Decreasing**
CL_(72:01)	0.251 ± 0.441	1.36 ± 1.4	0.758 ± 0.648	0.825 ± 1.56	$y = 0.0415x + 0.481$	0.10	

Table 1. *Cont.*

	Diet 1	Diet 2	Diet 3	Diet 4	Trendline Equation	R^2	Successive Change Across Groups
CL_(72:02)	0.874 ± 1.29	10.3 ± 12.3	4.1 ± 2.73	4.24 ± 7.52	$y = 0.144x + 3.77$	0.02	
CL_(72:03)	22.6 ± 21.7	55.2 ± 54.7	23.4 ± 12.2	17.9 ± 25	$y = -1.7x + 42.8$	0.12	
CL_(72:04)	246 ± 202	374 ± 291	228 ± 110	237 ± 229	$y = -6.41x + 320$	0.11	
CL_(72:05)	2310 ± 1300	3190 ± 1750	2650 ± 1080	2000 ± 1500	$y = -54.4x + 2950$	0.14	
CL_(72:06)	10,800 ± 5640	12,200 ± 4950	9770 ± 3660	5820 ± 3810	$y = -643x + 14,600$	0.67	
CL_(72:07)	**28,500 ± 15,500**	**25,400 ± 6760**	**17,100 ± 6180**	**7210 ± 4430**	$y = -2670x + 40,000 *$	**0.96**	**Decreasing**
CL_(72:08)	**31,100 ± 17,500**	**24,200 ± 5090**	**13,800 ± 4820**	**4420 ± 2860**	$y = -3350x + 44,000 *$	**0.99**	**Decreasing**
CL_(72:09)	**1280 ± 774**	**1150 ± 378**	**798 ± 339**	**456 ± 284**	$y = -105x + 1720 *$	**0.97**	**Decreasing**
CL_(72:10)	66.9 ± 45.2	62.8 ± 21.4	62 ± 40.5	70.8 ± 37.4	$y = 0.404x + 62.5$	0.12	
CL_(74:06)	**1080 ± 745**	**930 ± 448**	**489 ± 203**	**395 ± 366**	$y = -92.4x + 1430 *$	**0.94**	**Decreasing**
CL_(74:07)	**4550 ± 2900**	**3350 ± 1190**	**2000 ± 712**	**1650 ± 1280**	$y = -372x + 5740 *$	**0.95**	**Decreasing**
CL_(74:08)	**8200 ± 4930**	**6210 ± 1800**	**4510 ± 1460**	**3490 ± 2360**	$y = -586x + 10,100 *$	**0.98**	**Decreasing**
CL_(74:09)	**7180 ± 3960**	**6480 ± 1830**	**5090 ± 1530**	**3460 ± 2210**	$y = -465x + 9110 *$	**0.97**	**Decreasing**
CL_(74:10)	**3980 ± 2390**	**3050 ± 652**	**1860 ± 595**	**1030 ± 673**	$y = -372x + 5320 *$	**1.00**	**Decreasing**
CL_(74:11)	**1240 ± 707**	**784 ± 177**	**454 ± 136**	**284 ± 181**	$y = -118x + 1600 *$	**0.96**	**Decreasing**
CL_(76:09)	802 ± 495	560 ± 202	416 ± 152	633 ± 566	$y = -24.1x + 787$	0.27	
CL_(76:10)	1300 ± 724	974 ± 335	723 ± 242	960 ± 768	$y = -47.1x + 1350$	0.48	
CL_(76:11)	1580 ± 894	1140 ± 399	678 ± 217	710 ± 530	$y = -114x + 1900$	0.87	
CL_(76:12)	**1200 ± 704**	**925 ± 337**	**513 ± 163**	**441 ± 328**	$y = -99.6x + 1530 *$	**0.94**	**Decreasing**
DG_(32:0)	194 ± 108	64.5 ± 72.4	75.8 ± 122	99.9 ± 108	$y = -10x + 185$	0.35	
DG_(34:0)	84.3 ± 167	48.3 ± 145	ND ± ND	ND ± ND	$y = -13.3x + 132$	1.00	
DG_(34:1)	6950 ± 4570	3290 ± 1420	3300 ± 1600	4690 ± 4060	$y = -251x + 6480$	0.26	
GM1_(34:0)	1.47 ± 1.39	5.46 ± 4.45	4.11 ± 4.43	3.26 ± 2.36	$y = 0.149x + 2.44$	0.10	
GM1_(34:1)	4.89 ± 3.76	15.3 ± 11.1	10.6 ± 9.26	9.27 ± 6.24	$y = 0.313x + 7.62$	0.06	
GM1_(34:1-OH)	0.054 ± 0.0451	0.0988 ± 0.0944	0.0491 ± 0.0531	0.0338 ± 0.0588	$y = -0.00409x + 0.0902$	0.26	
GM1_(36:0)	ND ± ND	0.0285 ± 0.0603	ND ± ND	ND ± ND			
GM1_(36:1)	ND ± ND	0.157 ± 0.356	0.125 ± 0.158	0.152 ± 0.285	$y = -0.000926x + 0.153$	0.02	
Hex-Cer_(32:0)	0.578 ± 0.44	0.383 ± 0.558	0.199 ± 0.291	0.332 ± 0.324	$y = -0.0341x + 0.634$	0.57	
Hex-Cer_(32:1)	ND ± ND	0.0486 ± 0.104	0.0269 ± 0.076	0.0396 ± 0.112	$y = -0.00167x + 0.0534$	0.17	
Hex-Cer_(34:0-OH)	0.0882 ± 0.264	0.0548 ± 0.164	0.0611 ± 0.173	ND ± ND	$y = -0.00502x + 0.0997$	0.58	
Hex-Cer_(34:1)	0.565 ± 0.791	1.27 ± 1.2	1.74 ± 1.87	4.68 ± 1.85	$y = 0.475x - 1.57$	0.84	Increasing
Hex-Cer_(34:1-OH)	0.295 ± 0.719	ND ± ND	0.0372 ± 0.105	0.208 ± 0.587	$y = -0.016x + 0.31$	0.25	
Hex-Cer_(34:2)	0.32 ± 0.418	0.318 ± 0.598	ND ± ND	0.424 ± 1.2	$y = 0.0138x + 0.255$	0.88	
Hex-Cer_(34:2-OH)	**0.614 ± 0.37**	**0.518 ± 0.375**	**0.39 ± 0.512**	**0.306 ± 0.488**	$y = -0.039x + 0.755 *$	**0.99**	**Decreasing**
Hex-Cer_(35:0)	0.0425 ± 0.127	ND ± ND	0.663 ± 0.389	1.37 ± 1.56	$y = 0.157x - 0.579$	0.95	
Hex-Cer_(35:1)	**0.551 ± 0.625**	**0.649 ± 0.567**	**1.03 ± 0.58**	**1.66 ± 1.57**	$y = 0.137x - 0.0781 *$	**0.91**	**Increasing**
Hex-Cer_(36:0-OH)	**2.03 ± 1.06**	**2.58 ± 2.01**	**3.51 ± 1.07**	**4.75 ± 2.01**	$y = 0.337x + 0.642 *$	**0.97**	**Increasing**
Hex-Cer_(36:1)	0.235 ± 0.413	ND ± ND	0.688 ± 0.661	0.264 ± 0.491	$y = 0.0151x + 0.274$	0.06	

Table 1. *Cont.*

	Diet 1	Diet 2	Diet 3	Diet 4	Trendline Equation	R^2	Successive Change Across Groups
Hex-Cer_(36:2)	ND ± ND	0.0816 ± 0.245	0.289 ± 0.574	0.638 ± 0.942	$y = 0.103x - 0.591$	0.98	
Hex-Cer_(37:0)	ND ± ND	0.0835 ± 0.167	0.0639 ± 0.181	0.151 ± 0.295	$y = 0.0125x - 0.013$	0.55	
Hex-Cer_(37:0-OH)	9.56 ± 9.47	14.3 ± 11.8	38 ± 31.6	80.1 ± 67.3	$y = 8.72x - 31.2$	0.89	Increasing
Hex-Cer_(37:1)	0.0494 ± 0.148	0.167 ± 0.252	0.0497 ± 0.141	ND ± ND	$y = 0.0000556x + 0.0884$	0.00	
Hex-Cer_(37:2)	2.12 ± 2.98	3.57 ± 5.06	0.717 ± 2.03	0.957 ± 2.02	$y = -0.235x + 3.64$	0.39	
Hex-Cer_(38:0-OH)	ND ± ND	3.91 ± 5.51	2.8 ± 4.29	7.46 ± 8.35	$y = 0.657x - 1.19$	0.53	
Hex-Cer_(38:1)	14.8 ± 16	13.4 ± 12.3	22.6 ± 25.1	32.2 ± 25.2	$y = 2.27x + 3.35$	0.84	
Hex-Cer_(38:2)	0.569 ± 0.899	2.09 ± 1.23	1.81 ± 1.58	0.813 ± 0.906	$y = 0.0167x + 1.19$	0.01	
Hex-Cer_(39:0-OH)	5.62 ± 3.39	2.48 ± 3.2	4.86 ± 13.8	19.3 ± 21.8	$y = 1.61x - 4.24$	0.54	
Hex-Cer_(39:2)	20.7 ± 4.71	21.6 ± 7.5	24.1 ± 7.77	32.5 ± 7.01	$y = 1.4x + 14$	0.83	Increasing
Hex-Cer_(40:0)	10.1 ± 12.2	9.47 ± 10	15.5 ± 15	19.7 ± 9.5	$y = 1.29x + 3.82$	0.87	
Hex-Cer_(40:0-OH)	ND ± ND	5 ± 15	ND ± ND	ND ± ND			
Hex-Cer_(40:1)	**174 ± 204**	**221 ± 120**	**275 ± 249**	**403 ± 206**	$y = 27.4x + 58.3$ *	**0.94**	**Increasing**
Hex-Cer_(40:1-OH)	7.04 ± 2.48	6.38 ± 2.94	7.95 ± 2.57	5.14 ± 1.51	$y = -0.153x + 7.8$	0.20	
Hex-Cer_(40:2)	0.266 ± 0.178	0.854 ± 0.804	0.806 ± 1.66	0.672 ± 1.21	$y = 0.0433x + 0.318$	0.32	
Hex-Cer_(40:2-OH)	2.1 ± 1.32	2.17 ± 0.891	2.76 ± 1.9	1.86 ± 1.91	$y = -0.00481x + 2.26$	0.00	
Hex-Cer_(41:0-OH)	23.9 ± 8.7	33 ± 10.5	26.8 ± 4.88	20.1 ± 7.48	$y = -0.652x + 30.9$	0.17	
Hex-Cer_(41:1)	4.01 ± 1.31	5.78 ± 1.62	7.62 ± 3.75	17.7 ± 5.68	$y = 1.59x - 3.38$	0.82	Increasing
Hex-Cer_(41:2)	4.6 ± 1.03	4.78 ± 1.62	3.34 ± 1.5	3.71 ± 1.89	$y = -0.152x + 5.27$	0.59	
Hex-Cer_(42:0)	0.394 ± 0.271	0.329 ± 0.289	0.466 ± 0.227	0.705 ± 0.432	$y = 0.0396x + 0.17$	0.71	
Hex-Cer_(42:0-OH)	21.8 ± 14.5	21.3 ± 6.34	30.3 ± 20.7	34.6 ± 14.3	$y = 1.76x + 13.6$	0.88	
Hex-Cer_(42:1)	**14.1 ± 4.31**	**20.5 ± 5.26**	**22.5 ± 10.5**	**34.5 ± 12.3**	$y = 2.34x + 4.99$ *	**0.92**	**Increasing**
Hex-Cer_(42:2)	**7.61 ± 2.29**	**11.3 ± 3.91**	**14.3 ± 7.05**	**23.1 ± 7.39**	$y = 1.83x + 0.061$ *	**0.93**	**Increasing**
Hex-Cer_(42:2-OH)	4.99 ± 1.66	5.86 ± 2.35	5.19 ± 1.86	7.37 ± 3.43	$y = 0.24x + 4.02$	0.60	
Hex-Cer_(43:0)	0.0149 ± 0.0296	0.0155 ± 0.0308	0.145 ± 0.175	0.129 ± 0.153	$y = 0.0175x - 0.0576$	0.74	
Hex-Cer_(43:0-OH)	0.416 ± 0.255	0.583 ± 0.36	0.338 ± 0.18	0.32 ± 0.277	$y = -0.0197x + 0.565$	0.33	
Hex-Cer_(43:1)	1.69 ± 0.642	3.13 ± 1.14	3.47 ± 1.46	7.28 ± 2.91	$y = 0.634x - 0.955$	0.86	Increasing
Hex-Cer_(43:2)	0.866 ± 0.212	0.858 ± 0.321	0.799 ± 0.314	1.24 ± 0.718	$y = 0.0394x + 0.64$	0.46	
LPC_(14:0)	**1.4 ± 0.645**	**1.58 ± 0.732**	**2.09 ± 0.96**	**2.96 ± 1.34**	$y = 0.192x + 0.537$ *	**0.92**	**Increasing**
LPC_(15:0)	6.02 ± 1.49	7.53 ± 2.48	7.69 ± 2.62	8.38 ± 2.62	$y = 0.268x + 5.35$	0.88	Increasing
LPC_(16:0)	**1430 ± 246**	**1560 ± 272**	**1690 ± 300**	**1900 ± 427**	$y = 57x + 1210$ *	**0.98**	**Increasing**
LPC_(16:1)	**8.56 ± 4.67**	**11.3 ± 6.3**	**19.6 ± 8.38**	**33.1 ± 10.5**	$y = 3.03x - 5.07$ *	**0.92**	**Increasing**
LPC_(17:0)	64.5 ± 10.8	72.2 ± 13.1	74.9 ± 11.2	75.6 ± 27.5	$y = 1.33x + 61.6$	0.84	Increasing
LPC_(17:1)	0.395 ± 0.314	0.576 ± 0.537	2.2 ± 1.09	6.08 ± 1.62	$y = 0.692x - 2.98$	0.83	Increasing
LPC_(18:0)	**3720 ± 504**	**3550 ± 634**	**3450 ± 619**	**3300 ± 520**	$y = -50.4x + 3890$ *	**0.99**	**Decreasing**
LPC_(18:1)	252 ± 47.9	247 ± 56.4	332 ± 51.2	549 ± 121	$y = 36.1x + 68.5$	0.79	
LPC_(18:2)	**266 ± 52.6**	**232 ± 46.8**	**217 ± 19.7**	**198 ± 38.2**	$y = -8.11x + 290$ *	**0.97**	**Decreasing**
LPC_(18:3)	**80 ± 15.2**	**88.6 ± 22.2**	**93 ± 19.4**	**102 ± 22.5**	$y = 2.61x + 71$ *	**0.98**	**Increasing**
LPC_(18:4)	0.515 ± 0.37	0.758 ± 0.481	1.45 ± 0.744	3.08 ± 1.42	$y = 0.311x - 0.926$	0.88	Increasing

Table 1. *Cont.*

	Diet 1	Diet 2	Diet 3	Diet 4	Trendline Equation	R^2	Successive Change Across Groups
LPC_(19:0)	47.8 ± 5.54	37.7 ± 12.1	36.2 ± 8.19	40.1 ± 12.9	$y = -0.911x + 47.4$	0.38	
LPC_(20:0)	**45.5 ± 12**	**29 ± 9.84**	**22.8 ± 4.24**	**18.5 ± 5.39**	$y = -3.23x + 53.7$ *	**0.90**	**Decreasing**
LPC_(20:4)	344 ± 121	289 ± 32.6	288 ± 82.8	282 ± 53.7	$y = -6.93x + 354$	0.69	Decreasing
LPC_(20:5)	26.9 ± 9.49	21.5 ± 4.39	20.2 ± 2.89	21 ± 5.62	$y = -0.704x + 27.8$	0.65	
LPC_(21:0)	1.11 ± 0.227	0.89 ± 0.305	0.75 ± 0.257	0.896 ± 0.324	$y = -0.029x + 1.13$	0.46	
LPC_(22:4)	5.12 ± 2.83	3.37 ± 0.794	3.21 ± 0.909	3.52 ± 1	$y = -0.184x + 5.21$	0.52	
LPC_(22:5)	12.3 ± 9.44	6.01 ± 2.43	5.97 ± 2.51	9.46 ± 4.97	$y = -0.317x + 10.9$	0.13	
LPC_(22:6)	21.1 ± 13	14.4 ± 3.33	15.4 ± 4.74	26.5 ± 5.75	$y = 0.637x + 14.5$	0.16	
LPE_(16:0)	**9.94 ± 2.69**	**11.4 ± 3.82**	**16.7 ± 4.93**	**25.8 ± 11.7**	$y = 1.96x + 0.977$ *	**0.91**	**Increasing**
LPE_(16:1)	ND ± ND	0.0351 ± 0.0709	0.0116 ± 0.0329	0.177 ± 0.163	$y = 0.0263x - 0.162$	0.63	
LPE_(17:0)	**0.637 ± 0.392**	**0.724 ± 0.401**	**1.1 ± 0.151**	**1.61 ± 0.982**	$y = 0.122x + 0.0842$ *	**0.92**	**Increasing**
LPE_(18:0)	50.7 ± 12.9	50.3 ± 14.6	57.1 ± 12.6	67.7 ± 29.3	$y = 2.14x + 40.1$	0.84	
LPE_(18:1)	5.94 ± 1.16	4.7 ± 1.21	6.6 ± 1.36	11 ± 3.41	$y = 0.633x + 2.22$	0.65	
LPE_(18:2)	5.04 ± 3.8	2.76 ± 1.33	2.93 ± 1.15	2.42 ± 1.28	$y = -0.285x + 5.47$	0.70	
LPE_(18:3)	**1.85 ± 0.528**	**2.14 ± 0.76**	**2.96 ± 0.944**	**4.22 ± 1.93**	$y = 0.294x + 0.546$ *	**0.93**	**Increasing**
LPE_(20:0)	**0.444 ± 0.114**	**0.358 ± 0.169**	**0.31 ± 0.121**	**0.247 ± 0.17**	$y = -0.0237x + 0.521$ *	**0.99**	**Decreasing**
LPE_(20:3)	**0.472 ± 0.645**	**0.556 ± 0.49**	**0.945 ± 0.831**	**1.49 ± 1.22**	$y = 0.128x - 0.11$ *	**0.92**	**Increasing**
LPE_(20:4)	22.9 ± 12.3	17.9 ± 4.53	20.8 ± 7.66	28.6 ± 11.8	$y = 0.741x + 16.9$	0.33	
LPE_(20:5)	1.53 ± 1.12	0.932 ± 0.388	0.827 ± 0.252	0.645 ± 0.248	$y = -0.102x + 1.77$	0.86	Decreasing
LPE_(22:4)	0.264 ± 0.791	ND ± ND	0.0209 ± 0.0389	0.35 ± 0.382	$y = 0.00267x + 0.19$	0.00	
LPI_(16:0)	64.1 ± 21.4	82.6 ± 36.1	113 ± 39.7	103 ± 45	$y = 5.45x + 49$	0.76	
LPI_(17:0)	5.04 ± 3.17	6.56 ± 3.13	7.63 ± 3.8	5.1 ± 4.85	$y = 0.0463x + 5.73$	0.02	
LPI_(18:0)	1210 ± 404	1510 ± 647	1510 ± 376	1200 ± 547	$y = -1.11x + 1370$	0.00	
LPI_(18:1)	**46.2 ± 22.3**	**47.6 ± 17.7**	**57.4 ± 21.6**	**59.5 ± 26.8**	$y = 1.84x + 38.6$ *	**0.90**	**Increasing**
LPI_(18:2)	18.9 ± 10.4	17.8 ± 8	16.9 ± 5.86	7.43 ± 4.45	$y = -1.31x + 25.3$	0.74	Decreasing
LPI_(20:0)	0.0835 ± 0.1	ND ± ND	0.18 ± 0.264	0.0895 ± 0.127	$y = 0.00319x + 0.0918$	0.06	
LPI_(20:2)	**1.7 ± 1.09**	**1.18 ± 1.36**	**1.09 ± 0.93**	**0.661 ± 1.15**	$y = -0.119x + 2.07$ *	**0.94**	**Decreasing**
LPI_(20:3)	**25.8 ± 14.7**	**47.1 ± 27.2**	**85.6 ± 35.3**	**125 ± 49.4**	$y = 12.4x - 24.4$ *	**0.98**	**Increasing**
LPI_(20:4)	288 ± 102	453 ± 238	447 ± 108	360 ± 145	$y = 7.78x + 328$	0.12	
LPI_(22:4)	4.72 ± 2.12	3.96 ± 2.88	4.26 ± 2.1	3.26 ± 1.91	$y = -0.151x + 5.21$	0.74	
LPI_(22:5)	1.15 ± 1.67	1.51 ± 1.76	0.418 ± 0.701	1.98 ± 1.41	$y = 0.0518x + 0.868$	0.08	
LPI_(22:6)	2.06 ± 1.55	2.04 ± 1.58	0.605 ± 0.968	2.65 ± 2.37	$y = 0.0124x + 1.74$	0.00	
Lyso_CL_(52:01)	0.698 ± 1.02	1.75 ± 1.42	1.51 ± 0.755	1.99 ± 1.82	$y = 0.135x + 0.457$	0.70	
Lyso_CL_(52:02)	11.4 ± 10.4	13.7 ± 7.67	10.7 ± 4.9	12.5 ± 11.6	$y = 0.0111x + 12$	0.00	
Lyso_CL_(52:03)	**50.5 ± 42.8**	**32.9 ± 12.7**	**18.7 ± 8.32**	**17.2 ± 14.1**	$y = -4.23x + 62.2$ *	**0.90**	**Decreasing**
Lyso_CL_(52:04)	30.3 ± 36.6	19.7 ± 9.87	9.23 ± 4.21	10.4 ± 9.01	$y = -2.6x + 37.3$	0.86	
Lyso_CL_(52:05)	ND ± ND	ND ± ND	ND ± ND	0.785 ± 1.84			
Lyso_CL_(53:04)	ND ± ND	0.154 ± 0.261	0.221 ± 0.404	1.55 ± 1.72	$y = 0.259x - 1.69$	0.79	

Table 1. *Cont.*

	Diet 1	Diet 2	Diet 3	Diet 4	Trendline Equation	R^2	Successive Change Across Groups
Lyso_CL_(54:02)	ND ± ND	0.0851 ± 0.178	0.0412 ± 0.117	0.655 ± 1.15	$y = 0.106x - 0.689$	0.69	
Lyso_CL_(54:03)	24.8 ± 26.6	13.2 ± 7.1	9.89 ± 3.88	19.5 ± 18.6	$y = -0.711x + 22.3$	0.14	
Lyso_CL_(54:04)	172 ± 166	94.9 ± 32.4	67.3 ± 25.2	87 ± 72.4	$y = -10.5x + 185$	0.63	
Lyso_CL_(54:05)	405 ± 411	234 ± 94	144 ± 60.4	133 ± 104	$y = -33.6x + 486$	0.87	Decreasing
Lyso_CL_(54:06)	548 ± 555	247 ± 117	122 ± 47.8	65.7 ± 51.8	$y = -58.2x + 691$	0.89	Decreasing
Lyso_CL_(56:05)	42.9 ± 55.3	13.1 ± 3.27	4.18 ± 2.41	13.5 ± 16.1	$y = -3.6x + 45.9$	0.55	
Lyso_CL_(56:06)	135 ± 153	42.7 ± 13.4	20.9 ± 8.33	40.2 ± 42.2	$y = -11.3x + 146$	0.60	
PA_(30:0)	1.19 ± 2.41	2.63 ± 2.77	6.15 ± 4.83	30 ± 23.7	$y = 3.33x - 15.5$	0.74	Increasing
PA_(30:1)	17.2 ± 17.8	17.1 ± 11.8	20.7 ± 9.98	73.9 ± 41.3	$y = 6.43x - 17$	0.65	
PA_(32:1)	43.2 ± 46.5	56.3 ± 17.7	84.6 ± 30.3	291 ± 143	$y = 28.6x - 99.9$	0.74	Increasing
PA_(32:2)	223 ± 216	180 ± 96	154 ± 86.4	166 ± 70.4	$y = -7.3x + 237$	0.71	
PA_(34:1)	387 ± 401	402 ± 135	538 ± 176	1010 ± 307	$y = 74.3x + 16.2$	0.79	Increasing
PA_(34:2)	**2330 ± 2170**	**1990 ± 756**	**1600 ± 787**	**1230 ± 358**	$y = -137x + 2830$ *	**1.00**	**Decreasing**
PA_(36:1)	1440 ± 924	1640 ± 435	1410 ± 877	1240 ± 323	$y = -30.7x + 1670$	0.43	
PA_(36:2)	**3310 ± 1750**	**2480 ± 918**	**2250 ± 1360**	**1910 ± 1020**	$y = -164x + 3740$ *	**0.92**	**Decreasing**
PA_(36:4)	10,300 ± 9970	6760 ± 2840	7130 ± 3790	9580 ± 4540	$y = -66.3x + 8950$	0.02	
PA_(38:3)	1390 ± 910	1150 ± 544	1590 ± 683	2590 ± 1230	$y = 150x + 535$	0.68	
PA_(38:4)	42,400 ± 40,100	27,100 ± 11,700	25,400 ± 17,000	23,000 ± 7630	$y = -2220x + 46,400$	0.78	Decreasing
PA_(38:5)	3280 ± 2800	2230 ± 911	2430 ± 1180	3310 ± 1300	$y = 10.7x + 2730$	0.00	
PA_(40:5)	1730 ± 1530	838 ± 336	871 ± 449	1110 ± 412	$y = -67.7x + 1650$	0.33	
PC_(30:0)	**6.27 ± 6.94**	**21.5 ± 23.9**	**39.3 ± 18.4**	**81.4 ± 23.1**	$y = 9.01x - 31.8$ *	**0.94**	**Increasing**
PC_(30:1)	8.63 ± 6.24	12.9 ± 8.52	40.6 ± 32.6	188 ± 120	$y = 21x - 97.8$	0.74	Increasing
PC_(31:0)	54.7 ± 17.4	78.7 ± 33.5	85.4 ± 22.6	179 ± 44.5	$y = 14.1x - 8.1$	0.80	Increasing
PC_(32:0)	1750 ± 251	1930 ± 493	1650 ± 237	1950 ± 234	$y = 11.9x + 1730$	0.08	
PC_(32:1)	222 ± 110	352 ± 101	1240 ± 835	4860 ± 1550	$y = 548x - 2530$	0.77	Increasing
PC_(32:2)	565 ± 249	496 ± 222	646 ± 223	1350 ± 659	$y = 92.8x + 54.5$	0.67	
PC_(33:0)	28.2 ± 8.55	39.9 ± 13	49.6 ± 7.5	102 ± 25.7	$y = 8.56x - 10.6$	0.84	Increasing
PC_(33:1)	148 ± 45.4	267 ± 71.4	540 ± 136	1700 ± 388	$y = 183x - 733$	0.80	Increasing
PC_(33:2)	470 ± 56.2	515 ± 187	574 ± 241	566 ± 180	$y = 12.9x + 433$	0.85	
PC_(34:0)	789 ± 267	732 ± 165	615 ± 45.6	644 ± 83.3	$y = -20.4x + 851$	0.79	
PC_(34:1)	4160 ± 763	5580 ± 931	8080 ± 829	15,100 ± 1050	$y = 1310x - 1780$	0.88	Increasing
PC_(34:2)	11,900 ± 2050	14,700 ± 1880	15,200 ± 1930	20,400 ± 2030	$y = 963x + 8180$	0.90	Increasing
PC_(34:3)	3270 ± 1090	3620 ± 834	5400 ± 998	9050 ± 1860	$y = 708x - 82.3$	0.87	Increasing
PC_(35:0)	28.3 ± 9.27	32.3 ± 9.11	38.3 ± 6.8	71.2 ± 15.2	$y = 4.99x + 4.36$	0.79	Increasing
PC_(35:1)	108 ± 32	166 ± 36.7	297 ± 52.6	891 ± 153	$y = 91.9x - 337$	0.79	Increasing
PC_(35:2)	1240 ± 238	1530 ± 305	1530 ± 433	1640 ± 369	$y = 44.4x + 1150$	0.82	
PC_(36:0)	76.2 ± 27.6	66.4 ± 16.6	55.2 ± 22.9	83.4 ± 16.5	$y = 0.385x + 67.4$	0.01	
PC_(36:1)	3060 ± 753	3610 ± 756	5450 ± 944	11,600 ± 974	$y = 1020x - 1850$	0.82	Increasing

Table 1. *Cont.*

	Diet 1	Diet 2	Diet 3	Diet 4	Trendline Equation	R^2	Successive Change Across Groups
PC_(36:2)	6650 ± 1180	7290 ± 1320	7890 ± 1490	10,700 ± 954	$y = 472x + 4520$	0.85	Increasing
PC_(36:3)	8480 ± 912	9060 ± 1570	11,800 ± 1920	23,900 ± 2150	$y = 1810x - 573$	0.77	Increasing
PC_(36:4)	18,900 ± 3120	20,200 ± 2700	19,500 ± 1190	27,500 ± 1900	$y = 930x + 14,400$	0.65	
PC_(37:0)	0.557 ± 0.4	0.542 ± 0.232	0.338 ± 0.231	0.689 ± 0.551	$y = 0.00711x + 0.477$	0.03	
PC_(37:1)	40.9 ± 20.8	38 ± 19.4	75.1 ± 38.1	283 ± 96.2	$y = 28.3x - 107$	0.71	
PC_(37:2)	301 ± 88.2	217 ± 62.9	210 ± 61.6	301 ± 56.6	$y = -0.259x + 259$	0.00	
PC_(37:3)	**83.4 ± 17.3**	**86.2 ± 25.1**	**131 ± 33.9**	**160 ± 31.9**	**$y = 10.2x + 37.3$ ***	**0.92**	**Increasing**
PC_(37:4)	2810 ± 648	3180 ± 738	2520 ± 674	2430 ± 461	$y = -66.7x + 3250$	0.47	
PC_(37:5)	157 ± 59.9	171 ± 36.4	270 ± 103	445 ± 77.3	$y = 35.7x - 12.1$	0.88	Increasing
PC_(37:6)	85 ± 34.3	76 ± 37.5	55.8 ± 29.9	101 ± 29.7	$y = 1.03x + 71.6$	0.04	
PC_(38:0)	1.83 ± 0.323	1.77 ± 0.432	1.79 ± 0.773	2.67 ± 0.91	$y = 0.0941x + 1.3$	0.56	
PC_(38:1)	39.4 ± 20.9	14.1 ± 12.8	28.9 ± 21.3	103 ± 42.3	$y = 7.61x - 11.9$	0.46	
PC_(38:2)	649 ± 137	521 ± 186	493 ± 137	871 ± 171	$y = 23.6x + 453$	0.23	
PC_(38:3)	3030 ± 346	3260 ± 708	5020 ± 1050	12,200 ± 1770	$y = 1080x - 2420$	0.77	Increasing
PC_(38:4)	10,300 ± 2010	11,700 ± 2240	10,200 ± 1090	12,100 ± 2290	$y = 144x + 9970$	0.27	
PC_(38:5)	6010 ± 2050	5500 ± 1900	5290 ± 916	8730 ± 1930	$y = 294x + 4130$	0.41	
PC_(38:6)	16,000 ± 3530	14,000 ± 1550	12,900 ± 1690	18,900 ± 2600	$y = 281x + 13,300$	0.14	
PC_(40:2)	19.1 ± 8.48	14.8 ± 4.37	13.2 ± 3.21	20.1 ± 5.31	$y = 0.0519x + 16.4$	0.00	
PC_(40:3)	52.8 ± 15	44 ± 15.1	52.7 ± 13.6	124 ± 22.6	$y = 8.23x + 5.39$	0.59	
PC_(40:4)	**1200 ± 402**	**884 ± 230**	**726 ± 182**	**665 ± 175**	**$y = -65.3x + 1370$ ***	**0.90**	**Decreasing**
PC_(40:5)	947 ± 577	704 ± 574	514 ± 208	980 ± 459	$y = -3.37x + 812$	0.00	
PC_(40:6)	3800 ± 936	3630 ± 913	3090 ± 720	4370 ± 960	$y = 43.3x + 3390$	0.08	
PC_C18(plas)-18:1	ND ± ND	ND ± ND	ND ± ND	0.766 ± 1.43			
PE_(30:0)	4.78 ± 10.7	2.19 ± 4.47	ND ± ND	1.82 ± 5.14	$y = -0.323x + 5.26$	0.68	
PE_(32:1)	12.4 ± 20.9	26 ± 38.5	468 ± 553	1920 ± 896	$y = 228x - 1140$	0.78	Increasing
PE_(34:0)	80.3 ± 34.4	106 ± 24	54.8 ± 20.8	30.5 ± 35.6	$y = -7.43x + 125$	0.63	
PE_(34:1)	4110 ± 2220	4930 ± 1800	9030 ± 1940	18,100 ± 6000	$y = 1710x - 4010$	0.86	Increasing
PE_(34:2)	**71,600 ± 27,000**	**69,300 ± 42,400**	**48,600 ± 14,200**	**29,500 ± 12,000**	**$y = -5440x + 96,400$ ***	**0.92**	**Decreasing**
PE_(35:1)	**4.9 ± 2**	**10.6 ± 3.84**	**23.9 ± 7.89**	**42.4 ± 13**	**$y = 4.66x - 15.2$ ***	**0.95**	**Increasing**
PE_(36:0)	24.4 ± 11.3	35 ± 15	22.8 ± 6.68	17 ± 12.8	$y = -1.27x + 34.5$	0.35	
PE_(36:1)	6560 ± 4060	5730 ± 2600	6590 ± 1730	11,300 ± 4870	$y = 559x + 3270$	0.59	
PE_(36:2)	**46,300 ± 18,900**	**36,900 ± 20,900**	**24,400 ± 6230**	**19,300 ± 7930**	**$y = -3460x + 58,200$ ***	**0.98**	**Decreasing**
PE_(36:3)	**39,000 ± 15,000**	**28,800 ± 17,100**	**20,000 ± 4520**	**14,300 ± 5490**	**$y = -3070x + 49,000$ ***	**0.99**	**Decreasing**
PE_(36:4)	110,000 ± 44,700	107,000 ± 49,200	96,600 ± 12,500	126,000 ± 34,500	$y = 1390x + 99,200$	0.16	
PE_(38:1)	219 ± 130	144 ± 84.5	115 ± 22	121 ± 46.7	$y = -12x + 241$	0.76	
PE_(38:2)	1530 ± 645	822 ± 467	539 ± 98.4	535 ± 140	$y = -121x + 1780$	0.81	Decreasing
PE_(38:3)	13400 ± 4930	13,000 ± 7700	14,400 ± 2710	17,700 ± 3860	$y = 530x + 10,600$	0.75	
PE_(38:4)	165,000 ± 74,800	148,000 ± 59,300	113,000 ± 20,200	118,000 ± 26,000	$y = -6520x + 186,000$	0.84	

Table 1. *Cont.*

	Diet 1	Diet 2	Diet 3	Diet 4	Trendline Equation	R^2	Successive Change Across Groups
PE_(38:5)	72,900 ± 38,400	52,700 ± 23,200	39,700 ± 7150	44,700 ± 11,600	$y = -3610x + 80{,}200$	0.74	
PE_(38:6)	25,000 ± 12,000	18,200 ± 8270	12,300 ± 2890	19,100 ± 5400	$y = -874x + 25{,}300$	0.34	
PG_(32:0)	1.64 ± 0.748	1.59 ± 0.834	5.29 ± 4.05	11 ± 3.46	$y = 1.18x - 4.12$	0.86	
PG_(33:0)	ND ± ND	ND ± ND	ND ± ND	0.0213 ± 0.0434			
PG_(34:1)	ND ± ND	ND ± ND	2.24 ± 3.19	65.1 ± 51.8	$y = 23.3x - 207$	1.00	
PG_(35:1)	**0.256 ± 0.325**	**0.467 ± 0.67**	**0.816 ± 0.888**	**1.62 ± 0.913**	$y = 0.164x - 0.469$ *	**0.91**	**Increasing**
PG_(36:0)	11.8 ± 3.75	25.6 ± 18	13.4 ± 4.49	7.6 ± 5.1	$y = -0.919x + 21.6$	0.17	
PG_(36:1)	0.0408 ± 0.122	1.84 ± 2.72	0.525 ± 0.62	0.294 ± 0.831	$y = -0.0206x + 0.832$	0.01	
PG_(36:2)	125 ± 25.3	169 ± 81.3	106 ± 43.1	53.5 ± 21	$y = -10.3x + 192$	0.56	
PG_(36:3)	48.8 ± 14.2	62.3 ± 30	32.3 ± 9.53	28.1 ± 20.5	$y = -3.41x + 69$	0.57	
PG_(36:4)	25.6 ± 13.6	45.9 ± 26.9	33.6 ± 15.7	29.1 ± 15.4	$y = -0.0667x + 34.1$	0.00	
PG_(38:3)	1.72 ± 1.42	1.04 ± 0.753	3.19 ± 3.43	7.34 ± 3.64	$y = 0.704x - 2.06$	0.75	
PG_(38:4)	11.7 ± 5.76	19.5 ± 9.9	13.3 ± 5.22	11.9 ± 5.31	$y = -0.207x + 15.7$	0.04	
PG_(38:5)	6.31 ± 2.82	9.55 ± 5.47	6.6 ± 2.34	7.16 ± 4.03	$y = -0.0148x + 7.52$	0.00	
PG_(38:6)	0.0292 ± 0.0875	0.0656 ± 0.0754	0.0529 ± 0.067	0.0457 ± 0.0527	$y = 0.00136x + 0.0379$	0.10	
PG_(40:6)	0.306 ± 0.378	0.211 ± 0.434	0.028 ± 0.0792	0.773 ± 0.913	$y = 0.0451x - 0.0156$	0.25	
PG_(42:07)	0.469 ± 0.411	0.424 ± 0.714	0.269 ± 0.332	0.305 ± 0.316	$y = -0.024x + 0.55$	0.77	
PG_(42:08)	0.0147 ± 0.0192	0.0893 ± 0.11	0.112 ± 0.116	ND ± ND	$y = 0.018x - 0.0415$	0.91	Increasing
PG_(42:10)	0.0638 ± 0.191	0.314 ± 0.943	ND ± ND	0.0417 ± 0.118	$y = -0.00954x + 0.209$	0.07	
PI_(34:0)	455 ± 385	482 ± 396	354 ± 237	318 ± 196	$y = -20x + 555$	0.78	
PI_(34:1)	5070 ± 1420	5070 ± 909	7620 ± 1620	10,100 ± 2670	$y = 653x + 1970$	0.89	
PI_(34:2)	6510 ± 3850	9740 ± 4140	9450 ± 3880	3260 ± 1640	$y = -372x + 10{,}100$	0.18	
PI_(35:0)	0.0814 ± 0.244	0.09_7 ± 0.275	ND ± ND	ND ± ND	$y = 0.00381x + 0.0677$	1.00	
PI_(35:2)	134 ± 112	319 ± 164	261 ± 137	82.5 ± 68.9	$y = -7.87x + 259$	0.06	
PI_(36:1)	**186 ± 165**	**450 ± 292**	**659 ± 339**	**814 ± 589**	$y = 77.5x - 65.8$ *	**0.99**	**Increasing**
PI_(36:2)	3680 ± 2670	5400 ± 2720	4920 ± 2300	2020 ± 1140	$y = -202x + 5550$	0.22	
PI_(36:4)	16,000 ± 6720	25,800 ± 10,600	27,900 ± 8740	21,000 ± 9500	$y = 633x + 17{,}800$	0.17	
PI_(38:3)	6340 ± 4020	10,500 ± 6710	14,200 ± 8060	6420 ± 3950	$y = 146x + 8250$	0.02	
PI_(38:4)	86,600 ± 41,000	124,000 ± 50,900	119,000 ± 34,000	79,600 ± 40,400	$y = -963x + 110{,}000$	0.02	
PI_(38:5)	11,100 ± 5400	15,000 ± 5750	16,600 ± 4890	12,400 ± 5840	$y = 204x + 12{,}200$	0.08	
PI_(40:3)	**120 ± 56.6**	**94 ± 51.8**	**49.8 ± 19**	**25.2 ± 8.08**	$y = -12.2x + 165$ *	**0.99**	**Decreasing**
PI_(40:4)	1040 ± 615	1830 ± 1050	1380 ± 608	885 ± 607	$y = -33.9x + 1540$	0.08	
PI_(40:6)	4080 ± 1020	5270 ± 1350	4820 ± 901	6870 ± 1270	$y = 293x + 3020$	0.75	
PI_(40:8)	42.9 ± 25	48.9 ± 23.4	47.4 ± 22.1	50.8 ± 21	$y = 0.822x + 41.2$	0.72	
PS_(32:0)	0.673 ± 0.587	0.661 ± 0.918	0.321 ± 0.135	0.802 ± 0.336	$y = 0.00174x + 0.601$	0.00	
PS_(32:1)	0.00219 ± 0.00658	0.0339 ± 0.0329	0.00448 ± 0.0127	0.00374 ± 0.0106	$y = -0.000917x + 0.0181$	0.04	
PS_(33:1)	1.25 ± 0.504	1.76 ± 0.541	1.8 ± 0.553	1.19 ± 1.06	$y = -0.00519x + 1.54$	0.00	
PS_(34:0)	0.0337 ± 0.101	0.291 ± 0.452	0.16 ± 0.211	0.0675 ± 0.096	$y = -0.0011x + 0.146$	0.00	

Table 1. *Cont.*

	Diet 1	Diet 2	Diet 3	Diet 4	Trendline Equation	R^2	Successive Change Across Groups
PS_(34:1)	ND ± ND	0.0315 ± 0.0945	ND ± ND	0.73 ± 0.911	$y = 0.129x - 0.783$	1.00	
PS_(34:2)	10.3 ± 4.05	13.8 ± 4.69	15.3 ± 4.34	11.7 ± 5.1	$y = 0.211x + 11.2$	0.11	
PS_(34:3)	0.051 ± 0.0821	0.0582 ± 0.115	0.202 ± 0.314	0.911 ± 0.888	$y = 0.101x - 0.466$	0.74	Increasing
PS_(35:2)	0.215 ± 0.514	0.379 ± 0.946	0.0732 ± 0.207	0.101 ± 0.286	$y = -0.024x + 0.376$	0.36	
PS_(36:0)	0.242 ± 0.479	0.34 ± 0.527	0.261 ± 0.489	0.485 ± 0.672	$y = 0.0241x + 0.148$	0.58	
PS_(36:1)	63.9 ± 28.7	135 ± 76.6	101 ± 23.5	75.4 ± 43	$y = 0.0185x + 93.7$	0.00	
PS_(36:2)	115 ± 47	136 ± 51.5	123 ± 33.2	83 ± 23.2	$y = -4.04x + 145$	0.39	
PS_(38:2)	0.434 ± 0.578	1.66 ± 1.78	0.712 ± 0.42	0.536 ± 0.84	$y = -0.0238x + 1.02$	0.02	
PS_(38:5)	ND ± ND	0.0745 ± 0.224	ND ± ND	ND ± ND			
PS_(38:6)	113 ± 23.8	126 ± 29.7	140 ± 34.3	201 ± 55.3	$y = 10.3x + 66.2$	0.85	Increasing
PS_(40:0)	0.352 ± 0.406	1.18 ± 0.599	0.755 ± 0.566	1.46 ± 0.838	$y = 0.107x + 0.115$	0.59	
PS_(40:2)	0.485 ± 1.31	1.23 ± 2.17	ND ± ND	ND ± ND	$y = 0.276x - 0.508$	1.00	
PS_(40:3)	215 ± 36.3	243 ± 49.1	206 ± 51.9	147 ± 28.2	$y = -8.93x + 271$	0.59	
PS_(42:6)	185 ± 75.9	262 ± 104	237 ± 95.8	87.8 ± 37.6	$y = -11.7x + 283$	0.28	
S_(32:0)	0.134 ± 0.0379	0.128 ± 0.0538	0.112 ± 0.0157	0.0075 ± 0.014	$y = -0.0146x + 0.207$	0.74	Decreasing
S_(32:1)	ND ± ND	ND ± ND	ND ± ND	0.000566 ± 0.0016			
S_(34:0)	1.04 ± 0.105	2.09 ± 1.77	1.54 ± 0.259	1.31 ± 0.586	$y = 0.00963x + 1.42$	0.01	

Table 1. *Cont.*

	Diet 1	Diet 2	Diet 3	Diet 4	Trendline Equation	R^2	Successive Change Across Groups
S_(34:1-OH)	1.47 ± 0.48	1.72 ± 0.795	1.08 ± 0.388	1.35 ± 0.9	$y = -0.037x + 1.69$	0.24	
S_(34:2)	0.592 ± 0.147	0.769 ± 0.36	0.402 ± 0.187	0.327 ± 0.128	$y = -0.043x + 0.852$	0.57	
S_(35:0)	0.0316 ± 0.0617	0.022 ± 0.0369	0.0626 ± 0.123	0.0338 ± 0.0371	$y = 0.00175x + 0.0241$	0.12	
S_(35:1)	0.0235 ± 0.041	0.021 ± 0.0344	0.0404 ± 0.0922	0.16 ± 0.307	$y = 0.0159x - 0.0603$	0.70	
S_(35:1-OH)	0.209 ± 0.281	0.127 ± 0.381	ND ± ND	ND ± ND	$y = -0.0304x + 0.318$	1.00	
S_(35:2)	ND ± ND	ND ± ND	0.00482 ± 0.00897	0.00385 ± 0.00881	$y = -0.000359x + 0.00805$	1.00	
S_(36:1)	0.999 ± 0.437	1 ± 0.661	0.824 ± 0.375	0.561 ± 0.391	$y = -0.0552x + 1.27$	0.86	
S_(37:2)	0.159 ± 0.092	0.296 ± 0.22	0.206 ± 0.125	0.08 ± 0.0753	$y = -0.0121x + 0.278$	0.22	
S_(38:0)	1.88 ± 0.631	3.39 ± 2.13	1.89 ± 0.684	1.03 ± 0.594	$y = -0.15x + 3.2$	0.28	
S_(38:1-OH)	0.153 ± 0.242	0.197 ± 0.308	0.116 ± 0.253	0.0765 ± 0.109	$y = -0.0115x + 0.224$	0.61	
S_(39:1)	0.111 ± 0.0948	0.41 ± 0.632	0.194 ± 0.249	0.0621 ± 0.12	$y = -0.0134x + 0.297$	0.09	
S_(40:0)	7.77 ± 1.43	17.9 ± 12.7	10.5 ± 4.51	7.67 ± 4.6	$y = -0.285x + 13.1$	0.04	
S_(40:1-OH)	1.48 ± 0.815	1.73 ± 1.37	1.57 ± 0.909	1.46 ± 1.39	$y = -0.01x + 1.65$	0.06	
S_(40:2)	1.04 ± 0.637	1.44 ± 0.841	1.03 ± 0.425	0.566 ± 0.349	$y = -0.0679x + 1.54$	0.44	
S_(41:0)	0.352 ± 0.195	1.1 ± 0.943	0.253 ± 0.162	0.0222 ± 0.0413	$y = -0.068x + 0.952$	0.26	
S_(41:2)	ND ± ND	0.0108 ± 0.0324	0.021 ± 0.0318	0.0543 ± 0.0655	$y = 0.00806x - 0.0438$	0.91	
S_(42:0)	0.322 ± 0.225	0.59 ± 0.876	0.491 ± 0.223	0.294 ± 0.162	$y = -0.00678x + 0.476$	0.03	
S_(42:2)	0.395 ± 0.385	1.1 ± 1.26	0.958 ± 0.837	0.983 ± 0.763	$y = 0.0601x + 0.399$	0.44	
S_(42:2-OH)	2.43 ± 0.843	1.55 ± 0.588	0.786 ± 0.406	0.789 ± 0.434	$y = -0.211x + 3$	0.88	
S_(46:2-OH)	0.0599 ± 0.0337	0.0149 ± 0.0269	0.0228 ± 0.018	0.0258 ± 0.031	$y = -0.0035x + 0.0576$	0.37	
S_(48:2-OH)	0.00958 ± 0.0163	ND ± ND	0.00395 ± 0.00731	0.00115 ± 0.00326	$y = -0.00104x + 0.0133$	1.00	
SM_(30:1)	1.61 ± 0.262	1.59 ± 0.611	1.89 ± 0.679	1.84 ± 0.667	$y = 0.0367x + 1.45$	0.68	
SM_(32:0)	**4.99 ± 0.941**	**4.66 ± 2.13**	**3.81 ± 1.08**	**2.85 ± 1.21**	$y = -0.269x + 6.14 *$	**0.96**	**Decreasing**
SM_(32:1)	**428 ± 88.3**	**368 ± 107**	**334 ± 87.3**	**256 ± 63.6**	$y = -20.4x + 502 *$	**0.98**	**Decreasing**
SM_(33:1)	**273 ± 37.3**	**233 ± 62.2**	**168 ± 32.2**	**135 ± 44.4**	$y = -17.9x + 341 *$	**0.98**	**Decreasing**
SM_(34:0)	21.8 ± 5.98	26 ± 6.49	22.2 ± 6.65	18 ± 4.85	$y = -0.563x + 26.3$	0.36	
SM_(34:0-OH)	**7.39 ± 2.3**	**5.26 ± 2.46**	**3.5 ± 1.7**	**1.4 ± 1.69**	$y = -0.731x + 9.98 *$	**1.00**	**Decreasing**
SM_(34:1)	3400 ± 495	3830 ± 796	3240 ± 427	2620 ± 346	$y = -109x + 4100$	0.57	
SM_(34:1-OH)	4.36 ± 1.63	10.1 ± 8	5.89 ± 3.22	2.74 ± 2.48	$y = -0.336x + 8.34$	0.14	
SM_(34:2)	397 ± 102	401 ± 106	320 ± 50.4	282 ± 94.6	$y = -15.8x + 471$	0.88	
SM_(34:2-OH)	**4.44 ± 1.37**	**4.15 ± 1.73**	**3.2 ± 0.88**	**2.68 ± 0.928**	$y = -0.231x + 5.38 *$	**0.96**	**Decreasing**
SM_(35:0)	0.957 ± 1.35	0.969 ± 0.848	0.457 ± 0.267	0.389 ± 0.404	$y = -0.0821x + 1.32$	0.84	
SM_(35:1)	36.3 ± 3.74	48.6 ± 6.53	40.9 ± 7.5	41.1 ± 8.38	$y = 0.248x + 39.8$	0.03	
SM_(35:2)	0.632 ± 0.456	0.335 ± 0.314	0.144 ± 0.149	0.137 ± 0.18	$y = -0.0621x + 0.787$	0.87	Decreasing
SM_(36:1)	449 ± 43.4	529 ± 103	446 ± 58.3	298 ± 50.5	$y = -19.9x + 582$	0.52	
SM_(36:2)	32 ± 6.24	38.6 ± 9.35	30.9 ± 4.52	23.1 ± 5.49	$y = -1.27x + 40.9$	0.49	
SM_(36:3)	**2.19 ± 0.784**	**1.71 ± 0.891**	**0.76 ± 0.384**	**0.406 ± 0.326**	$y = -0.233x + 3.05 *$	**0.97**	**Decreasing**
SM_(37:2)	0.0328 ± 0.0984	0.024 ± 0.0721	ND ± ND	0.0197 ± 0.0556	$y = -0.0015x + 0.0363$	0.86	

Table 1. *Cont.*

	Diet 1	Diet 2	Diet 3	Diet 4	Trendline Equation	R^2	Successive Change Across Groups
SM_(38:0)	7.73 ± 1.45	7.83 ± 3.24	6.35 ± 1.81	4.64 ± 1.6	$y = -0.398x + 9.68$	0.86	
SM_(38:1)	**342 ± 56.4**	**293 ± 59.3**	**200 ± 27.7**	**106 ± 19.4**	$y = -29.7x + 462 *$	**0.98**	**Decreasing**
SM_(39:0)	8.55 ± 3.43	8.71 ± 6.78	12.5 ± 7.79	ND ± ND	$y = 0.731x + 5.31$	0.78	Increasing
SM_(39:1)	**178 ± 28.6**	**142 ± 35.5**	**109 ± 28.1**	**85.4 ± 10.7**	$y = -11.5x + 217 *$	**0.99**	**Decreasing**
SM_(40:0)	**17.3 ± 5.88**	**14.2 ± 4.27**	**10.8 ± 2.35**	**4.03 ± 1.2**	$y = -1.6x + 23.8 *$	**0.96**	**Decreasing**
SM_(40:0-OH)	**5.96 ± 1.62**	**4.29 ± 1.17**	**3.53 ± 1.12**	**0.919 ± 0.868**	$y = -0.588x + 8.17 *$	**0.95**	**Decreasing**
SM_(40:1)	**989 ± 137**	**867 ± 181**	**704 ± 125**	**520 ± 71.7**	$y = -58.1x + 1210 *$	**0.99**	**Decreasing**
SM_(40:2)	**138 ± 21.3**	**123 ± 21.6**	**91.7 ± 16.9**	**55.3 ± 11.1**	$y = -10.3x + 181 *$	**0.97**	**Decreasing**
SM_(41:0)	8.36 ± 2.45	8.68 ± 2.13	7.03 ± 1.56	5.33 ± 1.63	$y = -0.398x + 10.4$	0.83	
SM_(41:1)	**893 ± 123**	**839 ± 144**	**772 ± 135**	**695 ± 133**	$y = -24.5x + 987 *$	**0.99**	**Decreasing**
SM_(42:0)	8.67 ± 2.47	10.5 ± 3.94	7.13 ± 2.87	3.17 ± 1.49	$y = -0.736x + 13$	0.68	
SM_(42:0-OH)	**12.7 ± 2.47**	**11.1 ± 1.98**	**9.01 ± 2.88**	**4.92 ± 1.53**	$y = -0.942x + 16.6 *$	**0.95**	**Decreasing**
SM_(42:1)	**1610 ± 185**	**1540 ± 227**	**1280 ± 206**	**1020 ± 194**	$y = -75.2x + 1940 *$	**0.95**	**Decreasing**
SM_(42:2)	1160 ± 180	1180 ± 185	1010 ± 197	880 ± 109	$y = -37.4x + 1340$	0.86	
SM_(43:0)	0.202 ± 0.119	0.527 ± 0.275	0.368 ± 0.199	0.115 ± 0.152	$y = -0.0156x + 0.422$	0.09	
SM_(43:1)	158 ± 43.1	200 ± 35.8	163 ± 44.2	153 ± 57	$y = -1.93x + 183$	0.10	
SM_(44:1)	13.5 ± 4.24	17.9 ± 3.6	14.1 ± 4.99	9.32 ± 3.95	$y = -0.605x + 18.3$	0.36	
SM_(44:2)	27 ± 6.17	29.6 ± 5.31	27 ± 7.84	23.1 ± 5.83	$y = -0.53x + 30.7$	0.47	
TG_(18:0)	11 ± 8.91	11.3 ± 4.51	6.54 ± 4.04	2.47 ± 2.6	$y = -1.12x + 16.4$	0.88	
TG_(24:0)	920 ± 2500	39.9 ± 59.6	84.9 ± 195	177 ± 372	$y = -80.9x + 924$	0.46	
TG_(36:0)	170 ± 214	52.4 ± 36.9	93.4 ± 83.8	441 ± 288	$y = 31.6x - 52.8$	0.40	
TG_(44:1)	970 ± 941	325 ± 175	1550 ± 2130	2250 ± 1250	$y = 188x - 161$	0.63	
TG_(45:1)	14.3 ± 19	7.72 ± 5.87	32.3 ± 32.3	83.7 ± 49.3	$y = 8.62x - 31.4$	0.76	
TG_(45:2)	92.6 ± 86.2	31.6 ± 14.8	64.1 ± 58.7	84.2 ± 53.8	$y = 0.27x + 66.1$	0.00	
TG_(46:1)	85.4 ± 104	22.5 ± 16.1	87.6 ± 112	269 ± 189	$y = 22.8x - 58.4$	0.56	
TG_(46:2)	1060 ± 1040	293 ± 167	412 ± 311	787 ± 609	$y = -25.9x + 836$	0.07	
TG_(46:4)	3530 ± 3550	700 ± 350	464 ± 357	316 ± 241	$y = -366x + 4050$	0.70	Decreasing
TG_(47:0)	7.11 ± 7	2.59 ± 2.23	2.86 ± 4.1	6.91 ± 2.88	$y = -0.0122x + 4.96$	0.00	
TG_(47:1)	1.39 ± 3.19	0.253 ± 0.76	6.65 ± 10.2	19.6 ± 9.01	$y = 2.26x - 10.3$	0.79	
TG_(47:2)	11.5 ± 11.3	3.39 ± 2.87	10.9 ± 6.94	22.3 ± 17.7	$y = 1.48x + 0.715$	0.44	
TG_(48:0)	133 ± 69.6	63 ± 10.9	75.7 ± 98.4	92.4 ± 43.9	$y = -4.04x + 122$	0.21	
TG_(48:1)	142 ± 100	83 ± 29.2	279 ± 343	435 ± 172	$y = 39.8x - 69.8$	0.78	
TG_(48:2)	241 ± 196	115 ± 38.3	319 ± 401	440 ± 165	$y = 29.7x + 51.8$	0.57	
TG_(48:3)	250 ± 185	83.7 ± 27	112 ± 80.1	102 ± 49.9	$y = -15.4x + 255$	0.49	
TG_(49:0)	18 ± 9.16	10.7 ± 2.12	11 ± 5.69	10.2 ± 3.69	$y = -0.856x + 19$	0.65	
TG_(49:1)	35.7 ± 18.6	28.3 ± 9.67	67.9 ± 55.5	96.3 ± 30.3	$y = 8.2x - 5.68$	0.83	
TG_(49:2)	70.2 ± 29.8	50.1 ± 12.1	77.6 ± 55.4	86.3 ± 30.3	$y = 2.81x + 49.6$	0.40	
TG_(49:3)	27.3 ± 12.9	18.2 ± 5.03	26.8 ± 15.4	19.3 ± 7.36	$y = -0.57x + 27.3$	0.17	

Table 1. *Cont.*

	Diet 1	Diet 2	Diet 3	Diet 4	Trendline Equation	R^2	Successive Change Across Groups
TG_(50:0)	178 ± 68.6	118 ± 20.3	125 ± 57.6	102 ± 37.6	$y = -8.19x + 193$	0.75	
TG_(50:1)	1440 ± 548	1030 ± 339	1870 ± 1370	1870 ± 858	$y = 78.9x + 949$	0.47	
TG_(50:2)	2750 ± 1210	1830 ± 458	2920 ± 2200	2920 ± 1010	$y = 59.3x + 2150$	0.16	
TG_(50:3)	1500 ± 858	1050 ± 283	1760 ± 1370	1040 ± 452	$y = -24.8x + 1530$	0.06	
TG_(51:1)	79 ± 30.9	78.4 ± 31.8	138 ± 102	124 ± 39.6	$y = 7.21x + 49.7$	0.67	
TG_(51:2)	296 ± 163	274 ± 82.5	532 ± 293	509 ± 148	$y = 33.2x + 149$	0.72	
TG_(51:3)	**648 ± 459**	**467 ± 133**	**437 ± 175**	**186 ± 86.3**	$y = -52.4x + 836$ *	**0.92**	**Decreasing**
TG_(51:4)	**662 ± 568**	**330 ± 121**	**150 ± 55.5**	**37.2 ± 17.5**	$y = -76.1x + 877$ *	**0.94**	**Decreasing**
TG_(52:0)	**128 ± 70.4**	**80.4 ± 17.6**	**69.5 ± 28.8**	**47.4 ± 24.1**	$y = -9.36x + 153$ *	**0.92**	**Decreasing**
TG_(52:1)	849 ± 304	716 ± 311	1130 ± 904	823 ± 275	$y = 12.4x + 784$	0.06	
TG_(52:2)	12,300 ± 6460	9670 ± 3230	12,700 ± 6000	8610 ± 3300	$y = -298x + 13,100$	0.27	
TG_(52:3)	**36,800 ± 25,800**	**22,700 ± 5870**	**15,800 ± 5770**	**5160 ± 2670**	$y = -3770x + 49,000$ *	**0.98**	**Decreasing**
TG_(52:4)	**45,200 ± 36,700**	**21,100 ± 7240**	**9760 ± 3330**	**2220 ± 1880**	$y = -5200x + 59,300$ *	**0.93**	**Decreasing**
TG_(53:1)	28.5 ± 8.42	23.8 ± 10.3	36.4 ± 29.3	25.5 ± 8.57	$y = 0.133x + 27.5$	0.01	
TG_(53:2)	253 ± 129	240 ± 109	353 ± 259	214 ± 67.8	$y = -0.148x + 266$	0.00	
TG_(53:3)	**801 ± 527**	**518 ± 145**	**470 ± 175**	**183 ± 73.3**	$y = -70.4x + 1030$ *	**0.94**	**Decreasing**
TG_(53:4)	**877 ± 722**	**489 ± 180**	**277 ± 100**	**75.2 ± 38.5**	$y = -96.9x + 1170$ *	**0.97**	**Decreasing**
TG_(54:0)	20.1 ± 12.8	11.5 ± 2.27	9.51 ± 3.73	7.25 ± 5.07	$y = -1.5x + 23.6$	0.87	Decreasing
TG_(54:1)	117 ± 54.8	79 ± 24.1	96.6 ± 86.6	57.2 ± 25.8	$y = -5.99x + 133$	0.67	
TG_(54:2)	866 ± 326	637 ± 373	805 ± 731	408 ± 146	$y = -44.7x + 1020$	0.58	
TG_(54:3)	**5400 ± 3270**	**2880 ± 1140**	**2350 ± 1060**	**1000 ± 436**	$y = -509x + 6800$ *	**0.93**	**Decreasing**
TG_(54:4)	**14,200 ± 10,600**	**6230 ± 2180**	**3450 ± 1230**	**977 ± 444**	$y = -1570x + 18,200$ *	**0.91**	**Decreasing**
TG_(54:5)	26,100 ± 21,100	10,200 ± 4050	4300 ± 1350	1010 ± 595	$y = -3010x + 33,400$	0.89	Decreasing
TG_(54:6)	24,900 ± 20,300	9310 ± 3280	2980 ± 1090	554 ± 426	$y = -2940x + 31,900$	0.88	Decreasing
TG_(55:2)	13.8 ± 4.81	9.88 ± 6.38	16.2 ± 17.2	8.35 ± 3.23	$y = -0.371x + 14.9$	0.13	
TG_(55:3)	**77.5 ± 39.7**	**46.2 ± 18.2**	**44.1 ± 29.4**	**19.7 ± 8.29**	$y = -6.5x + 96.6$ *	**0.91**	**Decreasing**
TG_(55:4)	**142 ± 102**	**65.1 ± 25.8**	**44.9 ± 20.2**	**10.8 ± 6.06**	$y = -15.3x + 183$ *	**0.92**	**Decreasing**
TG_(55:5)	**161 ± 137**	**92.4 ± 44**	**49.8 ± 23.4**	**14.4 ± 11**	$y = -17.9x + 216$ *	**0.98**	**Decreasing**
TG_(55:6)	**122 ± 143**	**100 ± 46.1**	**50.3 ± 18.8**	**8.95 ± 8.19**	$y = -14.4x + 180$ *	**0.98**	**Decreasing**
TG_(56:2)	31.6 ± 11.6	20.8 ± 10.4	30.9 ± 40	11.7 ± 4.55	$y = -1.84x + 37.8$	0.46	
TG_(56:3)	**189 ± 87.4**	**100 ± 43.6**	**93.3 ± 74.8**	**31.4 ± 14.6**	$y = -17.8x + 239$ *	**0.91**	**Decreasing**
TG_(56:6)	**6970 ± 6060**	**3330 ± 1460**	**1200 ± 414**	**246 ± 173**	$y = -826x + 9260$ *	**0.93**	**Decreasing**
TG_(56:7)	**8570 ± 7750**	**3690 ± 1650**	**1190 ± 519**	**282 ± 194**	$y = -1010x + 11,200$ *	**0.90**	**Decreasing**
TG_(56:8)	7360 ± 7600	2990 ± 1460	893 ± 372	186 ± 119	$y = -875x + 9550$	0.89	Decreasing
TG_(57:2)	0.611 ± 0.405	0.434 ± 0.601	1 ± 1.89	0.371 ± 0.399	$y = -0.0057x + 0.648$	0.00	
TG_(57:6)	**83.1 ± 100**	**39.6 ± 24.4**	**16.6 ± 3.26**	**5.21 ± 5.03**	$y = -9.51x + 109$ *	**0.93**	**Decreasing**
TG_(58:10)	1370 ± 1470	564 ± 268	161 ± 77.3	29.9 ± 23.1	$y = -164x + 1780$	0.90	Decreasing
TG_(58:7)	1160 ± 1150	484 ± 241	123 ± 65.2	20 ± 18.1	$y = -140x + 1520$	0.90	Decreasing

Table 1. *Cont.*

	Diet 1	Diet 2	Diet 3	Diet 4	Trendline Equation	R^2	Successive Change Across Groups
TG_(58:8)	1600 ± 1440	654 ± 354	186 ± 77.9	42.6 ± 30.7	$y = -190x + 2080$	0.89	Decreasing
TG_(58:9)	1730 ± 1730	730 ± 418	206 ± 106	41.2 ± 28.3	$y = -207x + 2260$	0.90	Decreasing
TG_(59:3)	0.283 ± 0.273	0.333 ± 0.708	0.539 ± 1.26	0.0277 ± 0.0537	$y = -0.0207x + 0.454$	0.12	
TG_(59:4)	0.348 ± 0.474	0.22 ± 0.449	0.372 ± 0.807	0.00429 ± 0.0121	$y = -0.0326x + 0.485$	0.45	
TG_(59:5)	0.212 ± 0.241	0.104 ± 0.128	ND ± ND	ND ± ND	$y = -0.04x + 0.356$	1.00	
TG_(59:6)	0.382 ± 0.578	0.178 ± 0.227	0.137 ± 0.174	0.108 ± 0.197	$y = -0.032x + 0.446$	0.81	Decreasing
TG_(59:7)	0.214 ± 0.434	0.147 ± 0.232	ND ± ND	0.0326 ± 0.0607	$y = -0.0222x + 0.291$	1.00	
TG_(59:8)	**9.09 ± 14.6**	**6.77 ± 11.9**	**5.83 ± 9.99**	**1.75 ± 2.04**	$y = -0.85x + 12.4$ *	**0.94**	**Decreasing**
TG_(60:10)	194 ± 194	73.7 ± 56.6	4.21 ± 4.15	0.838 ± 1.23	$y = -24x + 252$	0.86	Decreasing
TG_(60:12)	**79.5 ± 74.4**	**39.6 ± 28**	**13 ± 12.1**	**1.42 ± 2.02**	$y = -9.66x + 107$ *	**0.94**	**Decreasing**
TG_(62:12)	**63.3 ± 43.7**	**31.3 ± 17.4**	**11 ± 5.05**	**0.863 ± 0.716**	$y = -7.69x + 85.4$ *	**0.95**	**Decreasing**
TG_(62:13)	**4.22 ± 2.92**	**2.86 ± 3.13**	**0.546 ± 1.15**	**0.0336 ± 0.0949**	$y = -0.551x + 6.13$ *	**0.95**	**Decreasing**

3. Discussion

Out of 472 lipids detected and semi-quantified, 100 showed a strong relationship with the dietary intake of ruminant fat with 35 species increasing and 65 species decreasing as the percentage of ruminant fat in the diet increased (NB. ruminant fat as a percentage composition with corn-oil and medium chain triacylglyceride oil). Interestingly, ceramides generally increased, whilst cardiolipins, sphingomyelins and triacylglycerides generally decreased as the dietary composition of ruminant fat rose. According to the literature, a rise in liver ceramides is typically associated with aggravated non-alcoholic fatty liver disease (NAFLD) and insulin resistance [23], this in conjunction with a decrease in cardiolipins (which are indicative of mitochondrial remodelling and dysfunction [24]) may suggest that the changes in the experimental diets here are detrimental for these pathologies. However, there was a clear decrease in the triacylglycerides (particularly evident in the unsaturated odd chain triacylglycerides), which is explicitly representative of an ameliorated pathology [25]. As previously published [26], many NAFLD and insulin resistance factors were mitigated as the ruminant fat increased in these diets, including: a reduction in the total body weight (g), total fat mass (%), serum ALT (U/mL) and degree of steatosis determined by Oil Red O staining, notably, the inflammatory marker TNFα did not change significantly (trend: *p*-value = 0.52). A key characteristic of NAFLD development is the accumulation of hepatic triacylglycerides [27]; therefore, the data here suggests that these dietary changes may be beneficial for NAFLD and insulin resistance by aiding in a reduced hepatic triacylglyceride load: possible mechanisms here include a lower saturated fatty acid composition resulting in a lower fatty acid incorporation into hepatic triacylglycerides and/or a higher pass-through of the medium chain triglyceride oil directly into the mitochondria stimulation fatty acid metabolism [26]. Work presented by Gonzalez-Cantero and colleagues [28] showed that hepatic triacylglyceride content were correlated with insulin resistance and these relationships were independently to the inflammatory marker TNFα. Therefore, it appears that the hepatic triacylglyceride load may be paramount in the development of NAFLD and insulin resistance, which is supported in the literature [29].

According to the literature, odd chain fatty acids are considered biomarkers of their dietary intake and particularly accredited as a biomarkers of ruminant fat intake (e.g., milk, butter and beef tallow, etc.); however, there is a vast amount of conflicting data [2]. Some studies have shown both positive correlations (either individual odd chain lipids or total odd chain lipids) and some studies have shown there were no significant correlations. Although these studies may conflict in their findings they all present their data as relative compositions (Mol%), which is the typical way lipid data appear in the literature [30]. By expressing the lipid data as relative compositions (Mol%), it normalises the data to the total fat in that sample; however, presenting the lipid data in this way confounds the results by interconnecting the individual data points. This interconnection can cause false positive and/or false negative conclusions (type 1 and 2 errors), i.e., if a single lipid increases it will artificially decrease the other(s) due to the Mol% calculation. As shown in the figure below (see Figure 2), the concentration of the total lipids containing either even chain or odd chain fatty acids and a combination are shown. Although lipids containing odd chain fatty acids did increase, albeit not statistically significantly; *p*-value: 0.197, it was also not proportionate to the increase in dietary ruminant fat. As shown, the lipids containing even chain fatty acids did significantly inversely decrease (slope *p*-value: 0.0189) as the dietary ruminant fat increased. Interestingly, due to both the decrease in the even chain lipids and the consistency of the odd chain lipids, if the relative composition (Mol%) of the lipids were calculated, there appears to be a statistically significant increase in the odd chain lipids (see Figure 3); however, this is an artefact of changes in the even chain lipids and a consequence of interconnecting the data.

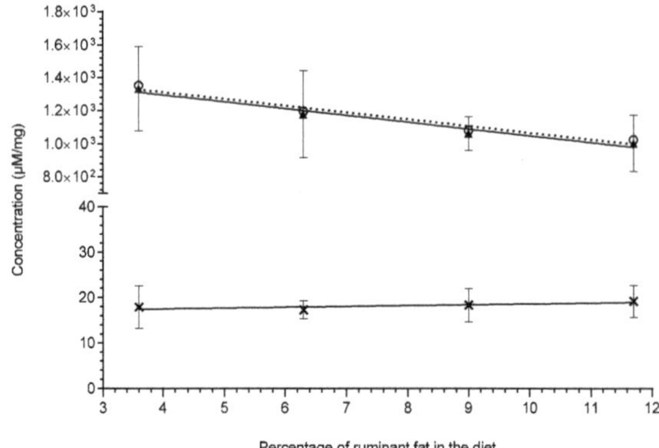

Figure 2. This figure shows the change in the liver lipid concentrations across the four high-fat diets fed to Sprague–Dawley rats (n = 8–9 per group): total odd chain lipids (symbol: **X**, trendline: ▬▬, gradient: 0.183 ± 0.0962, $R^2 = 0.64$, slope significance *p*-value: 0.197); total even chain lipids (symbol: ▲, trendline: ▬▬, gradient: -41.0 ± 5.71, $R^2 = 0.963$, slope significance *p*-value: 0.0189); total lipids containing both even and of odd chain (symbol: **O**, trendline: ▪▪▪▪▪, gradient: -40.8 ± 5.79, $R^2 = 0.961$, slope significance *p*-value: 0.0195. Lipid concentrations (μM/mg) are shown as means $\pm$ standard deviation and were extracted via the protein precipitation liquid extraction protocol (chloroform: methanol: acetone, ~7:3:4).

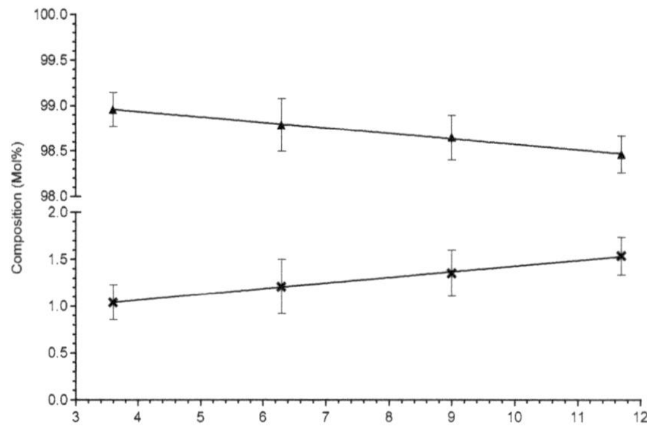

Figure 3. This figure shows the relative compositional (Mol%) change in the total odd chain lipids (symbol: **X**, trendline: ▬▬, gradient: 0.0604 ± 0.00203, $R^2 = 0.998$, slope significance *p*-value: 0.0011) and the total even chain lipids (symbol: ▲, trendline: ▬▬, gradient: -0.0604 ± 0.00203, $R^2 = 0.998$, slope significance *p*-value: 0.0011) across the four high-fat diets in Sprague–Dawley rats (n = 8–9 per group). Lipid compositions (Mol%) are shown as means $\pm$ standard deviation and were extracted via the protein precipitation liquid extraction protocol (chloroform: methanol: acetone, ~7:3:4). Diet one: 3.6% beef tallow; diet two: 6.3% beef tallow; diet three: 9.0% beef tallow; diet four: 11.7% beef tallow.

4. Materials and Methods

4.1. Chemicals and Standards

Stable isotope-labelled internal standards purchased from Sigma Aldrich (Haverhill, Suffolk, UK) include: N-palmitoyl-d31-D-erythro-sphingosine (abbreviated to IS_Cer_16:0-d31); order number: 868516P, 1-palmitoyl-d31-2-oleoyl-sn-glycero-3-phosphate (abbreviated to IS_PA_34:1-d31); order number: 860453P, 1-palmitoyl-d31-2-oleoyl-sn-glycero-3-phosphocholine (abbreviated to

IS_PC_34:1-d31); order number: 860399P, 1-palmitoyl-d31-2-oleoyl-sn-glycero-3-phosphoethanolamine (abbreviated to IS_PE_34:1-d31); order number: 860374P, 1-palmitoyl-d31-2-oleoyl-sn-glycero-3-[phospho-rac-(1-glycerol)] (abbreviated to IS_PG_34:1-d31); order number: 860384P, 1-palmitoyl-d31-2-oleoyl-sn-glycero-3-phosphoinositol (abbreviated to IS_PI_34:1-d31); order number: 860042P, 1,2-dimyristoyl-d54-sn-glycero-3-[phospho-L-serine] (abbreviated to IS_PS_28:0-d54); order number: 860401P, N-palmitoyl-d31-D-erythro-sphingosylphosphorylcholine (abbreviated to IS_SM_34:1-d31); order number: 868584P. Stable isotope-labelled internal standards purchased from QMX Laboratories Ltd. (QMX Laboratories Ltd., Thaxted, Essex, UK) include: Heptadecanoic-d33 acid (abbreviated to IS_FA_17:0-d33); order number: D-5261, N-tetradecylphosphocholine-d42 (abbreviated to IS_LPC_14:0-d42); order number: D-5885, Glyceryl tri(pentadecanoate-d29) (abbreviated to IS_TG_45:0-d87); order number: D-5265, Butyryl-d7-L-carnitine (abbreviated to IS_Car_4:0-d7); order number: D-7761, Hexadecanoyl-L-carnitine-d3 (abbreviated to IS_Car_16:0-d3); order number: D-6646.

Quality control standards (LIPID-QC) purchased from Cayman Chemical Company (Cambridge Bioscience, Cambridge, UK) include: Lysophosphatidylcholines (egg); order number: 24331, Phosphotidylcholines (egg); order number: 24343, Lysophosphatidylethanolamines (egg); order number: 25844, Phosphatidylethanolamines (bovine); order number: 16878, Phosphotidlethanolamine (soy); order number: 25845, Lysophosphatidyinositols (porcine liver); order number: 26016, Phosphatidylserines (soy); order number: 25847, Ceramides mixture; order number: 22853, Ceramides (non-hydroxy); order number: 24833, Ceramides (hydroxy); order number: 24834, Sphingomyelins (from bovine spinal cord); order number: 22674, Sphingomylins (egg); order number: 24345, Phosphatidylglycerols (egg); order number: 25846, Phosphatidic acid (egg); order number: 24344, Sulfatides (bovine); order number: 24323, Purified mixed gangliosides (bovine); order number: 24856, TLC Neutral Glycosphingolipid Mixture (bovine and porcine); order number: 1505, 2-Palmitoyl Glycerol; order number: CAY17882, 1,2-Dipalmitoyl-sn-glycerol; order number: CAY10008648. Quality control standards purchased from Sigma Aldrich include: Soy PC (95%); order number: 441601G, C18(Plasm)-18:1-PC; order number: 852467C, Brain CPE; order number: 860066P, Liver PI; order number: 840042P, Brain lyso PS; order number: 850092P, Milk SM Sphingomyelin (Milk, Bovine); order number: 860063P, Galactocerebrosides from bovine brain; order number: C4905, Glucosylceramide (Soy); order number: 131304P, Triglyceride mix, C2–C10; order number: 17810-1amp-s, Fish oil from menhaden; order number: F8020, Anhydrous butter fat, Cardiolipin solution from bovine heart; order number: C1649, Brain PI(4)P; order number: 840045P.

Commercially available blank human serum was purchased from BioIVT (Royston, Hertfordshire, UK; order number: HUMANSRMPNN. All solvents and additives were of HPLC grade or higher and purchased from Sigma Aldrich unless otherwise stated.

LIPID-IS: the lipid stable isotope-labelled internal standard was prepared by dissolving each of the individual lipid standards into chloroform: methanol (1:1) solution to produce a 1 mM primary stock solution. From each of these stock solutions, 1 mL was transferred into a volumetric flask and diluted with methanol to reach a final working solution concentration of 5 μM in methanol of IS_Cer_16:0-d31, IS_FA_17:0-d33, IS_LPC_14:0-d42, IS_PA_34:1-d31, IS_PC_34:1-d31, IS_PE_34:1-d31, IS_PG_34:1-d31, IS_PI_34:1-d31, IS_PS_28:0-d54, IS_SM_34:1-d31, IS_TG_45:0-d87.

ACYL-CARNITINE-IS: the acyl-carnitine stable isotope-labelled internal standard was prepared by dissolving each powdered stock into methanol to achieve a 5 mM stock solution. Taking 1 mL of the IS_Car_4:0-d7 and IS_Car_16:0-d3 stock solutions and diluting these into methanol until a final working solution of 5 μM was achieved for IS_Car_4:0-d7 and IS_Car_16:0-d3.

LIPID-QC: the lipid quality control standards were prepared by diluting each lipid mix to achieve a 50 μg/mL working stock solution in propan-2-ol: acetonitrile: water (2:1:1, respectively).

4.2. Extraction

Lipids were isolated comparing two methods; firstly, a novel protein-precipitation liquid extraction and secondly the liquid–liquid extraction previously described by Folch and colleagues [12] in an adapted version as we described previously [31]. Tissue quantities ranged from ~2–50 mg and fluid samples from 10–50 µL (e.g., plasma/serum) were tested (data not shown here).

4.2.1. Protein Precipitation Liquid Extraction Protocol (PPLE)

The protein-precipitation liquid extraction protocol was as follows: the tissue samples were weighed (NB. fluid samples were pipetted) and transferred into a 2 mL screw cap Eppendorf plastic tube (Eppendorf, Stevenage, UK) along with a single 5 mm stainless steel ball bearing. Immediately, 400 µL of chloroform: methanol (2:1, respectively) solution was added to each sample, followed by thorough mixing. The samples were then homogenised in the chloroform: methanol (2:1, respectively) using a Bioprep 24-1004 homogenizer (Allsheng, Hangzhou, China) run at speed; 4.5 m/s, time; 30 s for 2 cycles. Then, 400 µL of chloroform, 100 µL of the LIPID-IS (5 µM in methanol) and 100 µL of the CARNITINE-IS (5 µM in methanol) was added to each sample. The samples were homogenised again using a Bioprep 24-1004 homogenizer run at speed; 4.5 m/s, time; 30 s for 2 cycles. To ensure fibrous material was diminished, the samples were sonicated for 30 min in a water bath sonicator (Advantage-Lab, Menen, Belgium). Then, 400 µL of acetone was added to each sample. The samples were thoroughly vortexed and centrifuged for 10 min at ~20,000× g to pellet any insoluble material at the bottom of the vial. The single layer supernatant was pipetted into separate 2 mL screw cap amber-glass auto-sampler vials (Agilent Technologies, Cheadle, UK); being careful not to break up the solid pellet at the bottom of the tube. The organic extracts (chloroform, methanol, acetone composition, ~1.4 mL) were dried down to dryness using a Concentrator Plus system (Eppendorf, Stevenage, UK) run for 60 min at 60 °C. The samples were reconstituted in 100 µL of 2:1:1 (propan-2-ol, acetonitrile and water, respectively) then thoroughly vortex. The reconstituted sample was transferred into a 250 µL low-volume vial insert inside a 2 mL amber glass auto-sample vial ready for liquid chromatography with mass spectrometry detection (LC–MS) analysis.

4.2.2. Folch Liquid–Liquid Extraction Protocol (Folch LLE)

The Folch liquid–liquid extraction protocol is as follows: the tissue samples were weighed (NB. fluid samples were pipetted) and transferred into a 2 mL screw cap Eppendorf plastic tube (Eppendorf, Stevenage, UK) along with a single 5 mm stainless steel ball bearing. Immediately, 400 µL of chloroform: methanol (2:1, respectively) solution was added to each sample, followed by thorough mixing. The samples were then homogenised in the chloroform: methanol (2:1, respectively) using a Bioprep 24-1004 homogenizer (Allsheng, Hangzhou, China) run at speed; 4.5 m/s, time; 30 s for 2 cycles. Then, 400 µL of chloroform, 100 µL of the LIPID-IS (5 µM in methanol) and 100 µL of the ACYL-CARNITINE-IS (5 µM in methanol) was added to each sample. The samples were homogenised again using a Bioprep 24-1004 homogenizer run at speed; 4.5 m/s, time; 30 s for 2 cycles. To ensure fibrous material was diminished, the samples were sonicated for 30 min in a water bath sonicator. Then, 400 µL of HPLC water was added to each samples. The samples were thoroughly vortexed and centrifuged for 10 min at ~20,000 g to separate the two immiscible fractions. The organic fractions (the lower layer, mostly chloroform; ~700 µL) and aqueous fractions (the upper layer, methanol and water; ~700 µL) were pipetted into separate 2 mL screw cap amber-glass auto-sampler vials (Agilent Technologies, Cheadle, UK); being careful not to break up the solid pellet between the layers. To ensure complete lipid isolation a double extraction protocol was followed; 1 mL of chloroform: methanol (2:1, respectively) solution was added to each sample, along with 400 µL of HPLC water. The samples were thoroughly vortexed and centrifuged for 10 min at ~20,000× g. The organic fractions and aqueous fractions were pipetted into the corresponding 2 mL screw cap amber-glass auto-sampler vials containing the initial extracts (again being careful not to break up the solid pellet between the layers). The combined organic extracts (~1.4 mL) were dried down to dryness using a Concentrator Plus system (Eppendorf, Stevenage, UK) run for 60 min at 60 °C. The samples were reconstituted in 100 µL of 2:1:1 (propan-2-ol, acetonitrile and water, respectively) then thoroughly vortex. The reconstituted sample was transferred into a 250 µL low-volume vial insert inside a 2 mL amber glass auto-sample vial ready for liquid chromatography with mass spectrometry detection (LC–MS) lipidomics analysis.

4.3. LC–MS Method

Full chromatographic separation of intact lipids was achieved using a Shimadzu HPLC System (Shimadzu UK Limited, Milton Keynes, UK) with the injection of 10 µL onto a Waters Acquity UPLC® CSH C18 column (Waters, Hertfordshire, UK); 1.7 µm, I.D. 2.1 mm × 50 mm, maintained at 55 °C. Mobile phase A was 6:4, acetonitrile and water with 10 mM ammonium formate. Mobile phase B was 9:1, propan-2-ol and acetonitrile with 10 mM ammonium formate. The flow was maintained at 500 µL per minute through the following gradient: 0.00 min_40% mobile phase B; 0.40 min_43% mobile phase B; 0.45 min_50% mobile phase B; 2.40 min_54% mobile phase B; 2.45 min_70% mobile phase B; 7.00 min_99% mobile phase B; 8.00 min_99% mobile phase B; 8.3 min_40% mobile phase B; 10 min_40% mobile phase B. The sample injection needle was washed using 9:1, 2-propan-2-ol and acetonitrile. The mass spectrometer used was the Thermo Scientific Exactive Orbitrap with a heated electrospray ionisation source (Thermo Fisher Scientific, Hemel Hempstead, UK). The mass spectrometer was calibrated immediately before sample analysis using positive and negative ionisation calibration solution (recommended by Thermo Scientific). Additionally, the heated electrospray ionisation source was optimised at 50:50 mobile phase A to mobile phase B for spray stability (capillary temperature; 300 °C, source heater temperature; 420 °C, sheath gas flow; 40 (arbitrary), auxiliary gas flow; 15 (arbitrary), spare gas; 3 (arbitrary), source voltage; 4 kV. The mass spectrometer scan rate set at 4 Hz, giving a resolution of 25,000 (at 200 *m/z*) with a full-scan range of *m/z* 100 to 1800 with continuous switching between positive and negative mode.

4.4. Data Processing

Thermo Xcalibur Quan Browser (Thermo Fisher Scientific, Hemel Hempstead, UK) data processing involved the integration of the internal standard extracted ion chromatogram (EIC) peaks at the expected retention times (see Table 2). The EIC were selected from the ionisation mode for each analyte class; the ionisation mode is dependent on the molecular chemistry of the analytes, i.e., basic chemical groups ordinarily result in positive ionisation (e.g., $[M+H]^+$, $M+H-H_2O]^+$, $[M+Na]^+$, $[M+NH_4]^+$, $[M+K]^+$) whereas acidic chemical groups typically result in negative ionisation (e.g., $[M-H]^-$).

Table 2. This table shows the stable isotope-labelled internal standards with their ionisation products (i.e., $[M+H]^+$, $M+H-H_2O]^+$, $[M+Na]^+$, $[M+NH_4]^+$, $[M+K]^+$, $[M-H]^-$) and primary ionisation mode (positive; +ve or negative; −ve), along with their retention time (minutes). Butyryl-d7-L-carnitine (abbreviated to IS_Car_4:0-d7), N-tetradecylphosphocholine-d42 (abbreviated to IS_LPC_14:0-d42), hexadecanoyl-L-carnitine-d3 (abbreviated to IS_Car_16:0-d3), heptadecanoic-d33 acid (abbreviated to IS_FA_17:0-d33), 1,2-dimyristoyl-d54-sn-glycero-3-[phospho-L-serine] (abbreviated to IS_PS_28:0-d54), 1-palmitoyl-d31-2-oleoyl-sn-glycero-3-phosphoinositol (abbreviated to IS_PI_34:1-d31), N-palmitoyl-d31-D-erythro-sphingosylphosphorylcholine (abbreviated to IS_SM_34:1-d31), 1-palmitoyl-d31-2-oleoyl-sn-glycero-3-[phospho-rac-(1-glycerol)] (abbreviated to IS_PG_34:1-d31), 1-palmitoyl-d31-2-oleoyl-sn-glycero-3-phosphate (abbreviated to IS_PA_34:1-d31), N-palmitoyl-d31-D-erythro-sphingosine (abbreviated to IS_Cer_16:0-d31), 1-palmitoyl-d31-2-oleoyl-sn-glycero-3-phosphocholine (abbreviated to IS_PC_34:1-d31), 1-palmitoyl-d31-2-oleoyl-sn-glycero-3-phosphoethanolamine (abbreviated to IS_PE_34:1-d31), glyceryl tri(pentadecanoate-d29) (abbreviated to IS_TG_45:0-d87).

Internal Standard	Ionisation Product (*m/z*)	Ionisation Mode	Expected Retention Time (mins)
IS_Car_4:0-d7	239.1983	+ve	0.3
IS_LPC_14:0-d42	422.5560, 421.5498, 420.5435	+ve	0.4
IS_Car_16:0-d3	403.3610	+ve	0.5
IS_FA_17:0-d33	302.4557, 301.4495, 300.4432	−ve	1.1
IS_PS_28:0-d54	732.7741, 731.7678, 730.7615, 729.7553, 728.7490	−ve	1.4

Table 2. *Cont.*

Internal Standard	Ionisation Product (*m/z*)	Ionisation Mode	Expected Retention Time (mins)
IS_PI_34:1-d31	864.7162, 865.7225, 866.7288	−ve	2.9
IS_SM_34:1-d31	733.7632, 734.7670, 755.7451, 756.7514, 771.7190, 772.7253	+ve	3.0
IS_PG_34:1-d31	775.6939, 776.7002, 777.7065, 778.7127	−ve	3.0
IS_PA_34:1-d31	700.6509, 701.6571, 702.6634, 703.6697, 704.6760	−ve	3.4
IS_Cer_16:0-d31	548.6851, 549.6914, 550.6977, 551.7039, 566.6951, 567.7014, 568.7076, 569.7139, 590.6896, 591.6959, 606.6636, 607.6698	+ve	3.9
IS_PC_34:1-d31	790.7700, 791.7750, 812.7553, 813.7616, 828.7292, 829.7355	+ve	3.9
IS_PE_34:1-d31	747.7181, 748.7254, 749.7327, 769.7021, 770.7084, 771.7146, 785.6760, 786.6823, 787.6886	+ve	4.0
IS_TG_45:0-d87	850.2239, 851.2301, 852.2364, 853.2427, 867.2504, 868.2567, 869.2630, 870.2693, 872.2059, 873.2121, 874.2184, 875.2247, 888.1798, 889.1861, 890.1923, 891.1986	+ve	5.8

As shown in the table above (see Table 2), the internal standards have multiple ionisation products, these are the result of numerous ionisation mechanism (for example IS_TG_45:0-d87 having different adducts: $[M+H]^+$, $[M+Na]^+$, $[M+K]^+$ and $[M+NH_4]^+$, present) as well as an isotopic distribution (e.g., IS_TG_45:0-d87 having either the expected eighty-seven or fewer deuterium atoms present) all reasonably expected ions were included into the EIC for each internal standard (see Figure 4).

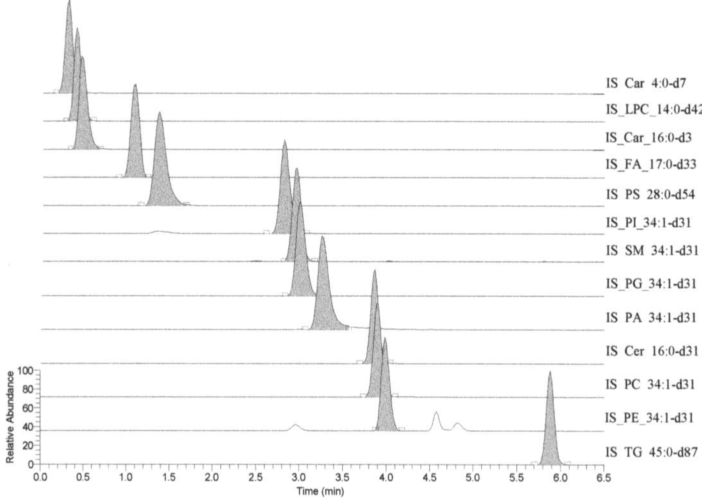

Figure 4. This figure shows a spiked (5 µM in methanol) commercial human plasma extracted ion chromatogram (EIC) for the stable isotope-labelled internal standards (lipids and acyl-carnitines): butyryl-d7-L-carnitine (abbreviated to IS_Car_4:0-d7); area ~5.9 × 10^6 counts, N-tetradecylphosphocholine-d42 (abbreviated to IS_LPC_14:0-d42); are ~2.1 × 10^8 counts, hexadecanoyl-L-carnitine-d3 (abbreviated to IS_Car_16:0-d3); area ~5.8 × 10^8 counts, heptadecanoic-d33 acid (abbreviated to IS_FA_17:0-d33); area ~1.1 × 10^4 counts, 1,2-dimyristoyl-d54-sn-glycero-3-[phospho-L-serine] (abbreviated to IS_PS_28:0-d54); area ~2.2 × 10^7 counts, 1-palmitoyl-d31-2-oleoyl-sn-glycero-3-phosphoinositol (abbreviated to IS_PI_34:1-d31): area ~1.1 × 10^7 counts, N-palmitoyl-d31-D-erythro-sphingosylphosphorylcholine (abbreviated to IS_SM_34:1-d31); 1.4 × 10^8 counts, 1-palmitoyl-d31-2-oleoyl-sn-glycero-3-[phospho-rac-(1-glycerol)] (abbreviated to IS_PG_34:1-d31); area ~6.1 × 10^7 counts, 1-palmitoyl-d31-2-oleoyl-sn-glycero-3-phosphate (abbreviated to IS_PA_34:1-d31); area ~1.3 × 10^7 counts, N-palmitoyl-d31-D-erythro-sphingosine (abbreviated to IS_Cer_16:0-d31); area ~2.2 × 10^8 counts, 1-palmitoyl-d31-2-oleoyl-sn-glycero-3-phosphocholine (abbreviated to IS_PC_34:1-d31); area ~2.5 × 10^8 counts, 1-palmitoyl-d31-2-oleoyl-sn-glycero-3-phosphoethanolamine (abbreviated to IS_PE_34:1-d31); area ~6.3 × 10^7 counts, glyceryl tri(pentadecanoate-d29) (abbreviated to IS_TG_45:0-d87); area ~3.2 × 10^7 counts.

The adduct composition of the total EIC produced from each of the ionisation mechanisms are shown in the figure below (see Figure 5).

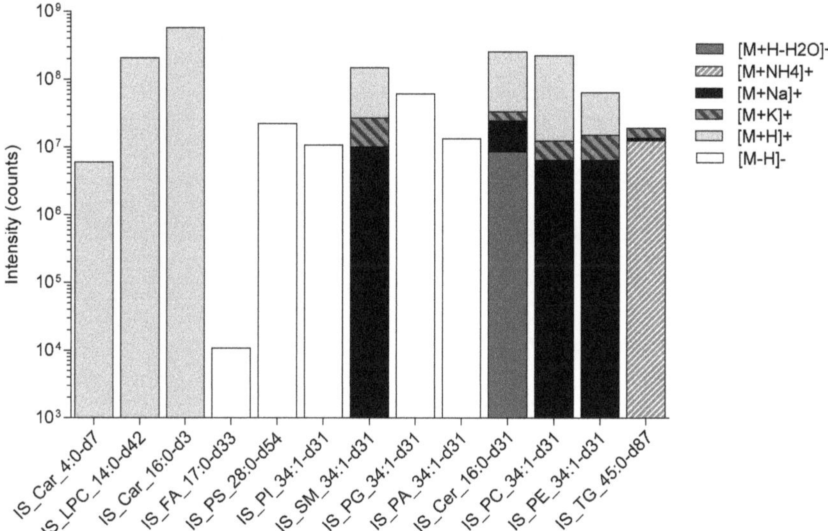

Figure 5. This figure shows the intensity of the extracted ion chromatogram for each stable isotope-labelled internal standard along with the ionisation adduct composition: $[M+H]^+$, $[M+H-H_2O]^+$, $[M+NH_4]^+$, $[M+Na]^+$, $[M+K]^+$ and $[M-H]^-$. Butyryl-d7-L-carnitine (abbreviated to IS_Car_4:0-d7), N-tetradecylphosphocholine-d42 (abbreviated to IS_LPC_14:0-d42), hexadecanoyl-L-carnitine-d3 (abbreviated to IS_Car_16:0-d3), heptadecanoic-d33 acid (abbreviated to IS_FA_17:0-d33), 1,2-dimyristoyl-d54-sn-glycero-3-[phospho-L-serine] (abbreviated to IS_PS_28:0-d54), 1-palmitoyl-d31-2-oleoyl-sn-glycero-3-phosphoinositol (abbreviated to IS_PI_34:1-d31), N-palmitoyl-d31-D-erythro-sphingosylphosphorylcholine (abbreviated to IS_SM_34:1-d31), 1-palmitoyl-d31-2-oleoyl-sn-glycero-3-[phospho-rac-(1-glycerol)] (abbreviated to IS_PG_34:1-d31), 1-palmitoyl-d31-2-oleoyl-sn-glycero-3-phosphate (abbreviated to IS_PA_34:1 d31), N palmitoyl d31 D erythro sphingosine (abbreviated to IS_Cer_16:0 d31), 1-palmitoyl-d31-2-oleoyl-sn-glycero-3-phosphocholine (abbreviated to IS_PC_34:1-d31), 1-palmitoyl-d31-2-oleoyl-sn-glycero-3-phosphoethanolamine (abbreviated to IS_PE_34:1-d31), glyceryl tri(pentadecanoate-d29) (abbreviated to IS_TG_45.0-d87).

The data processing also involved the integration of the individual lipid (and derivatives) species at their expected retention time (see Supplementary Table S2) allowing for a maximum of ±0.1 min of retention time drift: any retention time drift greater than ±0.1 min resulted in the exclusion of the analyte leading to a 'Not Found' result (i.e., zero concentration). A list of the analyte classes along with the number of species detected within each class are shown in the table below (see Table 3). The expected adducts for each analyte class and the internal standard used for semi-quantitation are also shown.

The lipid quality control (QC) standards were analysed with each batch of samples, these QC standards were used to check the retention times for the analytes ensuring that isobaric analytes were separated and expected analyte retention times remained robust.

Through the Thermo Xcalibur Quan Browser software, the responses of the analytes were normalised to the relevant internal standard response (producing area ratios) (see Supplementary Table S2), these area ratios corrected the intensity for any extraction and instrument variations. The area ratios were then blank corrected where intensities less than three times the blank samples were set to a 'Not Found' result (i.e., zero concentration). The accepted area ratios were then multiplied by the concentration of the internal standard to give the analyte concentrations. The results for fluid samples were expressed in molar concentrations (typically μM or nM). For tissue samples, the calculated concentrations of the analytes were then divided by the amount of tissue (in mg) used in the extraction protocol to give the final results in μM per mg of tissue extracted (μM/mg).

Table 3. This table shows the lipid classes detected with this LC–MS lipidomics method. The number of species per lipid class and the measured adducts (protonated: $[M+H]^+$, deprotonated: $[M-H]^-$, protonated with water loss: $[M+H-H_2O]^+$, sodiated: $[M+Na]^+$, potasiated: $[M+K]^+$, ammoniated: $[M+NH_4]^+$) are also shown. The internal standard used for semi-quantification are also shown: butyryl-d7-L-carnitine (abbreviated to IS_Car_4:0-d7), N-tetradecylphosphocholine-d42 (abbreviated to IS_LPC_14:0-d42), hexadecanoyl-L-carnitine-d3 (abbreviated to IS_Car_16:0-d3), heptadecanoic-d33 acid (abbreviated to IS_FA_17:0-d33), 1,2-dimyristoyl-d54-sn-glycero-3-[phospho-L-serine] (abbreviated to IS_PS_28:0-d54), 1-palmitoyl-d31-2-oleoyl-sn-glycero-3-phosphoinositol (abbreviated to IS_PI_34:1-d31), N-palmitoyl-d31-D-erythro-sphingosylphosphorylcholine (abbreviated to IS_SM_34:1-d31), 1-palmitoyl-d31-2-oleoyl-sn-glycero-3-[phospho-rac-(1-glycerol)] (abbreviated to IS_PG_34:1-d31), 1-palmitoyl-d31-2-oleoyl-sn-glycero-3-phosphate (abbreviated to IS_PA_34:1-d31), N-palmitoyl-d31-D-erythro-sphingosine (abbreviated to IS_Cer_16:0-d31), 1-palmitoyl-d31-2-oleoyl-sn-glycero-3-phosphocholine (abbreviated to IS_PC_34:1-d31), 1-palmitoyl-d31-2-oleoyl-sn-glycero-3-phosphoethanolamine (abbreviated to IS_PE_34:1-d31), glyceryl tri(pentadecanoate-d29) (abbreviated to IS_TG_45:0-d87).

Analyte Class	No. of Species	Adducts	Internal Standard
Acyl-carnitines	48	$[M+H]^+$	IS_Car_4:0-d7, IS_Car_16:0-d3
Ceramides	85	$[M+H]^+$, $[M+H-H_2O]^+$	IS_Cer_16:0-d31
Cardiolipins	56	$[M-H]^-$	IS_TG_45:0-d87
Diacylglycerols	6	$[M+H-H_2O]^+$, $[M+Na]^+$, $[M+K]^+$	IS_TG_45:0-d87
Gangliosides (GM1)	24	$[M-H]^-$	IS_PG_34:1-d31
Hexosylceramides	56	$[M+H]^+$, $[M+H-H_2O]^+$	IS_Cer_16:0-d31
Lyso-phosphatidylcholines	23	$[M+H]^+$	IS_LPC_14:0-d42
Lyso-phosphatidyethanolamines	19	$[M+H]^+$	IS_LPC_14:0-d42
Lyso-phosphatidylinositols	19	$[M-H]^-$	IS_PI_34:1-d31
Lyso-phosphoserines	20	$[M-H]^-$	IS_PS_28:0-d54
Lyso-cardiolipins	23	$[M-H]^-$	IS_TG_45:0-d87
Monoacylglycerols	1	$[M+H-H_2O]^+$, $[M+Na]^+$, $[M+K]^+$	IS_TG_45:0-d87
Phosphatidic acids	26	$[M-H]^-$	IS_PA_34:1-d31
Phosphatidylcholines	43	$[M+H]^+$	IS_PC_34:1-d31
Phosphatidylethanolamines	19	$[M+H]^+$	IS_PE_34:1-d31
Phosphatidylglycerol	34	$[M-H]^-$	IS_PG_34:1-d31
Phosphatidylinositols	21	$[M-H]^-$	IS_PI_34:1-d31
Phosphatidylserines	36	$[M-H]^-$	IS_PS_28:0-d54
Sulfatides	72	$[M-H]^-$	IS_PG_34:1-d31
Sphingomyelins	54	$[M+H]^+$, $[M+Na]^+$, $[M+K]^+$	IS_SM_34:1-d31
Triacylglycerides	89	$[M+H]^+$, $[M+NH_4]^+$, $[M+Na]^+$, $[M+K]^+$	IS_TG_45:0-d87

4.5. Animal Intervention

Sprague–Dawley rats (Harlan, IN, USA) were overfed using one of four experimental diets (n = 6–9 per group) at 17% above matched growth via an intragastric cannula surgically inserted as previously described [26]. Animals had ad libitum access to water throughout the experiments. The four experimental diets were 70% fat (% energy) including different amounts of medium chain triacylglycerides oil (MCT), beef tallow and corn oil; the fat composition of each diet are shown in the table below (see Table 4).

Protein (19% whey protein), vitamin and mineral contents were the same in all diets. Diets were formulated to meet the caloric and nutritional recommendations established by the National Research Council (NRC), but were fed at a level that exceeded the recommended caloric intake by 17% to increase weight gain and adiposity and produce steatohepatitis.

Liver tissue was collected after 21 days. All experimental procedures were ethically approved by the Institutional Animal Care and Use Committee at the University of Arkansas for Medical Science.

Table 4. This table shows the dietary fat composition of each of the four experimental diets fed to Sprague–Dawley rats (n = 8–9 per group). MCT: medium chain triglyceride oil.

Diet	Corn Oil	MCT Oil	Beef Tallow
1	50%	16.4%	3.6%
2	35%	28.7%	6.3%
3	20%	41.0%	9.0%
4	5%	53.3%	11.7%

5. Conclusions

This lipidomics protocol has been developed to quantify lipids across a broad range of hydrophobicities, from acyl-carnitines through to long chain glycerolipids. The extraction method produces a single liquid supernatant phase ideal for high-throughput workflows with an increased extraction capability over the frequently published liquid–liquid extraction previously published by Folch and colleagues [12].

Following the establishment and validation of this method, we applied it to a ruminant fat dose response dietary intervention in Sprague–Dawley rats, where we found 100 lipid species correlated strongly with the composition of ruminant fat within the diet.

It has been previously suggested that dietary ruminant fat is beneficial/protective in type 2 diabetes [32], the results presented in this manuscript suggest possible target mechanisms that need to be examined could include ceramide fatty acid compositions, cardiolipin remodeling, sphingomyelins and/or triacylglycerides concentration (particularly unsaturated odd chain species) and their associated fatty acid compositions, as well as the liver total lipid content.

Supplementary Materials: The following are available online at http://www.mdpi.com/2218-1989/10/7/296/s1, Figure S1: This figure shows the comparison between the two lipid extraction techniques regarding their extraction efficiency on each lipid class detected in the rat liver samples (Folch liquid–liquid extraction with a compositions of chloroform: methanol: water, ~7:3:4, and Protein precipitation liquid extraction with a composition of chloroform: methanol: acetone, ~7:3:4). n = 34 rat liver samples per extraction method. The intensity of the lipids were measured by liquid chromatography with mass spectrometry. The significance of the difference between the two extraction protocols are shown by the p-value star system; where $p \leq 0.05$ was considered statistically significant (* $p < 0.05$, ** $p < 0.01$, *** $p < 0.001$). Error bars represent ± standard deviation. Table S1: This table shows the comparison between the two lipid extraction techniques regarding their extraction efficiency on the total intensity of each lipid class detected in the rat liver samples by each sample extraction method (Folch-LLE: Folch liquid–liquid extraction with a compositions of chloroform: methanol: water, ~7:3:4, and PPLE: protein precipitation liquid extraction with a composition of chloroform: methanol: acetone, ~7:3:4). n = 34 rat liver samples per extraction method. The percentage increase the PPLE method is over the Folch-LLE method is shown (% diff.) along with the p-value resulting from a t-test ($p < 0.05$ designates statistical significance are in bold & shaded). The total number of lipid species detected and pass the quality control process are also shown. Table S2: This table shows the lipids quantified in this LC–MS method, along with the ionisation mode (either positive; +ve, or negative; −ve), the detected ion (m/z), the expected retention time (minutes) and the internal standard used for normalisation and quantification. Lipid are shown in their shorthand notations with the number of carbons and unsaturated bonds in the fatty acid moiety separated by a colon; acyl-carnitines (Carn), ceramides (Cer), cardiolipins (CL), diacylglycerols (DG), gangliosides (GM1), hexosylceramides (Hex-Cer), lyso-phosphatidylcholines (LPC), lyso-phosphatidyethanolamines (LPE), lyso-phosphatidylinositols (LPI), lyso-cardiolipins (Lyso_CL), phosphatidic acids (PA), phosphatidylcholines (PC), phosphatidylethanolamines (PE), phosphatidylglycerol (PG), phosphatidylinositols (PI), phosphatidylserines (PS), sulfatides (S), sphingomyelins (SM), triacylglycerides (TG).

Author Contributions: Conceptualization, B.J.; methodology, B.J.; validation, B.J.; formal analysis, B.J.; investigation, B.J. and M.R.; resources, M.R. and A.K.; data curation, B.J.; writing—original draft preparation, B.J., M.R. and A.K.; writing—review and editing, B.J., M.R. and A.K.; visualization, B.J.; supervision, M.R. and A.K.; project administration, B.J.; funding acquisition, M.R. and A.K. All authors have read and agreed to the published version of the manuscript.

Funding: This research was funded by BBSRC, grant number BB/M027252/1, USDA, grant number ACNC-USDA-CRIS 6251-51000-005-03S, and COST Action, grant number CM0603.

Conflicts of Interest: The authors declare no conflict of interest.

References

1. LIPID MAPS® Lipidomics Gateway LIPID MAPS® Lipidomics Gateway. Available online: https://www.lipidmaps.org/ (accessed on 20 April 2020).
2. Jenkins, B.; West, J.A.; Koulman, A. A Review of Odd-Chain Fatty Acid Metabolism and the Role of Pentadecanoic Acid (C15:0) and Heptadecanoic Acid (C17:0) in Health and Disease. *Molecules* **2015**, *20*, 2425–2444. [CrossRef] [PubMed]
3. Berg, J.M.; Tymoczko, J.L.; Stryer, L. *Biochemistry*, 5th ed.; W.H. Freeman: New York, NY, USA, 2002.
4. Nakamura, M.T.; Yudell, B.E.; Loor, J.J. Regulation of energy metabolism by long-chain fatty acids. *Progress Lipid Res.* **2014**, *53*, 124–144. [CrossRef] [PubMed]
5. Wymann, M.P.; Schneiter, R. Lipid signalling in disease. *Nat. Rev. Mol. Cell Biol.* **2008**, *9*, 162–176. [CrossRef] [PubMed]
6. Harayama, T.; Riezman, H. Understanding the diversity of membrane lipid composition. *Nat. Rev. Mol. Cell Biol.* **2018**, *19*, 281–296. [CrossRef]
7. Jenkins, B.J.; Seyssel, K.; Chiu, S.; Pan, P.-H.; Lin, S.-Y.; Stanley, E.; Ament, Z.; West, J.A.; Summerhill, K.; Griffin, J.L.; et al. Odd Chain Fatty Acids; New Insights of the Relationship Between the Gut Microbiota, Dietary Intake, Biosynthesis and Glucose Intolerance. *Sci. Rep.* **2017**, *7*, 44845. [CrossRef]
8. Jenkins, B.; de Schryver, E.; Van Veldhoven, P.P.; Koulman, A. Peroxisomal 2-Hydroxyacyl-CoA Lyase Is Involved in Endogenous Biosynthesis of Heptadecanoic Acid. *Molecules* **2017**, *22*, 1718. [CrossRef]
9. Zarrouk, A.; Debbabi, M.; Bezine, M.; Karym, E.M.; Badreddine, A.; Rouaud, O.; Moreau, T.; Cherkaoui-Malki, M.; El Ayeb, M.; Nasser, B.; et al. Lipid Biomarkers in Alzheimer's Disease. *Curr. Alzheimer Res.* **2018**, *15*, 303–312. [CrossRef]
10. Dorcely, B.; Katz, K.; Jagannathan, R.; Chiang, S.S.; Oluwadare, B.; Goldberg, I.J.; Bergman, M. Novel biomarkers for prediabetes, diabetes, and associated complications. *Diabetes Metab. Syndr. Obes.* **2017**, *10*, 345–361. [CrossRef]
11. Niki, E. Biomarkers of lipid peroxidation in clinical material. *Biochim. Biophys. Acta (BBA) Gen. Subj.* **2014**, *1840*, 809–817. [CrossRef]
12. Folch, J.; Lees, M.; Stanley, G.H.S. A Simple Method for the Isolation and Purification of Total Lipides from Animal Tissues. *J. Biol. Chem.* **1957**, *226*, 497–509.
13. Bligh, E.G.; Dyer, W.J. A Rapid Method of Total Lipid Extraction and Purification. *Can. J. Biochem. Physiol.* **1959**, *37*, 911–917. [CrossRef]
14. Matyash, V.; Liebisch, G.; Kurzchalia, T.V.; Shevchenko, A.; Schwudke, D. Lipid extraction by methyl-tert-butyl ether for high-throughput lipidomics. *J. Lipid Res.* **2008**, *49*, 1137–1146. [CrossRef] [PubMed]
15. Gil, A.; Zhang, W.; Wolters, J.C.; Permentier, H.; Boer, T.; Horvatovich, P.; Heiner-Fokkema, M.R.; Reijngoud, D.-J.; Bischoff, R. One- vs two-phase extraction: Re-evaluation of sample preparation procedures for untargeted lipidomics in plasma samples. *Anal. Bioanal. Chem.* **2018**, *410*, 5859–5870. [CrossRef] [PubMed]
16. Pellegrino, R.M.; Di Veroli, A.; Valeri, A.; Goracci, L.; Cruciani, G. LC/MS lipid profiling from human serum: A new method for global lipid extraction. *Anal. Bioanal. Chem.* **2014**, *406*, 7937–7948. [CrossRef] [PubMed]
17. Cajka, T.; Fiehn, O. Comprehensive analysis of lipids in biological systems by liquid chromatography-mass spectrometry. *Trends Anal. Chem.* **2014**, *61*, 192–206. [CrossRef]
18. Cajka, T.; Fiehn, O. Increasing lipidomic coverage by selecting optimal mobile-phase modifiers in LC–MS of blood plasma. *Metabolomics* **2016**, *12*, 34. [CrossRef]
19. Isaac, G.; McDonald, S.; Astarita, G. *Lipid Separation Using UPLC with Charged Surface Hybrid Technology*; Waters Corp: Milford, MA, USA, 2011; pp. 1–8.
20. Ulmer, C.Z.; Patterson, R.E.; Koelmel, J.P.; Garrett, T.J.; Yost, R.A. A Robust Lipidomics Workflow for Mammalian Cells, Plasma, and Tissue Using Liquid-Chromatography High- Resolution Tandem Mass Spectrometry. *Methods Mol. Biol.* **2017**, *1609*, 91–106. [CrossRef]
21. Haider, A.; Wei, Y.-C.; Lim, K.; Barbosa, A.D.; Liu, C.-H.; Weber, U.; Mlodzik, M.; Oras, K.; Collier, S.; Hussain, M.M.; et al. PCYT1A Regulates Phosphatidylcholine Homeostasis from the Inner Nuclear Membrane in Response to Membrane Stored Curvature Elastic Stress. *Dev. Cell* **2018**, *45*, 481–495.e8. [CrossRef]

22. Petkevicius, K.; Virtue, S.; Bidault, G.; Jenkins, B.; Çubuk, C.; Morgantini, C.; Aouadi, M.; Dopazo, J.; Serlie, M.; Koulman, A.; et al. Accelerated phosphatidylcholine turnover in macrophages promotes adipose tissue inflammation in obesity. *eLife* **2019**, *8*, e47990. [CrossRef]

23. Pagadala, M.; Kasumov, T.; McCullough, A.J.; Zein, N.N.; Kirwan, J.P. Role of ceramides in nonalcoholic fatty liver disease. *Trends Endocrinol. Metab.* **2012**, *23*, 365–371. [CrossRef]

24. Shi, Y. Emerging roles of cardiolipin remodeling in mitochondrial dysfunction associated with diabetes, obesity, and cardiovascular diseases. *J. Biomed. Res.* **2010**, *24*, 6–15. [CrossRef]

25. Kawano, Y.; Cohen, D.E. Mechanisms of hepatic triglyceride accumulation in non-alcoholic fatty liver disease. *J. Gastroenterol.* **2013**, *48*, 434–441. [CrossRef] [PubMed]

26. Ronis, M.J.J.; Baumgardner, J.N.; Sharma, N.; Vantrease, J.; Ferguson, M.; Tong, Y.; Wu, X.; Cleves, M.A.; Badger, T.M. Medium chain triglycerides dose-dependently prevent liver pathology in a rat model of non-alcoholic fatty liver disease. *Exp. Biol. Med.* **2013**, *238*, 151–162. [CrossRef] [PubMed]

27. Alkhouri, N.; Dixon, L.J.; Feldstein, A.E. Lipotoxicity in Nonalcoholic Fatty Liver Disease: Not All Lipids Are Created Equal. *Expert Rev. Gastroenterol. Hepatol.* **2009**, *3*, 445–451. [CrossRef] [PubMed]

28. Gonzalez-Cantero, J.; Martin-Rodriguez, J.L.; Gonzalez-Cantero, A.; Arrebola, J.P.; Gonzalez-Calvin, J.L. Insulin resistance in lean and overweight non-diabetic Caucasian adults: Study of its relationship with liver triglyceride content, waist circumference and BMI. *PLoS ONE* **2018**, *13*, e192663. [CrossRef]

29. Dongiovanni, P.; Stender, S.; Pietrelli, A.; Mancina, R.M.; Cespiati, A.; Petta, S.; Pelusi, S.; Pingitore, P.; Badiali, S.; Maggioni, M.; et al. Causal relationship of hepatic fat with liver damage and insulin resistance in nonalcoholic fatty liver. *J. Intern. Med.* **2018**, *283*, 356–370. [CrossRef]

30. Liebisch, G.; Ahrends, R.; Arita, M.; Arita, M.; Bowden, J.A.; Ejsing, C.S.; Griffiths, W.J.; Holčapek, M.; Köfeler, H.; Mitchell, T.W.; et al. Lipidomics needs more standardization. *Nat. Metab.* **2019**, *1*, 745–747. [CrossRef]

31. Virtue, S.; Petkevicius, K.; Moreno-Navarrete, J.M.; Jenkins, B.; Hart, D.; Dale, M.; Koulman, A.; Fernández-Real, J.M.; Vidal-Puig, A. Peroxisome Proliferator-Activated Receptor γ2 Controls the Rate of Adipose Tissue Lipid Storage and Determines Metabolic Flexibility. *Cell Rep.* **2018**, *24*, 2005–2012.e7. [CrossRef]

32. Forouhi, N.G.; Koulman, A.; Sharp, S.J.; Imamura, F.; Kröger, J.; Schulze, M.B.; Crowe, F.L.; Huerta, J.M.; Guevara, M.; Beulens, J.W.; et al. Differences in the prospective association between individual plasma phospholipid saturated fatty acids and incident type 2 diabetes: The EPIC-InterAct case-cohort study. *Lancet Diabetes Endocrinol.* **2014**, *2*, 810–818. [CrossRef]

 *metabolites*

Article

Lipid Annotation by Combination of UHPLC-HRMS (MS), Molecular Networking, and Retention Time Prediction: Application to a Lipidomic Study of in Vitro Models of Dry Eye Disease

Romain Magny [1,2], Anne Regazzetti [2], Karima Kessal [1,3], Gregory Genta-Jouve [2,4], Christophe Baudouin [1,3,5], Stéphane Mélik-Parsadaniantz [1], Françoise Brignole-Baudouin [1,2,3], Olivier Laprévote [2,6] and Nicolas Auzeil [2,*]

[1] Sorbonne Université UM80, INSERM UMR 968, CNRS UMR 7210, Institut de la Vision, IHU ForeSight, 75006 Paris, France; romain.magny@inserm.fr (R.M.); karima.kessal@inserm.fr (K.K.); cbaudouin@15-20.fr (C.B.); stephane.melik-parsadaniantz@inserm.fr (S.M.-P.); fbaudouin@15-20.fr (F.B.-B.)
[2] UMR CNRS 8038 CiTCoM, Chimie Toxicologie Analytique et Cellulaire, Université de Paris, Faculté de Pharmacie, 75006 Paris, France; anne.regazzetti@parisdescartes.fr (A.R.); gregory.genta-jouve@parisdescartes.fr (G.G.-J.); olivier.laprevote@aphp.fr (O.L.)
[3] Centre Hospitalier National d'Ophtalmologie des Quinze-Vingts, IHU ForeSight, 75006 Paris, France
[4] Laboratoire Ecologie, Evolution, Interactions des Systèmes Amazoniens (LEEISA), USR 3456, Université De Guyane, CNRS Guyane, 97300 Cayenne, French Guiana, France
[5] Hôpital Ambroise Paré, AP-HP, Université Versailles Saint-Quentin-en-Yvelines, 92100 Boulogne-Billancourt, France
[6] Hôpital Européen Georges Pompidou, AP-HP, Service de Biochimie, 75006 Paris, France
[*] Correspondence: nicolas.auzeil@parisdescartes.fr

Received: 10 April 2020; Accepted: 25 May 2020; Published: 29 May 2020

Abstract: Annotation of lipids in untargeted lipidomic analysis remains challenging and a systematic approach needs to be developed to organize important datasets with the help of bioinformatic tools. For this purpose, we combined tandem mass spectrometry-based molecular networking with retention time (t_R) prediction to annotate phospholipid and sphingolipid species. Sixty-five standard compounds were used to establish the fragmentation rules of each lipid class studied and to define the parameters governing their chromatographic behavior. Molecular networks (MNs) were generated through the GNPS platform using a lipid standards mixture and applied to lipidomic study of an *in vitro* model of dry eye disease, *i.e.*, human corneal epithelial (HCE) cells exposed to hyperosmolarity (HO). These MNs led to the annotation of more than 150 unique phospholipid and sphingolipid species in the HCE cells. This annotation was reinforced by comparing theoretical to experimental t_R values. This lipidomic study highlighted changes in 54 lipids following HO exposure of corneal cells, some of them being involved in inflammatory responses. The MN approach coupled to t_R prediction thus appears as a suitable and robust tool for the discovery of lipids involved in relevant biological processes.

Keywords: lipidomic; liquid chromatography; tandem mass spectrometry; molecular network; dry eye disease; hyperosmolarity

---

1. Introduction

Over the last decade, lipids have become a major research topic and are now recognized as key biological compounds displaying various roles in cell functions. They include the coordination of bio-membrane structures, intra- and extra-cellular communication, metabolic efficiency, and signaling cascades, all of which are critical for cell functionality [1]. Disruption of lipid homeostasis is now recognized to be involved in numerous pathologies such as cancer, diabetes, neurodegenerative disorders or chronic inflammatory diseases [2,3]. Lipids, especially phospholipid and sphingolipid classes, are central in both inflammatory and cell death processes [4,5]. Arachidonic acid, mainly originating from the cleavage of phospholipids, is, for example, widely recognized as a pro-inflammatory fatty acid [6]. Besides, ceramides, a sphingolipid subclass, mediate apoptosis through a caspase-3 dependent mechanism and inflammation through the release of cytokines such as IL-1ß or IL-6 [7].

Nevertheless, lipids encompass a tremendous number of molecular species exhibiting a wide variety of structures. Indeed, the cellular lipidome includes numerous subclasses of sphingolipids, phospholipids, glycerolipids, sterol lipids, and lipid metabolites. Lipidomics, the comprehensive analysis of lipids in biological systems, remains challenging and must involve not only efficient analytical techniques but also appropriate sample processing and integrative computational approaches [8]. Electrospray ionization (ESI) mass spectrometry (MS), either through a shotgun approach or hyphenated to liquid chromatography (LC), has become the gold standard of lipidome study [8,9]. Indeed, lipidomic analysis using an infusion approach represents a useful strategy to easily access a large part of the lipidome [10,11]. Nevertheless, despite fruitful applications, this approach still suffers from ion suppression which limits the analysis of low abundant lipid species [10,12]. In contrast, lipidomic analysis using liquid chromatography hyphenated to mass spectrometry includes a separation step reducing ion suppression and, therefore, improves detection of low abundant lipid species through an increase in sensitivity [9,11,13]. Recent advances highlight that ion mobility, combined to chromatography and mass spectrometry, represents an additional source of information in the case of lipid identification [14,15]. Nevertheless, whatever the benefit of these techniques, the fact is that, in the context of a lipidomic analysis, the large amount of data to be processed requires the development of new data processing approaches, in particular, to simplify the annotation of lipid species while maintaining a high degree of reliability.

In the course of a lipidomics study, the reliable identification of the numerous lipid species detected represents a rigorous and demanding task. Lipids identification is performed taking into account four analytical features: retention time (t_R), accurate precursor ion *m/z* value, isotopic ratio, and MS/MS data through comparison to reference compounds [16]. In the case of phospholipids, identification deals with the determination of the polar head group and the length of the acyl chains and of their sn_1/sn_2 location on the glycerol moiety. It must be emphasized that such identification is biologically strongly relevant inasmuch as the nature and position of acyl chains depend on the homeostatic balance between biosynthesis rates, remodeling, and degradation and also reflect the extent of the pool of fatty acids available.

To elucidate the structure of unknown compounds, bioinformatics tools are strongly valuable. Among them, molecular networks (MNs) have recently been proposed, historically in the field of plant secondary metabolites, to identify compounds of biotechnological interest or exhibiting promising pharmacological activity [17–20]. Molecular networks are a computational strategy aimed at organizing and visualizing hundreds of molecules using their MS/MS spectra in accordance to their similarities through the assumption that structurally related molecules display similar product ion spectra [21]. The structural similarity of a set of compounds is visualized in a network which may be generated through online platforms such as GNPS or MetGem [21,22].

Taking into account the LC retention time represents an important benefit for the identification of compounds, as it makes it possible to discriminate isobaric compounds [23]. For this purpose, a pre-processing step of the LC-MS/MS data is required; it may be performed using software such as MzMine 2 [24]. Furthermore, based on structural properties, retention time can easily be predicted with good reliability, and t_R values have previously been used to support the identification of numerous metabolite classes, especially lipids [25,26].

The aim of our study was to propose a new approach for the rapid and reliable structural annotation of phospholipids and sphingolipids species using molecular networking and t_R prediction. Using 65 commercial lipid standards, we first determined the collision energy conditions required to achieve the fragmentation patterns appropriate to the structural annotation of lipids. To support lipids annotation, commercial standards were further used to define the relationship between lipid structure and t_R. Unknown lipids were then identified based on their exact mass measurements and MS/MS fragmentation through molecular networking and t_R values.

This approach was used to perform a lipidomic analysis of human corneal epithelial cells exposed to hyperosmolarity (HO)—an *in vitro* model of dry eye disease (DED) [27,28]. Dry eye disease, a chronic multifactorial inflammatory pathology, is characterized by alteration of tear film, cell damage, and inflammation of the ocular surface [29,30]. This very common ocular pathology is also characterized by HO, one of the core mechanisms of DED [31]. Disruption of lipid homeostasis, known to be involved in inflammation and the cell death process, may also be a key feature in the pathophysiology of DED [32]. Thanks to MNs and t_R prediction, our lipidomic approach allowed annotation of 150 unique lipid species and highlights homeostasis disruption of 54 lipid species. Several of them are involved in inflammation and cell death.

2. Results and Discussion

Reliable annotation of unknown compounds using MN highly depends on the quality of the acquired MS/MS spectra [33]. For this purpose, we performed a set of MS/MS experiments for which collision energy was increased step by step in order to optimize the diagnostic fragment ion intensities. Phospholipid annotation needs the presence on the MS/MS spectra of product ions corresponding to the polar head group, the fatty acyl side chains and of the precursor ion. For sphingolipid annotation, fragments corresponding to the sphinganine base moiety and fatty acyl side chain must be detected on MS/MS spectra. Figures 1 and 2 exhibit the main MS characteristics related to PC (16:0/18:1) and Cer (d18:1/16:0), respectively.

2.1. Fragmentation Patterns of Phospholipids

In the negative ion mode, phosphatidylcholine (PC) are mainly detected as $[M-CH_3]^-$ ions corresponding to an in-source loss of a methenium and, to a less extent, as formiate ($[M+HCOO]^-$) and acetate ($[M+CH_3COO]^-$) adducts (Figure 1A). In our study, annotation of a PC was based on the detection of six diagnostic product ion peaks in the MS/MS spectrum of $[M-CH_3]^-$.

In the example shown in Figure 1 related to PC (16:0/18:1), the precursor ion was observed at *m/z* 744.5540. At low mass, a peak at *m/z* 168.0423 corresponded to the deprotonated demethylated phosphocholine ion formed at a 25 eV collision energy (Figure 1B). At higher mass, the peaks at *m/z* 255.2334 and *m/z* 281.2479 were assigned to oleate and palmitate and exhibited an increased intensity from 20 to 50 eV collision energy (Figure 1C). Two other key product ions at *m/z* 480.3098 and *m/z* 506.3256 corresponded to demethylated lysophosphatidylcholine LPC (16:0) and LPC (18:1) ions, respectively. They were detected from 20 to 40 eV collision energy with a maximum intensity at 30 eV (Figure 1D). A collision energy ramp between 20 and 40 eV thus appeared to be suitable to obtain the six diagnostic ions with sufficient sensitivity and mass accuracy ($\Delta < 10$ ppm) (Figure 1E). The ions used to identify 10 standard PC species are compiled in Table S3.

Interestingly, while in the positive ion mode, the phosphocholine product ion at *m/z* 184.0733 allows for the highly sensitive detection of PC, and the negative ion mode is essential to perform fatty acyl chains identification [34]. This ionization mode was also successfully applied to identify the fatty acyl chains of other phospholipid subclasses, namely, phosphatidylethanolamine (PE), phosphatidylinositol (PI), phosphatidylserine (PS), phosphatidylglycerol (PG), and phosphatidic acid (PA). It also proved useful to locate the fatty acyl chains on the sn_1 and sn_2 positions of the glycerol core. Indeed, for PC, PE, and PG species, the intensity of the carboxylate ion peak corresponding to the fatty acid at the sn_2 position was always significantly higher than that at sn_1. This is in agreement with previously published data [35]. For example, Figure 1C shows an oleate peak at sn_2 more intense than the sn_1 palmitate in the whole collision energy range. Similarly, the fragment corresponding to demethylated LPC (16:0) formed by the loss of the sn_2 oleate from the precursor ion was more intense than the demethylated LPC (18:1) arising from the loss of the sn_1 palmitate (Figure 1D). For the lipids belonging to the PS, PI, and PA subclasses, the acyl group at sn_1 always led to the more intense product ion peak (Figures S2 and S3) [35]. However, it is noteworthy that the relative intensity of carboxylate product ions displayed in the MS/MS spectra only provided information on fatty acids sn_1 and sn_2 locations regarding the major regio-isomer. Indeed, the presence in the mixture of a minor amount of the other regio-isomer cannot be excluded. Ensuring it, would need to build a calibration curve using the two pure regio-isomers [36]. In this study, we thus report what is likely to be the major regio-isomer.

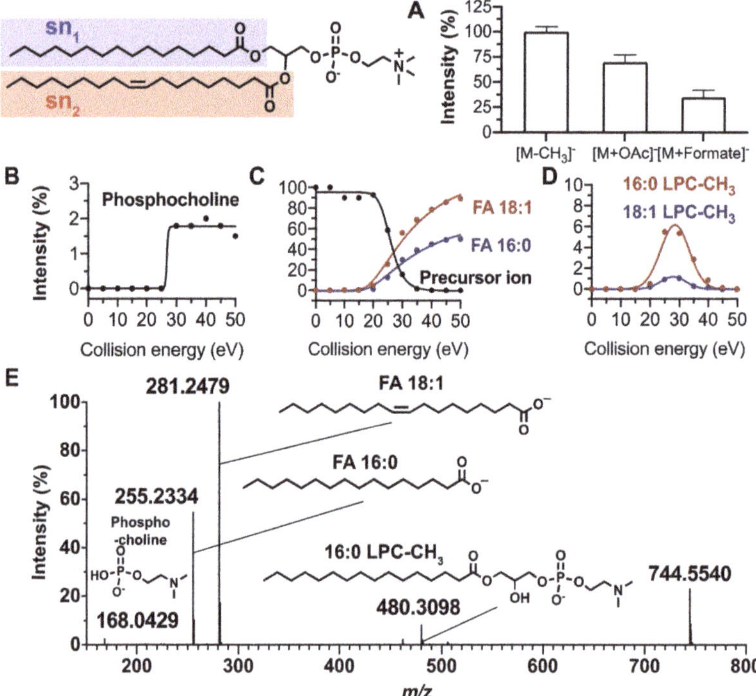

Figure 1. MS characteristics of PC (16:0/18:1). (**A**) Relative intensities of [M−CH₃]⁻, [M+OAc]⁻, and [M+HCOO]⁻ ions. (**B–D**) Fragmentation patterns of the [M−CH₃]⁻ ion formed from PC (16:0/18:1) under negative ionization conditions: relative intensities versus collision energy of (**B**) polar head group, (**C**) fatty acyl side chains, and (**D**) demethylated lysophosphatidylcholine ions. (**E**) Tandem mass spectrum of PC (16:0/18:1) [M−CH₃]⁻ ion obtained by collision energy ramping from 20 to 40 eV. Blue and red colors relate to the sn_1 and sn_2 positions, respectively.

The polar head groups were identified owing to the presence of specific fragment ions such as *m/z* 168.0428 and *m/z* 224.0694 for PC, *m/z* 140.0113 and *m/z* 196.0380 for PE, *m/z* 227.0326 for PG, and *m/z* 241.0119 for PI. For PS, an abundant and specific serine loss (87.0326 Da) was observed in the MS/MS spectra of the deprotonated molecules [M−H]⁻ together with a glycerophosphate ion at *m/z* 152.9958, whereas PA only led to a glycerophosphate ion. Table S3 in the Supplementary Materials lists the diagnostic ions for the 65 phospholipid standard species included in our study.

2.2. Fragmentation Patterns of Sphingolipids

Ceramides (Cer) are based on a sphinganine backbone amidated by a fatty acyl side chain. Ceramides are also building blocks of more complex sphingolipids such as hexosyl ceramide (HexCer) or sphingomyelins (SM). Under negative ionization conditions, Cer were mainly detected as [M−H]⁻ and, to a lesser extent, as acetate ([M+CH₃COO]⁻) and formate ([M+HCOO]⁻) adducts (Figure 2A). As for phospholipids, annotation of individual ceramides was based on the detection of six diagnostic ions which made it possible to determine the double bound number of the sphinganine backbone as well as the FA side chain length. Fatty acyl chain identification was performed thanks to the MS/MS spectra of the [M−H]⁻ ion at collision energies of 20 eV and more (Figure 2B,C).

In the example shown at Figure 2, product ions labelled T (*m/z* 280.2646), U (*m/z* 254.2486) and S (*m/z* 296.259) are indicative of the fatty acyl chain (Figure 2B,C). The sphingosine moiety is characterized by the product ions Q at *m/z* 263.2379 and P at *m/z* 237.2225 formed at collision energies higher than 20 eV (Figure 2D). The [M−H]⁻ precursor ion and its product ions [M−H−H₂O]⁻ and [M−H−2H₂O]⁻ are detected in the high mass region (Figure 2E). As exemplified by Cer (d18:1/16:0), a collision energy ramping from 20 to 40 eV proved suitable to detect the six diagnostic ions with good sensitivity and mass accuracy (Δ < 2 ppm), allowing an easy annotation of Cer species. MS/MS spectra of Cer displayed more intense product ion peaks than PC therefore resulting in a better accuracy of *m/z* values.

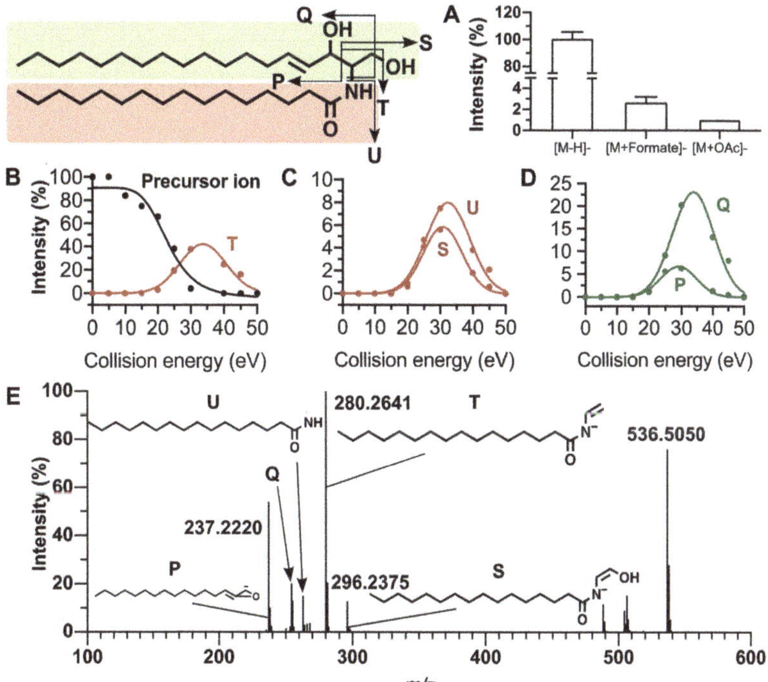

Figure 2. MS characteristics of Cer (d18:1/16:0). (**A**) Relative intensities of [M−H]⁻, [M+OAc]⁻, and [M+HCOO]⁻ ions. (**B–D**) Fragmentation patterns of the [M−H]⁻ ion formed from Cer (d18:1/16:0) under negative ionization conditions: relative intensities versus collision energy of (**B**) precursor ion, (**C**) fatty acyl side chains, and (**D**) sphinganine ions. (**E**) Tandem mass spectrum of Cer (d18:1/16:0) [M−H]⁻ ion obtained by collision energy ramping from 20 to 40 eV. Product ions: T (*m/z* 280.2646), U (*m/z* 254.2486), and S (*m/z* 296.259) are indicative of the fatty acyl chains; Q (*m/z* 263.2379) and P (*m/z* 237.2225) are indicative of the sphingosine moiety. Product ions are labelled according to Reference [37]. Green and red colors relate to sphingosine and fatty acyl position, respectively.

2.3. Retention Time Prediction

A typical UHPLC-ESI-MS negative ion mode chromatogram of a mixture of 65 lipid standards representative of the nine studied subclasses is displayed in Figure 3A. In reversed-phase liquid chromatography, an elution of lipids is closely related to the fatty acyl chain lengths, and this property has been widely used in the frame of lipidomic analyses [38,39]. Under our conditions, the FA and lysophospholipids were firstly eluted for 6 min followed by Cer and phospholipids (*i.e.*, PE, PI, PG, and PS) between 6 and 9 min (Figure 3A). Furthermore, in the case of phospholipids, the chromatographic behavior was also dependent on the polar head group, the elution order being for a given fatty acyl chain pattern as follows: PI, PG, PS, PC, and PE (Figure S1). Retention time may thus be considered as a valuable analytical feature helpful in confirming annotation or in highlighting misannotation. However, this requires a robust chromatographic system able to deliver stable and reliable retention times. In our hands, RP-UHPLC operating with a reduced particle size (1.7 μm), associated to a column temperature of 50 °C, and optimized elution conditions provided chromatograms with peak widths lower than 20 s and highly reproducible retention times (Table S2).

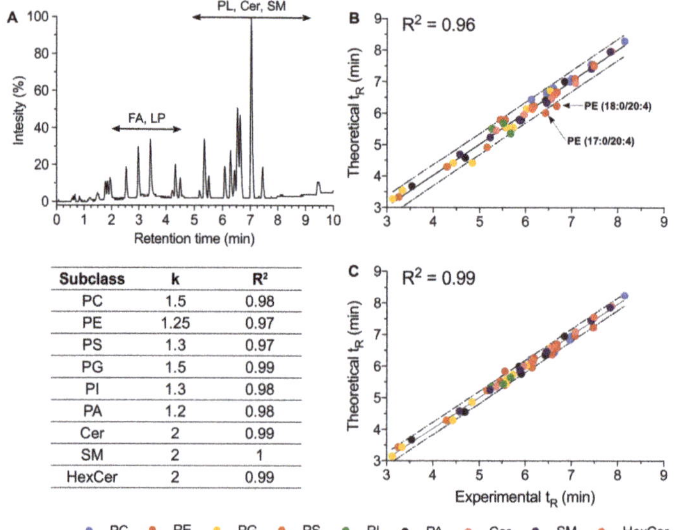

Subclass	k	R²
PC	1.5	0.98
PE	1.25	0.97
PS	1.3	0.97
PG	1.5	0.99
PI	1.3	0.98
PA	1.2	0.98
Cer	2	0.99
SM	2	1
HexCer	2	0.99

Figure 3. Retention time prediction model based on equivalent carbon number (ECN) of nine subclasses of phospholipids and sphingolipids. (**A**) UHPLC-HRMS chromatogram of a commercial standard lipid mixture in the negative ion mode. (**B**) Theoretical t_R plotted against experimental t_R for standard lipid mixture before and (**C**) after k value fitting. A linear (phospholipids) and polynomial (sphingolipids) regression model was used. Dotted lines represent the 95% confidence interval displaying a relative error which reached 15% on the linear regression graph (B) (k = 2) and did not exceed 5% on the linear regression graph (C). Table: Parameters of relationship between ECN and experimental t_R determined using a standard lipid mixture. The general expression of ECN is ECN = NC - k × DB where NC and DB are the total number of carbons and the number of double bonds, respectively. PC: phosphatidylcholine, PE: phosphatidylethanolamine, PG: phosphatidylglycerol, PS: phosphatidylserine, PI: phosphatidylinositol, PA: phosphatidic acid, Cer: ceramide, SM: sphingomyelin, HexCer: hexosyl ceramide.

In order to reinforce lipid annotation using t_R, we first used the commercial lipid standard to build a t_R predictive model based on the equivalent carbon numbers (ECNs). The general expression of ECNs is ECN = NC – k × DB, where NC and DB are the total number of carbons and the number of double bonds, respectively. For lipid species, k = 2 is usually applied [40].

In the case of phospholipids, especially when containing polyunsaturated fatty acid (PUFA), the linear t_R prediction model using k = 2 was fairly poor (Figure 3B). To improve the t_R prediction model, we plotted ECN values with experimental t_R allowing, next to an appropriate fitting step, to determine more accurate k values. New k values calculated for the different investigated phospholipid subclasses thus improved the linear correlation between ECN and t_R (Figure 3C). For example, the difference between the theoretical and experimental t_R values for PE (17:0/20:4) was 13% for k = 2 and decreased to 0% using k = 1.25.

In the case of sphingolipids, the linear t_R prediction model with k = 2 was suitable for species whose fatty acyl chains did not exceed 22 carbon atoms. However, regarding this lipid class, a polynomial curve led to a better t_R prediction model than a linear one. The selection of the polynomial degree was based on the value of the correlation coefficient using a tolerance of $1/100$ ($R^2 > 0.99$). For the three investigated sphingolipid subclasses, a quadratic function was finally retained.

Based on this improved fitting step, we predicted more accurately the theoretical t_R for all the commercial standards lipids including six phospholipid subclasses and three sphingolipid subclasses (Figure 3C,D). Indeed, the difference between the experimental and theoretical t_R values did not exceed 5%, whatever the commercial standard lipid. Consequently, an uncertainty of 5% was retained when t_R was used in the purpose to support the annotation of unknown lipids from HCE cell extract and also to discriminate two isobaric lipid species as exemplified later in the text (see Section 2.6).

2.4. Instrument Stability

The stability of the UHPLC-ESI-MS/MS system was assessed for both MS and chromatography. For this purpose, exact mass measurements were performed on the standard lipid species. High-resolution mass measurements led to a mass accuracy better than 5 ppm for whichever standard lipid species under consideration (Table S2) and the chromatographic system delivered t_R values with a deviation within a 3-day period not exceeding 3% (Table S2).

2.5. Lipidic Networking of Human Corneal Epithelial Cells

Although untargeted analysis by MS constitutes a relevant and powerful tool to characterize a cell lipidome, the annotation of the lipids of interest remains a real challenge. Fortunately, phospholipids and sphingolipids display structural characteristics which make their identification suitable through molecular networking.

In our case, MNs were used as part of a study aimed at assessing the impact on the HCE cell lipidome exposed to HO and were implemented as described hereafter. Tandem mass spectrometry in the DDA mode (see Section 3.3) was used to acquired MS and MS/MS data for 65 commercial standard lipids and for the whole lipids contained in HCE cell extracts. Next to a preprocessing step performed with MzMine 2 (see Section 3.4), preprocessed data were subsequently used to build MNs through the GNPS platform. The MNs thus included three types of nodes corresponding first, to commercial standard lipids, second, to lipids available both as commercial references and identified in HCE cell extracts, and, finally, to lipids only detected in HCE cell extracts (Figures 4 and 5, Table S3).

In such a MN, commercial lipid standards of known structure were thus used to anchor the molecular network within which lipids from HCE cell extracts were clustered according to the similarities of structures that they shared with reference lipids. Thanks to the standard lipids, the key parameters were optimized to provide a reliable and relevant MN. The nodes of the network, corresponding to the MS/MS spectra, were only linked to others if they displayed a common fragmentation pattern, *i.e.*, a minimum number of six identical product ions and/or neutral losses. Moreover, the similarity score between a pair of MS/MS spectra, also called "cosine score" (cos) had to be greater than 0.6. Values selected for the aforementioned parameters were widely used for molecular networking [21].

Figure 4 corresponds to the network of phospholipids; it includes PC, PE, PA, and PG. It was mainly built from the fatty acyl chains located at the sn_1 and sn_2 positions, and, to a less extent, to the polar head group. For instance, various phospholipid species (*i.e.*, PC, PE, PC-P, PE-P, PG, PA) containing palmitate in the sn_1 and sn_2 positions were clustered (cluster MN-C1 in Figure 4). The PC (16:0/18:1) and PC (18:0/18:1) were also clustered, as they both contained oleate in the sn$_2$ position and a phosphocholine polar head group (Figure 4). Similarly, PC (18:0/18:1) and PC (18:0/18:2) were clustered, as they contained stearate in the sn$_1$ position (Figure 4). In some cases, depending on the precursor ion intensity, the MN could also include several adducts corresponding to only one PC. This is especially the case for the lipid standard PC (17:0/17:0) observed as [M−CH$_3$]$^-$, [M+HCOO]$^-$ and [M+CH$_3$COO]$^-$ ion species (cluster MN–C2 in Figure 4). Phospholipids containing ether (PC–O and PE–O) or vinyl ether (PC–P and PE–P) bonds in the sn$_1$ position are displayed on the same MN as the diacylphospholipids. For instance, PE–P (16:0/16:1) included in MN–C1 was connected to PE–P (16:0/16:0), as they both contained a phosphoethanolamine head group and a sn$_1$ palmitoyl moiety. The PE–P (16:0/16:0) was also connected to PC–O (16:0/16:0) and PE (16:0/16:0), as they all contained the same two palmitoyl moieties. In contrast, PE–P (16:0/16:1) was not connected to PE (16:0/16:1), indeed, their MS/MS spectra displayed only four common product ions, this being insufficient to connect them in the MN. To improve data visualization, the MN was organized under two orthogonal axes; the abscissa and the ordinate corresponding, respectively, to the sn$_1$ and sn$_2$ fatty acyl chains of glycerophospholipids. For instance, lipids containing an arachidonate side chain in the sn$_2$ position were displayed on the same line (cluster MN–C3 in Figure 4).

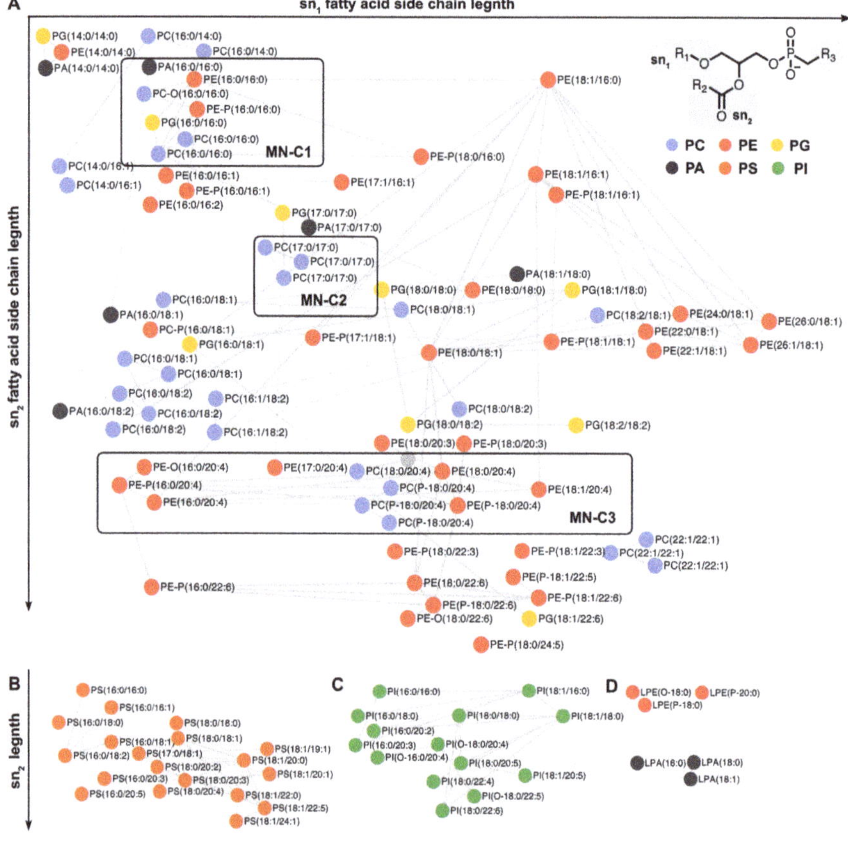

Figure 4. Lipidic molecular network of phospholipids including commercial standards and epithelial corneal cell lipids. The MN between (**A**) PC, PE, PA, and PG; (**B**) PS; (**C**) PI; and (**D**) LP. Networking was based on both fatty acyl side chains and polar head groups. The MN was organized along two orthogonal axes: abscissa and ordinate correspond to the sn$_1$ and sn$_2$ fatty acids, respectively. See the text for the explanation of the MN–C1, MN–C2, and MN–C2.

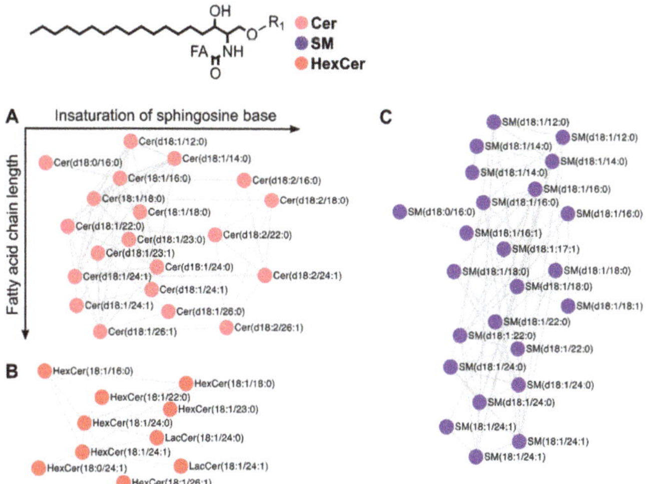

Figure 5. Lipidic molecular network of sphingolipids including commercial standards and epithelial corneal cell lipids. MN of (**A**) ceramide, (**B**) sphingomyelin, and (**C**) hexosylceramide. The MN was organized along two orthogonal axes: abscissa and ordinate corresponding to insaturation of the sphinganine base and fatty acid side chain length, respectively.

It must be emphasized that, in contrast to other phospholipids, PS and PI clustering was mainly based on their polar head group because of abundant characteristic fragment ions (Figure 4B,C). The [M−H]⁻ precursor ions of PS were dissociated by the loss of the serine moiety leading to an intense [M−H−87.0326]⁻ product ion [35] (Figure S2). The MS/MS spectra of the deprotonated PI displayed an inositol phosphate fragment at *m/z* 259.0225 and two product ions corresponding to the consecutive loss of one (*m/z* 241.0119) and two (*m/z* 223.0013) water molecules (Figure S3).

The networks connecting sphingolipids were all organized by subclasses (Figure 5). Ceramides (Figure 5A) were thus clustered separately from sphingomyelins (Figure 5B) and hexosylceramides (Figure 5C). Furthermore, networks of each sphingolipid subclasses displayed clusters depending on sphinganine (d18:0), sphingosine (sphing-4-enine, d18:1) or a sphingadienine (sphing-4,14-dienine, d18:2) moiety. In the ceramide subclass, Cer (d18:1/26:0) was connected to Cer (d18:1/26:1), as they both contained a sphingosine (d18:1) moiety. Clustering also depended on the nature of fatty acyl chain as exemplified by Cer (d18:0/16:0) which was connected to Cer (d18:1/16:0), as they both contained a palmitoyl moiety but displayed no connection with Cer (d18:1/18:0). To simplify the data visualization in Figure 6, the MN was organized on two orthogonal axes: the abscissa and the ordinate correspond to the number of insaturation of the sphinganine base moiety and to the fatty acyl chain length, respectively.

2.6. Use of Retention Time Prediction for Lipid Annotation

Applied to HCE cell lipidome, the MN approach was helpful to annotate more than 150 phospholipid and sphingolipid species (Table S3). Annotation was based on tandem mass spectrometry and confirmed by retention time with a maximum tolerance of 5% between theoretical and experimental t_R values. In some cases, a co-elution and a co-selection of precursor ions in MS/MS was encountered leading to difficult mass spectra interpretation which did not readily permit to decide between two different lipid structures. In such a case, the t_R prediction made it possible to annotate unequivocal lipid species.

For instance, the MS/MS spectrum of the ion at *m/z* 800.619 displayed the characteristic fragment ions at *m/z* 140.0123 and 196.0369 of a phosphoethanolamine headgroup suggesting a PE (40:1) structure. The spectrum also exhibited intense peaks corresponding to oleate (*m/z* 281.2480) and behenate (*m/z* 339.3254) but also two small peaks at *m/z* 253.2175 and 367.3488 indicative of palmitoleate and lignocerate, respectively (Figure 6A,B). The selected precursor ion thus corresponded to a mixture of an abundant PE (22:0/18:1) and a less abundant PE (24:0/16:1). However, the presence of an ion at *m/z* 168.0444 could correspond to the headgroup of an isobaric [M−CH$_3$]$^-$ precursor ion from PC (16:1/22:0). Thanks to the t$_R$ prediction, PC (16:1/22:0) was excluded as the difference between experimental and theoretical t$_R$ was 7% (Figure 6C). The origin of the phosphocholine ion at *m/z* 168.0444 was explained by the co-elution and co-selection by the Q1 quadrupole of the demethylated SM (d42:1) at *m/z* 799.668, a very scarce species in the precursor ion beam. This annotation was confirmed by a tiny difference of experimental and theoretical t$_R$ (0.3%).

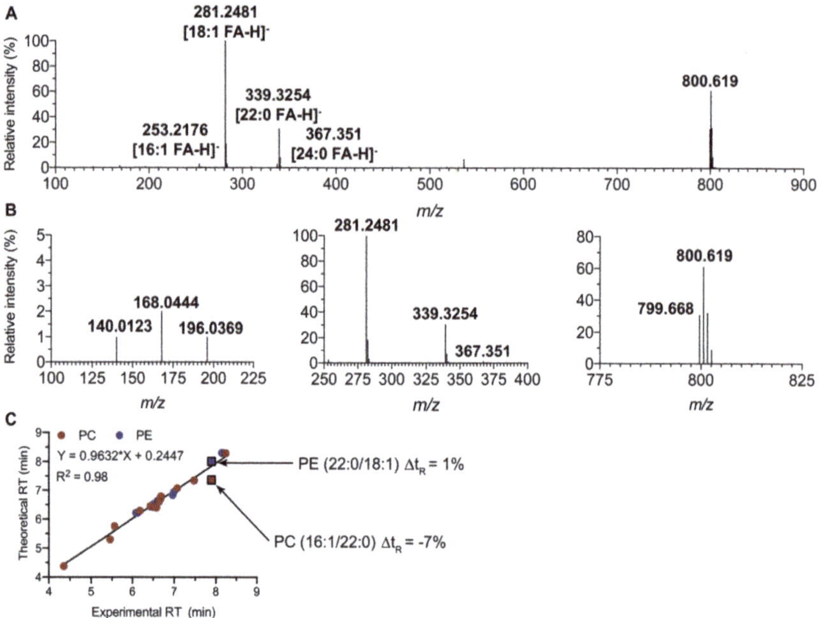

Figure 6. Retention time prediction use to discriminate two isobaric PE and PC species. (**A**) MS/MS spectrum of PE (40:1) [M−H]$^-$ ion (*m/z* 800,619). (**B**) Low, middle, and high mass region of the spectrum A. (**C**) Theoretical t$_R$ plotted against experimental t$_R$ for standard PC and PE species (circles) and for PE (22:0/18:1) and PC (16:1/22:0) (squares).

2.7. Use of Existing Lipid Library Database

Although the proposed annotation procedure is performed through an individual inspection of each MS/MS spectrum, an automatization of the annotation may, at least in part, be considered using currently available lipid library especially LipidBlast or Lipidex [41–43].

For example, using LipidBlast, we performed annotation of the commercial standard lipids available in this study. Among the 65 lipids species studied, 50 were successfully annotated using LipidBlast and one misannotation was reported (Figure S4). None of the six SM species were annotated as SM subclass is not supported by the LipidBlast library using negative ion mode LC-ESI-MS/MS data. Moreover, among the 13 PC species included in the commercial standard lipid mixture, three PC failed to be annotated using LipidBlast.

In the frame of a lipidomic analysis, MN in combination with lipid databases may be regarded as a valuable and saving time approach giving a strong insight in the structural elucidation of lipid. Nevertheless, we believe that it remains important to use at least several known standard lipids to perform annotation of unknown lipid species with a high degree of confidence.

2.8. Effect of HO on HCE Cells

The lipid annotation through molecular networking and the retention time prediction approach proposed in this study was applied to assess lipid perturbations in human corneal epithelial cells exposed to hyperosmolarity (HO)—an *in vitro* model of dry eye disease [27,28]. Dry eye disease is a chronic inflammatory pathology of the ocular surface. It is characterized by alteration of tear film, HO, cell damage, and inflammation of the ocular surface, all contributing to a vicious circle [30,31,44,45]. Therapeutic strategies targeting inflammatory processes, such as cyclosporine or other anti-inflammatory agents, have been proposed to break this deleterious cycle [30,46]. To better understand the mechanism underlying DED and to find new marker of this pathology, especially to improve patient monitoring and to develop new targeted treatments, further investigations are still needed.

Hyperosmolarity is a key feature of DED [31]. Indeed, it induces significant stress targeting the corneal cell membrane [28,47,48]. Studying the modification of lipid homeostasis related to HO is relevant as this stressor induces the disruption of processes closely associated to cell membranes. Indeed, HO activates pro-inflammatory and pro-apoptotic processes, initiated at the cell membrane level. Hyperosmolarity stimulates downstream signaling pathways mediated by multiple membrane-bound proteins and enzymes [27,49–51]. The interest to study cell lipid disruption is also reinforced as HO favors ROS production which, in turn, targets lipids through peroxidation [52–54].

Lipidomic analysis associated to molecular networking and t_R calculation led to the annotation of 150 lipid species and revealed that among them, 54 phospholipids and sphingolipids were significantly up- or downregulated in human corneal epithelial cells exposed to HO (Figure 7).

Regarding phospholipids, an increase of the cell concentration was observed for 4 PC, 10 PE, 3 PS, and 3 PI species under HO exposure. These phospholipids mainly contained oleate (18:1) located in the sn_2 position. On the contrary, abundances of six ether-phospholipid species were strikingly decreased. Thanks to MN, they were identified as 2 PC-P, 2 PI-O, and 2 PE-P species containing polyunsaturated fatty acids at position sn_2, especially arachidonic (20:4) and docosahexaenoic (22:6) acids (Figure 7). Ether phospholipids are known to be pools of FA (22:6) and FA (20:4) [55]. The FA (20:4) is the preferential substrate of cyclooxygenase (COX) and lipoxygenase (LOX), enzymes leading to pro-inflammatory eicosanoids. Therefore, our results suggest that HO, a key feature in DED, may favor the release of arachidonic acid from ether phospholipids to promote inflammatory process. Ether phospholipid, especially PE-P and PC-P species, are also known to be targets of oxidative stress through vinyl ether bonds. Beside inflammation, oxidative stress is also a key feature of DED pathophysiology [44,56]. Therefore, the significant decrease in PC-P and PE-P species observed in HCE cells under HO exposure may proceed through oxidation of these lipids which is known to generate toxic malondialdehyde (MDA) and 4-hydroxynonenal (4-HNE). Of note, MDA and 4-HNE release have previously been described in DED *in vitro* models as well as in conjunctival imprints of DED suffering patients [53,57].

Regarding sphingolipids, an increase of ceramide abundance was observed in HCE cells submitted to HO (Figure 7). Ceramides are bioactive lipid species known to promote inflammation through IL-1β release and to induce apoptosis through caspase 3 activation [7,58]. Dysregulation of ceramide metabolism is involved in many inflammatory diseases such as atherosclerosis, inflammatory bowel disease or multiple sclerosis [2,3]. In the frame of DED, apoptosis and inflammatory processes induced by ceramides may thus be considered as important mediators of the deleterious effects of HO in accordance with previous published data [47,59]. Thanks to MN, Cer (d18:0/16:0) and Cer (d18:1/16:0) were successfully identified. Under HO exposure, these two lipids species are increased in HCE cells. Because Cer (d18:0/16:0) and Cer (d18:1/16:0) are respectively substrate and a product of dehydroceramide desaturase—a key enzyme in *de novo* synthesis of ceramide—this result suggests that HO promotes *de novo* synthesis of ceramides making this ceramide biosynthetic pathway a putative therapeutic target in the frame of DED.

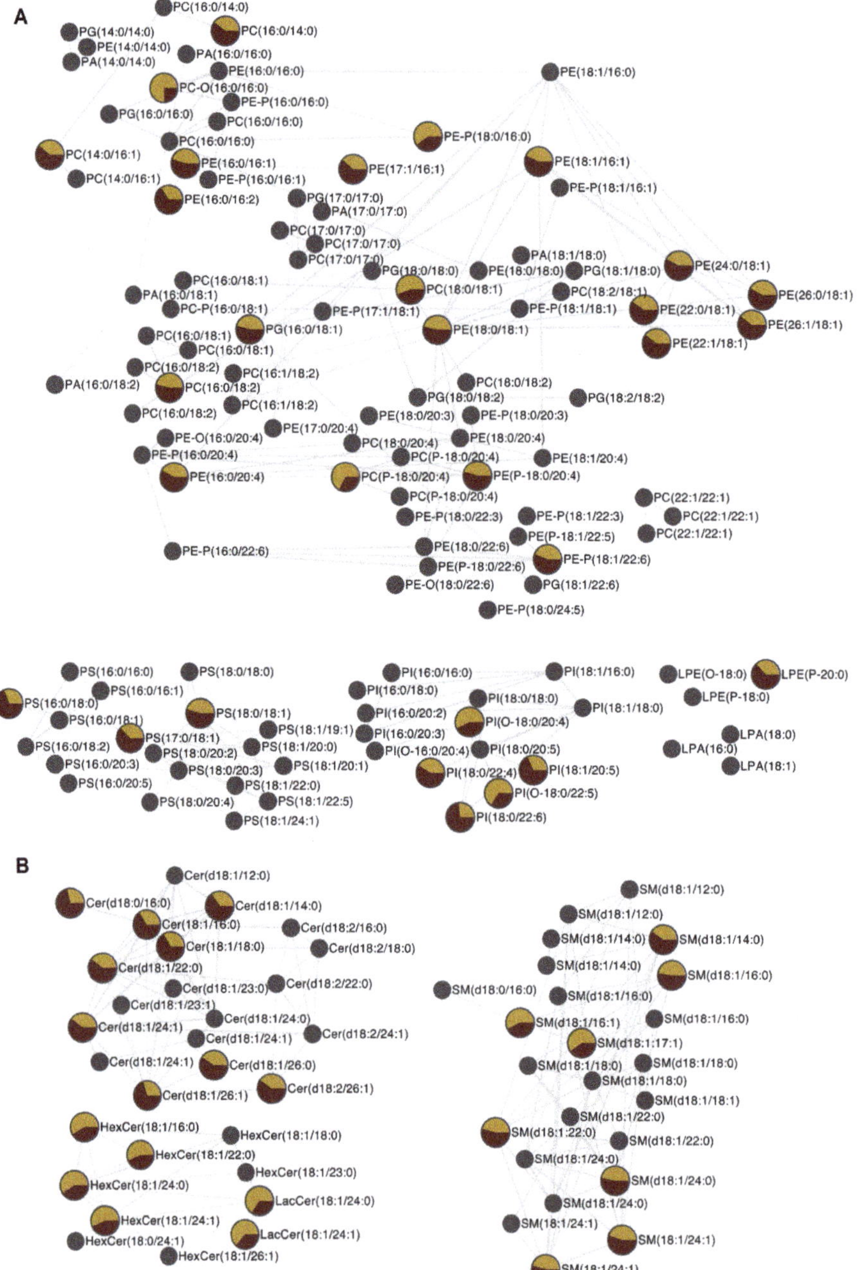

Figure 7. MN of lipids up- or downregulated in human corneal epithelial cells exposed to HO. MN displays (**A**) phospholipids and (**B**) sphingolipids species up- or downregulated in human corneal epithelial cells exposed to HO. The pie chart represents mean ion intensity of a lipid molecule from the control cell (yellow) and HO exposed cell (red) lipidomes. Of note, networking, based on the structural characteristics of lipid species, does not provide information regarding biosynthesis and modeling/remodeling pathways of the lipids found within HCE cells, especially under the effect of hyperosmolarity.

3. Materials and Methods

3.1. Chemicals and Reagents

Chloroform (Carlo Erba Reactifs SDS, Val-de-Reuil, France), acetonitrile, methanol, isopropanol, water of LC-MS grade (J.T. Baker, Phillipsburg, NJ, USA) and 3,5-di-*tert*-4-butylhydroxytoluene (Sigma–Aldrich, Saint-Quentin Fallavier, France) were used to perform cell lipid extraction and to prepare mobile phase for liquid chromatography. All commercial lipid standards were purchased from Avanti Polar Lipids, Inc. (Alabaster, AL, USA) and are listed in Table S1 of the Supplementary Materials.

3.2. Sample Preparation

The HCE cells were exposed to HO (500 mOsM) for 24 h and then were washed with Dulbecco's Phosphate-Buffered Saline (DPBS). Cells were harvested using trypsin-EDTA 0.05%, washed with DPBS, centrifuged at 2000 rpm for 10 min. Dry cell pellets were adjusted to 3 million cells and stored at −80 °C until analysis. After thawing, the cell pellets were resuspended in ultra-pure water (1 mL) and were sonicated for 5 min. Lipids were extracted using a chloroform/methanol/water (5:5:2, *v/v/v*) mixture containing 3,5-di-*tert*-4-butylhydroxytoluene 0.01% (*w/v*) as an antioxidant agent. Samples were subsequently centrifuged at 3000 rpm for 10 min, organic phases were collected, and solvents were evaporated under reduced pressure at 45 °C. Dry residues were dissolved in a 100 µL mixture containing acetonitrile/isopropanol/chloroform/water (35:35:20:10, *v/v/v/v*) before injection into the UHPLC-MS system.

3.3. Data-Dependent LC-ESI-HRMS/MS Analysis

Liquid chromatography-negative electrospray ionization mass spectrometry analysis of lipid extracts was performed on a UHPLC system (Waters®, Manchester, UK) combined with a Synapt®G2 High Definition MS™ (Manchester, UK) (Q-TOF) mass spectrometer (Waters®). Chromatographic separation was achieved on an Acquity® (Manchester, UK) CSH C18 column (100 × 2.1 mm; 1.7 µm). Lipids were eluted using a binary gradient system consisting in 10 mM ammonium acetate in acetonitrile/water mixture (40:60, *v/v*) as solvent A and 10 mM ammonium acetate in acetonitrile/isopropanol mixture (10:90, *v/v*) as solvent B. The eluent increased from 40% B to 100% B in 10 min, was held at 100% B for 2 min before returning to 40% B. The flow rate was kept at 0.4 mL.min^{-1}. The column oven was set at 50 °C and the injection volume was 5 µL. The source parameters were as follows: capillary voltage 2400 V, cone voltage 45 V, source temperature 120 °C, desolvation temperature 550 °C, cone gas flow 20 L h^{-1}, and desolvation gas flow 1000 L h^{-1}. Leucine enkephalin (2 ng mL^{-1}) was used as an external reference compound (Lock-Spray™, Manchester, UK) for mass correction. In a data-dependent acquisition mode (DDA), MS full scans were followed by MS/MS scans performed on the five most intense ions above an absolute threshold of 1000 counts. Selected parent ions were fragmented at collision energy ramp 20–40 eV and a selection window size of 1.0 Th. Scan durations for both MS and MS/MS were 0.2 s. In the full scan mode, the data were acquired between *m/z* 50 and 1200 using a resolution of 20,000 FWHM at *m/z* 500. Data acquisition was managed using Waters MassLynx™ software (version 4.1; Waters MS Technologies, Manchester, UK). A mixture of 65 standard lipids belonging to 9 of the main lipid classes (*i.e.*, phosphatidic acid (PA), phosphatidylethanolamine (PE), phosphatidylserine (PS), phosphatidylcholine (PC), phosphatidylglycerol (PG), ceramide (Cer), sphingomyelin (SM), HexosylCeramide (HexCer)) at a final individual concentration of 1 µM was also periodically injected throughout the analytical batch.

3.4. Data-Preprocessing Parameters

Raw data files were converted into universal open source mzXML file with MSConvert 3.0 and were then processed using MZmine 2.51 software. The MS and MS/MS spectra were extracted using MZmine 2.51 with a mass detection noise level set at 2E2 and 0E0, respectively. Chromatograms were then built with the ADAP algorithms [60] using a minimum group size of 5 scans, a group intensity threshold of 5000, and an *m/z* tolerance of 0.005 Da (about 10 ppm). The ADAP wavelets chromatogram deconvolution algorithm was used with the following settings: signal-to-noise ratio = 10, coefficient/area ratio = 50, peak duration range = 0.05–0.4 min, retention time wavelet range = 0.02–0.1, *m/z* range for MS/MS scan pairing of 0.01, and t_R range for MS/MS scan pairing of 0.15 min. Chromatograms were de-isotoped using the isotopic peaks grouper algorithm with a *m/z* tolerance set at 0.005 (*m/z* < 500) and 10 ppm (*m/z* > 500), and a t_R tolerance of 0.1 min. Peak alignment was performed using the join aligner method using the following parameters: *m/z* tolerance at 0.005 (*m/z* < 500) and 10 ppm (*m/z* > 500) and an absolute t_R tolerance 0.15 min. Each MS/MS scans were associated with the corresponding MS scans using a t_R tolerance of 0.1 min and a *m/z* tolerance of 0.005 (*m/z* < 500) and 10 ppm (*m/z* > 500). The peak list was finally gap-filled using the so-called module "same RT and *m/z* range gap filler" with *m/z* tolerance 0.005 (*m/z* < 500) and 5 ppm (*m/z* > 500).

3.5. Molecular Network Analysis

The MNs were created using the feature based molecular networking workflow of the Global Natural Products Social (GNPS) platform [61]. The following settings were used to build the network: minimum pairs Cos > 0.60, parent ion mass tolerance = 0.02 Da, fragment ion mass tolerance = 0.02, network topK < 100, minimum matched peaks = 6, and minimum cluster size = 2. The library spectra inquiries were performed using the same parameter values as those define for the network building. The MNs were finally visualized and annotated using Cytoscape 3.4.0 software (San Diego, California, USA) [62].

3.6. Lipid Structure Assignment

The structural annotation of unknown lipid species was based on the MNs generated on the GNPS platform using MS and MS/MS data as follow: (*i*) nodes associated to lipid standards were indexed using MzMine 2 thanks to MS, MS/MS data, and t_R value, (*ii*) nodes associated to unknown lipids were subsequently indexed based on MS data, using online data base LIPIDMAPS and METLIN, and MS/MS data, through manual inspection on MzMine 2, (*iii*) annotation was finally supported by t_R values by comparison of experimental t_R values to the calculated one using t_R prediction models. Based on these three criteria, lipids already annotated were used to create a database valuable for later annotation.

In order to demonstrate the relevance of the MNs in lipid species annotation, MS/MS spectra were individually inspected to select diagnostic product ions essential for annotation and the differences between theoretical and experimental *m/z* values were calculated using Excel software (see Supplementary Materials, compilation of experimental and theoretical *m/z* values of diagnostic product ions for PL and SL).

In accordance with the guidelines provided by the minimum reporting standards of the Metabolomics Standards Initiative [16], Table S3 includes the level of identification for all annotated lipids. Indeed, thanks to accurate *m/z* measurement, the MS/MS data inspection and retention time analysis, lipids annotated in HCE cells were assigned to group 1 or 2. Lipids for which we had the corresponding commercial standards were assigned to group 1. In addition, lipids for which we did not have the corresponding standards, the annotation was performed on the adequacy of *m/z* value, MS/MS data and retention time analysis and were thus assigned to group 2.

3.7. Statistical Analysis

A false discovery rate ((FDR)-adjusted $p < 0.01$) controlling procedure was performed to assess the statistical significance of the concentration differences of the identified lipids from cell extracts of HO-treated cells versus control cells. Each experiment was performed independently at least five times. The ANOVA, Dunnett test, and Student *t*-test were performed using GraphPad Prism 8 software (version 8; GraphPad Software, La Jolla, CA, USA) with a risk set at 0.05 (* $p < 0.05$, ** $p < 0.01$, *** $p < 0.001$).

4. Conclusions

For the first time, the present study showed that the use of molecular networks makes it possible to facilitate and increase the reliability of lipid annotation in the course of lipidomic analysis. This approach, based on the use of the tandem mass spectrometry DDA mode in negative ionization conditions, allowed characterizing the fatty acyl chains of phospholipids and sphingolipids. In addition, if an ambiguity in the annotation of a lipid persists, the prediction of the retention time makes it possible to remove the latter. This new strategy makes it possible to cover the entire lipidome despite the limited number of standard lipids to which it is possible to have access commercially. The present study was limited to lipid subclasses which had structural characteristics which were clearly depicted by the MN under negative ionization conditions. Nevertheless, through this approach, we were able, in the context of a differential lipidomic analysis of an *in vitro* model of DED (*i.e.*, HCE cells exposed to HO) to annotate many lipids potentially involved in cell death and inflammation. Regarding others lipid subclasses such as glycerolipids, the positive ionization mode appears suitable to highlight their structural characteristics using molecular networks. It is currently in progress and will be the subject of a separate study.

Supplementary Materials: The following are available online at http://www.mdpi.com/2218-1989/10/6/225/s1, Figure S1: Extracted ion chromatograms of m/z values corresponding to precursor ions of standard phospholipids containing two palmitoyl moieties, Figure S2: MN of PS subclass, Figure S3: MN in PI subclass, Figure S4: Annotation of MS/MS spectra of commercial standard lipids using LipidBlast library, Table S1: Lipid composition of standard mix, Table S2: Repeatability and method precision (within-day, between-day, and intermediate precision), Table S3: Annotation of lipid species by MS/MS experiment. Supplementary Materials include the Excel table used for tR prediction and for diagnostic ion checking.

Author Contributions: Conceptualization, N.A., F.B.B., C.B., O.L., G.G.J., R.M., K.K., A.R., S.M.P.; methodology, R.M., N.A., G.G.J., F.B.B., K.K., A.R.; validation, N.A., G.G.J., F.B.B., K.K., A.R., S.M.P., O.L., C.B., R.M.; formal analysis, N.A., G.G.J., F.B.B., R.M., K.K., A.R.; writing—original draft preparation, R.M., N.A., F.B.B., K.K., A.R.; supervision, O.L., C.B., S.M.P., N.A.; All authors have read and agreed to the published version of the manuscript.

Funding: This study was funded by Sorbonne Université, the Institut National de la Santé et de la Recherche Médicale and Centre National de la Recherche Scientifique.

Acknowledgments: This work was completed with the support of the Programme Investissements d'Avenir IHU FOReSIGHT (ANR-18-IAHU-01). The authors thank the core facilities of the Institut de la vision, the Centre Hospitalier National d'Ophtalmologie des Quinze-Vingts, Région Ile-de-France and Ville de Paris.

Conflicts of Interest: The authors declare no conflict of interest.

References

1. Zhang, C.; Wang, K.; Yang, L.; Liu, R.; Chu, Y.; Qin, X.; Yang, P.; Yu, H. Lipid metabolism in inflammation-related diseases. *Analyst* **2018**, *143*, 4526–4536. [CrossRef] [PubMed]

2. Stith, J.L.; Velazquez, F.N.; Obeid, L.M. Advances in determining signaling mechanisms of ceramide and role in disease. *J. Lipid Res.* **2019**, *60*, 913–918. [CrossRef] [PubMed]

3. Meikle, P.J.; Summers, S.A. Sphingolipids and phospholipids in insulin resistance and related metabolic disorders. *Nat. Rev. Endocrinol.* **2017**, *13*, 79–91. [CrossRef] [PubMed]

4. Magtanong, L.; Ko, P.J.; Dixon, S.J. Emerging roles for lipids in non-apoptotic cell death. *Cell Death Differ.* **2016**, *23*, 1099–1109. [CrossRef]

5. Agmon, E.; Stockwell, B.R. Lipid homeostasis and regulated cell death. *Curr. Opin. Chem. Biol.* **2017**, *39*, 83–89. [CrossRef]

6. Serhan, C.N.; Savill, J. Resolution of inflammation: The beginning programs the end. *Nat. Immunol.* **2005**, *6*, 1191–1197. [CrossRef]

7. Maceyka, M.; Spiegel, S. Sphingolipid metabolites in inflammatory disease. *Nature* **2014**, *510*, 58–67. [CrossRef]

8. Holčapek, M.; Liebisch, G.; Ekroos, K. Lipidomic Analysis. *Anal. Chem.* **2018**, *90*, 4249–4257. [CrossRef]

9. Ivanisevic, J.; Want, E.J. From samples to insights into metabolism: Uncovering biologically relevant information in LC- HRMS metabolomics data. *Metabolites* **2019**, *9*, 308. [CrossRef]

10. Hsu, F.F. Mass spectrometry-based shotgun lipidomics – a critical review from the technical point of view. *Anal. Bioanal. Chem.* **2018**, *410*, 6387–6409. [CrossRef]

11. Wang, J.; Wang, C.; Han, X. Tutorial on lipidomics. *Anal. Chim. Acta* **2019**, *1061*, 28–41. [CrossRef] [PubMed]

12. Han, X.; Yang, K.; Gross, R.W. Multi-dimensional mass spectrometry-based shotgun lipidomics and novel strategies for lipidomic analyses. *Mass Spectrom. Rev.* **2012**, *31*, 134–178. [PubMed]

13. Züllig, T.; Trötzmüller, M.; Köfeler, H.C. Lipidomics from sample preparation to data analysis: A primer. *Anal. Bioanal. Chem.* **2019**, *412*, 2191–2209. [PubMed]

14. Hinz, C.; Liggi, S.; Griffin, J.L. The potential of Ion Mobility Mass Spectrometry for high-throughput and high-resolution lipidomics. *Curr. Opin. Chem. Biol.* **2018**, *42*, 42–50.

15. Leaptrot, K.L.; May, J.C.; Dodds, J.N.; McLean, J.A. Ion mobility conformational lipid atlas for high confidence lipidomics. *Nat. Commun.* **2019**, *10*, 985.

16. Sumner, L.W.; Amberg, A.; Barrett, D.; Beale, M.H.; Beger, R.; Daykin, C.A.; Fan, T.W.M.; Fiehn, O.; Goodacre, R.; Griffin, J.L.; et al. Proposed minimum reporting standards for chemical analysis: Chemical Analysis Working Group (CAWG) Metabolomics Standards Initiative (MSI). *Metabolomics* **2007**, *3*, 211–221.

17. Allard, P.-M.; Genta-Jouve, G.; Wolfender, J.-L. Deep metabolome annotation in natural products research: Towards a virtuous cycle in metabolite identification. *Curr. Opin. Chem. Biol.* **2017**, *36*, 40–49.

18. Allard, P.M.; Péresse, T.; Bisson, J.; Gindro, K.; Marcourt, L.; Pham, V.C.; Roussi, F.; Litaudon, M.; Wolfender, J.L. Integration of Molecular Networking and In-Silico MS/MS Fragmentation for Natural Products Dereplication. *Anal. Chem.* **2016**, *88*, 3317–3323. [CrossRef]

19. Quinn, R.A.; Nothias, L.F.; Vining, O.; Meehan, M.; Esquenazi, E.; Dorrestein, P.C. Molecular Networking As a Drug Discovery, Drug Metabolism, and Precision Medicine Strategy. *Trends Pharmacol. Sci.* **2017**, *38*, 143–154. [CrossRef]

20. Bauvais, C.; Bonneau, N.; Blond, A.; Pérez, T.; Bourguet-Kondracki, M.-L.; Zirah, S. Furanoterpene Diversity and Variability in the Marine Sponge Spongia officinalis, from Untargeted LC–MS/MS Metabolomic Profiling to Furanolactam Derivatives. *Metabolites* **2017**, *7*, 27. [CrossRef]

21. Wang, M.; Carver, J.J.; Phelan, V.V.; Sanchez, L.M.; Garg, N.; Peng, Y.; Nguyen, D.D.; Watrous, J.; Kapono, C.A.; Luzzatto-Knaan, T.; et al. Sharing and community curation of mass spectrometry data with Global Natural Products Social Molecular Networking. *Nat. Biotechnol.* **2016**, *34*, 828–837. [PubMed]

22. Olivon, F.; Elie, N.; Grelier, G.; Roussi, F.; Litaudon, M.; Touboul, D. MetGem Software for the Generation of Molecular Networks Based on the t-SNE Algorithm. *Anal. Chem.* **2018**, *90*, 13900–13908. [PubMed]

23. Tada, I.; Tsugawa, H.; Meister, I.; Zhang, P.; Shu, R.; Katsumi, R.; Wheelock, C.E.; Arita, M.; Chaleckis, R. Creating a Reliable Mass Spectral–Retention Time Library for All Ion Fragmentation-Based Metabolomics. *Metabolites* **2019**, *9*, 251. [CrossRef] [PubMed]

24. Pluskal, T.; Castillo, S.; Villar-Briones, A.; Orešič, M. MZmine 2: Modular framework for processing, visualizing, and analyzing mass spectrometry-based molecular profile data. *BMC Bioinform.* **2010**, *11*, 395. [CrossRef] [PubMed]

25. Aicheler, F.; Li, J.; Hoene, M.; Lehmann, R.; Xu, G.; Kohlbacher, O. Retention Time Prediction Improves Identification in Nontargeted Lipidomics Approaches. *Anal. Chem.* **2015**, *87*, 7698–7704.

26. Randazzo, G.M.; Tonoli, D.; Hambye, S.; Guillarme, D.; Jeanneret, F.; Nurisso, A.; Goracci, L.; Boccard, J.; Rudaz, S. Prediction of retention time in reversed-phase liquid chromatography as a tool for steroid identification. *Anal. Chim. Acta* **2016**, *916*, 8–16.

27. Clouzeau, C.; Godefroy, D.; Riancho, L.; Rostène, W.; Baudouin, C.; Brignole-Baudouin, F. Hyperosmolarity potentiates toxic effects of benzalkonium chloride on conjunctival epithelial cells in vitro. *Mol. Vis.* **2012**, *18*, 851–863.

28. Warcoin, E.; Clouzeau, C.; Roubeix, C.; Raveu, A.L.; Godefroy, D.; Riancho, L.; Baudouin, C.; Brignole-Baudouin, F. Hyperosmolarity and Benzalkonium Chloride Differently Stimulate Inflammatory Markers in Conjunctiva-Derived Epithelial Cells in vitro. *Ophthalmic Res.* **2017**, *58*, 40–48.

29. Craig, J.P.; Nichols, K.K.; Akpek, E.K.; Caffery, B.; Dua, H.S.; Joo, C.; Liu, Z.; Nelson, J.D.; Nichols, J.J.; Tsubota, K.; et al. The Ocular Surface TFOS DEWS II Definition and Classification Report. *Ocul. Surf.* **2017**, *15*, 276–283.

30. Pflugfelder, S.C.; de Paiva, C.S. The Pathophysiology of Dry Eye Disease: What We Know and Future Directions for Research. *Ophthalmology* **2017**, *124*, S4–S13.

31. Baudouin, C.; Aragona, P.; Messmer, E.M.; Tomlinson, A.; Calonge, M.; Boboridis, K.G.; Akova, Y.A.; Geerling, G.; Labetoulle, M.; Rolando, M. Role of hyperosmolarity in the pathogenesis and management of dry eye disease: Proceedings of the ocean group meeting. *Ocul. Surf.* **2013**, *11*, 246–258. [PubMed]

32. Robciuc, A.; Hyo, T. Ceramides in the Pathophysiology of the Anterior Segment of the Eye. *Curr. Eye Res.* **2013**, *38*, 1006–1016. [PubMed]

33. Olivon, F.; Roussi, F.; Litaudon, M.; Touboul, D. Optimized experimental workflow for tandem mass spectrometry molecular networking in metabolomics. *Anal. Bioanal. Chem.* **2017**, *409*, 5767–5778. [CrossRef] [PubMed]

34. Godzien, J.; Ciborowski, M.; Martínez-Alcázar, M.P.; Samczuk, P.; Kretowski, A.; Barbas, C. Rapid and Reliable Identification of Phospholipids for Untargeted Metabolomics with LC-ESI-QTOF-MS/MS. *J. Proteome Res.* **2015**, *14*, 3204–3216.

35. Hsu, F.F.; Turk, J. Electrospray ionization with low-energy collisionally activated dissociation tandem mass spectrometry of glycerophospholipids: Mechanisms of fragmentation and structural characterization. *J. Chromatogr. B Anal. Technol. Biomed. Life Sci.* **2009**, *877*, 2673–2695. [CrossRef]

36. Maccarone, A.T.; Duldig, J.; Mitchell, T.W.; Blanksby, S.J.; Duchoslav, E.; Campbell, J.L. Characterization of acyl chain position in unsaturated phosphatidylcholines using differential mobility-mass spectrometry. *J. Lipid Res.* **2014**, *55*, 1668–1677. [CrossRef]

37. Ann, Q.; Adams, J. Structure-Specific Collision-Induced Fragmentations of Ceramides Cationized with Alkali-Metal Ions. *Anal. Chem.* **1993**, *65*, 7–13. [CrossRef]

38. Lanzini, J.; Dargère, D.; Regazzetti, A.; Tebani, A.; Laprévote, O.; Auzeil, N. Changing in lipid profile induced by the mutation of Foxn1 gene: A lipidomic analysis of Nude mice skin. *Biochimie* **2015**, *118*, 234–243.

39. Petit, J.; Wakx, A.; Gil, S.; Fournier, T.; Auzeil, N.; Rat, P.; Laprévote, O. Lipidome-wide disturbances of human placental JEG-3 cells by the presence of MEHP. *Biochimie* **2018**, *149*, 1–8. [CrossRef]

40. Ovčačíková, M.; Lísa, M.; Cífková, E.; Holčapek, M. Retention behavior of lipids in reversed-phase ultrahigh-performance liquid chromatography-electrospray ionization mass spectrometry. *J. Chromatogr. A* **2016**, *1450*, 76–85. [CrossRef]

41. Kind, T.; Liu, K.H.; Lee, D.Y.; Defelice, B.; Meissen, J.K.; Fiehn, O. LipidBlast in silico tandem mass spectrometry database for lipid identification. *Nat. Methods* **2013**, *10*, 755–758. [CrossRef] [PubMed]

42. Cajka, T.; Fiehn, O. LC–MS-based lipidomics and automated identification of lipids using the LipidBlast in-silico MS/MS library. *Methods Mol. Biol.* **2017**, *1609*, 149–170. [PubMed]

43. Hutchins, P.D.; Russell, J.D.; Coon, J.J. LipiDex: An Integrated Software Package for High-Confidence Lipid Identification. *Cell Syst.* **2018**, *6*, 621–625.e5. [CrossRef] [PubMed]

44. Bron, A.J.; de Paiva, C.S.; Chauhan, S.K.; Bonini, S.; Gabison, E.E.; Jain, S.; Knop, E.; Markoulli, M.; Ogawa, Y.; Perez, V.; et al. TFOS DEWS II pathophysiology report. *Ocul. Surf.* **2017**, *15*, 438–510. [CrossRef]

45. Baudouin, C.; Messmer, E.M.; Aragona, P.; Geerling, G.; Akova, Y.A.; Benítez-Del-Castillo, J.; Boboridis, K.G.; Merayo-Lloves, J.; Rolando, M.; Labetoulle, M. Revisiting the vicious circle of dry eye disease: A focus on the pathophysiology of meibomian gland dysfunction. *Br. J. Ophthalmol.* **2016**, *100*, 300–306. [CrossRef]

46. Baudouin, C.; Irkeç, M.; Messmer, E.M.; Benítez-del-Castillo, J.M.; Bonini, S.; Figueiredo, F.C.; Geerling, G.; Labetoulle, M.; Lemp, M.; Rolando, M.; et al. Clinical impact of inflammation in dry eye disease: Proceedings of the ODISSEY group meeting. *Acta Ophthalmol.* **2018**, *96*, 111–119. [CrossRef]

47. Robciuc, A.; Hyötyläinen, T.; Jauhiainen, M.; Holopainen, J.M. Hyperosmolarity-induced lipid droplet formation depends on ceramide production by neutral sphingomyelinase 2. *J. Lipid Res.* **2012**, *53*, 2286–2295. [CrossRef]

48. Warcoin, E.; Clouzeau, C.; Brignole-Baudouin, F.; Baudouin, C. Hyperosmolarité: Effets intracellulaires et implication dans la sécheresse oculaire. *J. Fr. Ophtalmol.* **2016**, *39*, 641–651. [CrossRef]

49. Warcoin, E.; Baudouin, C.; Gard, C.; Brignole-Baudouin, F. In Vitro Inhibition of NFAT5-Mediated Induction of CCL2 in Hyperosmotic Conditions by Cyclosporine and Dexamethasone on Human HeLa-Modified Conjunctiva-Derived Cells. *PLoS ONE* **2016**, *11*, e0159983. [CrossRef]

50. Cavet, M.E.; Harrington, K.L.; Vollmer, T.R.; Ward, K.W.; Zhang, J.-Z. Anti-inflammatory and anti-oxidative effects of the green tea polyphenol epigallocatechin gallate in human corneal epithelial cells. *Mol. Vis.* **2011**, *17*, 533–542.

51. Ren, Y.; Lu, H.; Reinach, P.S.; Zheng, Q.; Li, J.; Tan, Q.; Zhu, H.; Chen, W. Hyperosmolarity-induced AQP5 upregulation promotes inflammation and cell death via JNK1/2 Activation in human corneal epithelial cells. *Sci. Rep.* **2017**, *7*, 4727. [PubMed]

52. Hua, X.; Deng, R.; Li, J.; Chi, W.; Su, Z.; Lin, J.; Pflugfelder, S.; Li, D. Protective Effects of L-Carnitine against Oxidative Injury by Hyperosmolarity in Human Corneal Epithelial Cells. *Investig. Ophthalmol. Vis. Sci.* **2015**, *56*, 5503–5511. [CrossRef]

53. Deng, R.; Hua, X.; Li, J.; Chi, W.; Zhang, Z.; Lu, F.; Zhang, L.; Pflugfelder, S.C.; Li, D.-Q. Oxidative stress markers induced by hyperosmolarity in primary human corneal epithelial cells. *PLoS ONE* **2015**, *10*, e0126561. [CrossRef]

54. Wang, P.; Sheng, M.; Li, B.; Jiang, Y.; Chen, Y. High osmotic pressure increases reactive oxygen species generation in rabbit corneal epithelial cells by endoplasmic reticulum. *Am. J. Transl. Res.* **2016**, *8*, 860.

55. Braverman, N.E.; Moser, A.B. Functions of plasmalogen lipids in health and disease. *Biochim. Biophys. Acta Mol. Basis Dis.* **2012**, *1822*, 1442–1452.

56. Dogru, M.; Kojima, T.; Simsek, C.; Tsubota, K. Potential role of oxidative stress in ocular surface inflammation and dry eye disease. *Investig. Ophthalmol. Vis. Sci.* **2018**, *59*, DES163–DES168.

57. Choi, W.; Lian, C.; Ying, L.; Kim, G.E.; You, I.C.; Park, S.H.; Yoon, K.C. Expression of Lipid Peroxidation Markers in the Tear Film and Ocular Surface of Patients with Non-Sjogren Syndrome: Potential Biomarkers for Dry Eye Disease. *Curr. Eye Res.* **2016**, *41*, 1143–1149.

58. Chaurasia, B.; Summers, S.A. Ceramides—Lipotoxic Inducers of Metabolic Disorders. *Trends Endocrinol. Metab.* **2015**, *26*, 538–550. [CrossRef]

59. Magny, R.; Kessal, K.; Regazzetti, A.; Yedder, A. Ben; Baudouin, C.; Parsadaniantz, S.M.; Brignole-Baudouin, F.; Laprévote, O.; Auzeil, N. Lipidomic analysis of epithelial corneal cells following hyperosmolarity and benzalkonium chloride exposure: New insights in dry eye disease. *Biochim. Biophys. Acta - Mol. Cell Biol. Lipids* **2020**, 158728.

60. Myers, O.D.; Sumner, S.J.; Li, S.; Barnes, S.; Du, X. One Step Forward for Reducing False Positive and False Negative Compound Identifications from Mass Spectrometry Metabolomics Data: New Algorithms for Constructing Extracted Ion Chromatograms and Detecting Chromatographic Peaks. *Anal. Chem.* **2017**, *89*, 8696–8703.

61. GNPS Home Page. Available online: http://gnps.ucsd.edu (accessed on 26 May 2020).

62. Shannon, P.; Markiel, A.; Ozier, O.; Baliga, N.S.; Wang, J.T.; Ramage, D.; Amin, N.; Schwikowski, B.; Ideker, T. Cytoscape: A software Environment for integrated models of biomolecular interaction networks. *Genome Res.* **2003**, *13*, 2498–2504. [CrossRef] [PubMed]

 *metabolites*

Article

Oxylipin Profiles in Plasma of Patients with Wilson's Disease

Nadezhda V. Azbukina [1], Alexander V. Lopachev [2], Dmitry V. Chistyakov [3,*], Sergei V. Goriainov [4], Alina A. Astakhova [3], Vsevolod V. Poleshuk [5], Rogneda B. Kazanskaya [6], Tatiana N. Fedorova [2,*] and Marina G. Sergeeva [3,*]

[1]   Faculty of Bioengineering and Bioinformatics, Moscow Lomonosov State University, Moscow 119234, Russia; ridernadya@gmail.com
[2]   Laboratory of Clinical and Experimental neurochemistry, Research Center of Neurology, Moscow 125367, Russia; lopachev@neurology.ru
[3]   Belozersky Institute of Physico-Chemical Biology, Lomonosov Moscow State University, Moscow 119992, Russia; alina_astakhova@yahoo.com
[4]   SREC PFUR Peoples' Friendship University of Russia (RUDN University), Moscow 117198, Russia; goryainovs@list.ru
[5]   Research Center of Neurology, Moscow 125367, Russia; pol82@yandex.ru
[6]   Biological Department, Saint Petersburg State University, Universitetskaya Emb. 7/9, St Petersburg 199034, Russia; magnolia@kwakwa.org
*   Correspondence: Chistyakof@gmail.com (D.V.C.); tnf51@bk.ru (T.N.F.); mg.sergeeva@gmail.com (M.G.S.)

Received: 17 April 2020; Accepted: 25 May 2020; Published: 29 May 2020

Abstract: Wilson's disease (WD) is a rare autosomal recessive metabolic disorder resulting from mutations in the copper-transporting, P-type ATPase gene ATP7B gene, but influences of epigenetics, environment, age, and sex-related factors on the WD phenotype complicate diagnosis and clinical manifestations. Oxylipins, derivatives of omega-3, and omega-6 polyunsaturated fatty acids (PUFAs) are signaling mediators that are deeply involved in innate immunity responses; the regulation of inflammatory responses, including acute and chronic inflammation; and other disturbances related to any system diseases. Therefore, oxylipin profile tests are attractive for the diagnosis of WD. With UPLC-MS/MS lipidomics analysis, we detected 43 oxylipins in the plasma profiles of 39 patients with various clinical manifestations of WD compared with 16 healthy controls (HCs). Analyzing the similarity matrix of oxylipin profiles allowed us to cluster patients into three groups. Analysis of the data by VolcanoPlot and partial least square discriminant analysis (PLS-DA) showed that eight oxylipins and lipids stand for the variance between WD and HCs: eicosapentaenoic acid EPA, oleoylethanolamide OEA, octadecadienoic acids 9-HODE, 9-KODE, 12-hydroxyheptadecatrenoic acid 12-HHT, prostaglandins PGD2, PGE2, and 14,15-dihydroxyeicosatrienoic acids 14,15-DHET. The compounds indicate the involvement of oxidative stress damage, inflammatory processes, and peroxisome proliferator-activated receptor (PPAR) signaling pathways in this disease. The data reveal novel possible therapeutic targets and intervention strategies for treating WD.

Keywords: COX; CYP450; LOX; oxylipins; PUFAs; lipidomics; UPLC-MS/MS; copper; Wilson's disease

---

1. Introduction

Wilson's disease (WD) is a rare autosomal recessive metabolic disorder resulting from mutations in the copper-transporting, P-type ATPase gene, ATP7B gene, which encodes a copper-transporting P-type ATPase [1]. The enzyme is responsible for the transport of copper into bile from hepatocytes, facilitating its incorporation into apoceruloplasmin to form ceruloplasmin, a major copper-transporting protein in the blood. The mutations lead to copper accumulation in the affected tissues [1–3], which causes biochemical deviations, followed by the fluctuating occurrence of hepatic and extrapyramidal symptoms (EPSs), accompanied by the impairment of other organs (for more detail, see recent reviews [4–6]). The worldwide prevalence of WD is 1 in 30,000; it is even higher in populations with a high frequency of consanguinity [5]. The manifestations of WD are variable, and, in addition to liver diseases, may include neurological and/or psychiatric symptoms, as well as abnormalities in the blood or kidneys. It is therefore hypothesized that other genetic and/or environmental factors could influence the phenotypes of WD [5,7]. Besides different ATP7B mutations [8], other genetic variations may influence the variability in WD manifestation, such as apolipoprotein E (APOE), human prion protein (PRNP), 5,10-methylenetetrahydrofolate reductase (MTHFR), the interleukin-1 receptor antagonist (IL1RN), peroxisomal catalase, and other genes [7,9]. The influence of epigenetics, environment, age, and sex-related factors on the WD phenotype further complicates diagnosis [3,7,10]. The clear–phenotype correlations for WD are still unclear, although recent epigenetic whole genome screening data may shed light on in-depth pathogenic mechanisms [11]. Obtained from liver and blood samples from patients with WD, the data have shown specific sets of modified genes, enriched for functions in lipid metabolism and inflammatory responses [11]. Such changes can manifest themselves on the level of the organism as a whole in a variety of ways and be the cause of the observed differences in manifestations of the disease. This attracted attention to studying the disease on the level of the metabolome. Understanding the variations in the metabolome may help in identifying variations in the biochemical pathways leading to different manifestations of the disease. High-throughput techniques, such as metabolomic profiling, can deepen our understanding of the disease's pathogenesis and biology of WD manifestation [12,13], and therefore lead to new therapeutic approaches.

A promising type of metabolomic profiling is oxylipin measurement in the blood or other tissues. Oxylipins, derivatives of omega-3, and omega-6 polyunsaturated fatty acids (PUFAs) are signaling mediators that are deeply involved in innate immunity responses; the regulation of inflammatory responses, including acute and chronic inflammation; and other disturbances related to any system disease [14–17]. The conversion of PUFAs into oxylipins occurs via three major pathways, named according to their respective key pathway enzymes, such as the cyclooxygenase (COX), lipoxygenase (LOX), and cytochrome P450 monooxygenase (CYP450) branches of metabolism. Besides this, there are non-enzymatic conversions of PUFAs [15]. Due to the diversity of the individual oxylipin functions, it is difficult to predict the general direction of their action. For example, eicosapentaenoic (EPA) and docosahexaenoic (DHA) omega-3 PUFAs, as well as their derivative oxylipins, hydroxyeicosapentaenoic acids (HEPEs) and hydroxydocosahexaenoic acids (HDoHEs), are regarded as anti-inflammatory mediators [14,15]. Arachidonic acid (AA), an omega-6 PUFA, is mainly the source of prostaglandins (PGs), thromboxane (TX), leukotrienes (LTs), and hydroxyeicosatetraenoic acids (HETEs), attributed to groups of proinflammatory oxylipins. Meanwhile, cyclopentenone PGs, non-enzymatic metabolites of PGE2 and PGD2, possess anti-inflammatory features [18]. Oxidative derivatives of α-linolenic acid (ALA) can be transformed into hydroxyoctadecatrienoic (HOTrEs) acids or others [15]. Linoleic acid (LA)-derived oxylipins, such as hydroxyoctadecadienoic (HODEs) acids, agonists of PPARγ [19], or dihydroxyoctadecamonoenoic (DiHOMEs) acids, which are cytotoxic [20], exhibit both pro- and anti-inflammatory features [15]. Taken together, these data show that oxylipin synthesis should not be studied in groups of separate substances, but in terms of oxylipin profiles, which can characterize the different states of the studied organisms.

Indirect data indicate the possibility of oxylipin profile changes in WD. The roles of oxidative stress in the pathogenesis of WD, and dietary omega-3 PUFAs' usefulness in an animal model of WD suggest that oxylipins may be involved in the clinical manifestation of WD. Moreover, the number of some lipid-related nuclear receptors, such as retinoid X receptor (RXR), peroxisome proliferator-activated receptor α (PPARα), and hepatocyte nuclear factor 4 alpha (HNF4A), is generally decreased in WD animal models and humans [9,21–25]. It is important to note that although oxylipins exhibit multiple effects, they somehow change the state of the innate immunity system [13–15]. Oxylipins are important markers of activation of the system, including the regulation of inflammatory resolution processes [13,16]. Despite these facts, the role of the innate immunity system, and oxylipins as parts of the system, is still underestimated in many diseases.

The development of mass spectrometric methods for oxylipin detection made it possible to obtain oxylipin profiles from the plasma of patients with diseases, such as Alzheimer's disease [24], alcohol-related liver disease [25], or cancer [26]. However, no such studies have been conducted to characterize WD. Therefore, in the present study, UPLC-MS/MS lipidomics analyses were performed to characterize the plasma profiles of patients with various clinical manifestations of WD, compared to healthy subjects in order to identify the oxylipin characteristics of this disease.

2. Results

2.1. Clinical Characteristics

The study involved 39 WD patients and 16 healthy controls. The anthropometric, demographic, and blood biochemical parameters of the enrolled individuals are presented in Table 1. In total, 25 patients had the akinetic-rigid form, 10 had the trembling form, and 4 had other forms. Biochemical profiling of the blood of WS patients was conducted; data for ceruloplasmin and serum Cu concentrations are shown in Table 1.

Table 1. Demographic parameters of the patients, disease characteristics, and medications.

Wilson Disease Patients						
Sex		F (n = 22)			M (n = 17)	
	mean	sd	n	mean	sd	n
Age	35.68	13.17		32.18	12.36	
Serum Cu, mkM	8.5	4.3		9.62	4.36	
Shvab scale, %	80.95	18.41		67.5	22.36	
Leipzig score	7.32	2.19		6.42	2.57	
Ceruloplasmin, mg/dL	9.61	7.2		12.78	8.02	
Height, cm	169.95	5.55		180.8	7.44	
Longevity illness, years	13.86	11.34		9.81	9.32	
Longevity treatment, years	12.86	11.25		8.21	8.75	
Weight, kg	61.2	13.27		75.67	12.21	
Form (akinetic-rigid/trembling/others)			2014/6/2			2011/4/2
Nephropathy			5			8
Portal hypertension			5			11
Psycoproductive somatic			8			6
Healthy Donors						
		F (11)			M (5)	
	mean	sd	n	mean	sd	n
Age	37.88	15.96		49.2	12.19	

2.2. Metabolomic Profiling

Using UPLC-MS/MS, we detected a total of 43 metabolites in human plasma (Table S1). Metabolites were from different lipid classes: 3 PUFA (AA, DHA and EPA), 19 AA derivatives, one DGLA derivate, 7 DHA derivatives, 3 EPA derivatives, 7 LA derivatives, and 3 non-PUFA-derived compounds (OEA, AEA, Lyso-PAF).

2.3. Volcano Plot Analysis

To evaluate the separate metabolites that differ among WD and HC groups, we performed pairwise comparisons of age and gender-adjusted metabolite concentrations. The results were then illustrated using a volcano plot with Holm–Bonferroni correction (Figure 1). The four metabolites whose concentrations were changed significantly are indicated in red (12-HHT, EPA, PGE2, and PGD2). Barplots of the indicated compounds' relative concentrations are presented in Figure 1B.

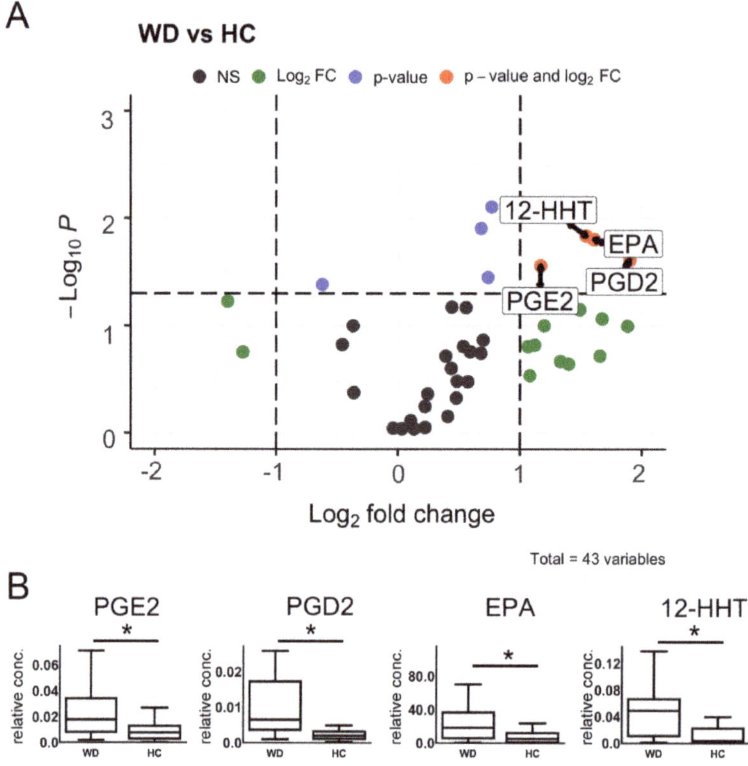

Figure 1. (**A**) Volcano plot indicating significantly changed compounds. The X-axis indicates a log2 fold change of wilson disease WD to HC (healthy control) patients. Y-axis indicates −log10 *p*-values (adjusted). The cut-off for *p*-values is indicated based on Bonferroni correction. Compounds that changed insignificantly are indicated in gray, compounds whose means changed in WD (relative to HCs) more than twofold or less than twofold but insignificantly are indicated in green. Red dots stand for compounds, which changed more than twofold and had a *p*-value (adjusted < 0.05). (**B**) Relative concentrations of separate metabolites that changed significantly in WD patients in comparison with HCs. Pairwise comparison of adjusted means was conducted taking into account the age and sex of patients. * *p* < 0.05 (adjusted for multiple testing).

2.4. PLS-DA Model

For data analysis, we used normalized concentrations of metabolites (see Section 2.6). The presence of outliers was identified by performing principal component analysis (PCA) to prevent their effects on the model. Hotelling's T2 test indicated three outliers in the healthy control group. A total of 52 samples, which were placed inside a 95% confidence interval ellipse red bounds (Figure 2A), were used for further analyses. For testing whether WD (Wilson disease) and HC (healthy control) patients could be distinguished based on oxylipin concentrations, the partial least square discriminant analysis (PLS-DA) was performed. The model was evaluated via cross-validation based on the overall error, balanced error rate (BER), and area under curve (AUC) values (Figure S1, Table S2). The optimal number of components was three. Projections on the first two components are presented in Figure 2B, and on the first three components in Figure 2C. Studied groups were separated with a small overlap. For each metabolite, the VIP score was estimated (as described in Section 2.6). The value of this parameter addresses the explained variation between classes in each projection. A total of seven metabolites, including 12-HHT, EPA, 14,15-DHET, 9-HODE, OEA, PGE2, and 9-KODE, with VIP score values > 1.5 are shown in Table 2.

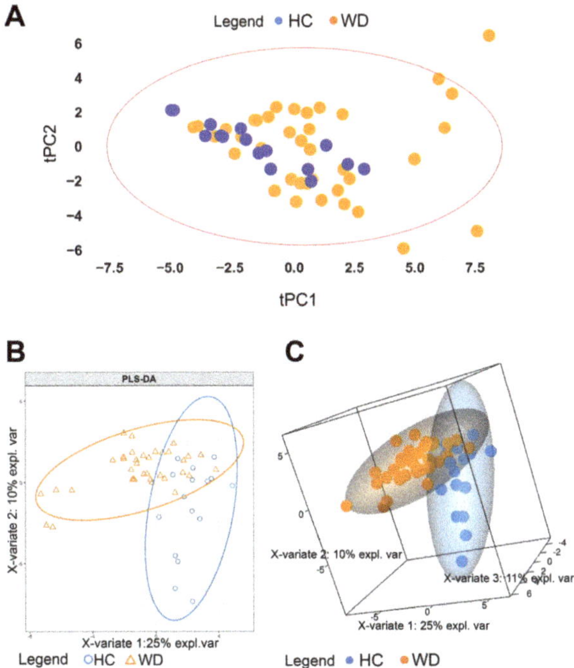

Figure 2. (**A**) The principal component analysis (PCA) performed to verify outliers. The 95% Hotelling T2 confidence interval is indicated as an ellipse. (**B**) The partial least square discriminant analysis (PLS-DA) model discriminating healthy control (HC) and Wilson disease patients (WD). The explained variance of each component is indicated in brackets on the corresponding axis. (**C**) PLS2-DA model represented in 3-D showing separation among the HC and WD patients.

Table 2. Variable importance in projection (VIP) scores are shown for 7 metabolites. A cutoff value of 1.5 is established for VIP selection.

Name	12-HHT *	EPA *	14,15-DHET	9-HODE	OEA	PGE2 *	9-KODE
VIP-scores	1.899456	1.741633	1.739218	1.624940	1.617023	1.594828	1.837164

* volcano plot indicating significantly changed compounds, $p < 0.05$ (adjusted for multiple testing).

2.5. Similarity Matrix

Since oxylipins represent different branches of metabolic pathways [14], we decided to estimate possible interconnections among compounds by calculating the pairwise association matrix, using data obtained using PLS-DA. A clustered image map (CIM), based on a hierarchical clustering of both the rows and the columns, was built using the Euclidean distance and complete linkage clustering algorithm (Figure 3). In the figure, each entry of the matrix is colored according to the association between metabolite concentrations and illness status (X and Y-variables in the model). The red color indicates positive correlation, whereas yellow/green indicates a weaker correlation. Dendrograms are shown on the left side (for metabolites) and on top (for patients). Color bar A indicates whether the patient belongs to WD (black) or HC (red). Based on the dendrogram and illness status, we subdivided the subjects into four groups (bar on the top of heatmap). WD patients can be subdivided into three groups: Not distinguished from healthy donors (mix), with enrichment of HdOHE and HETE compounds, and with DiHETE, DiHOME enrichment (Figure 3).

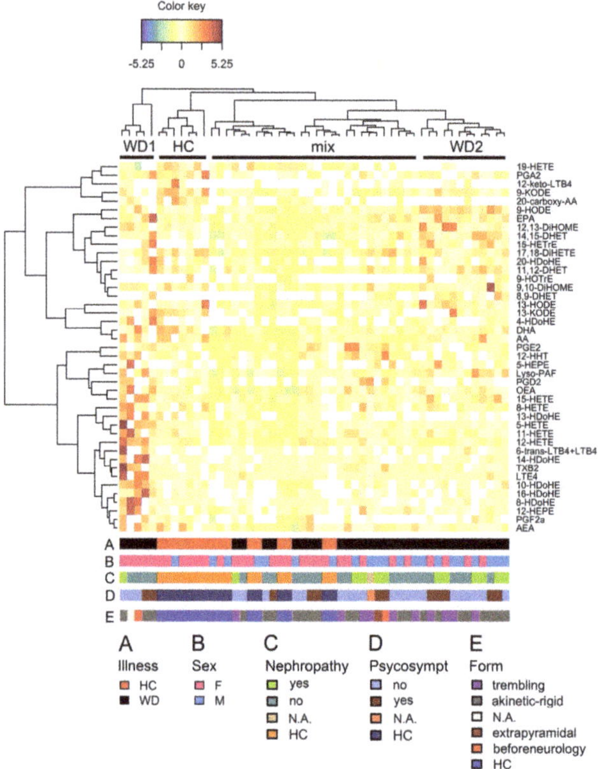

Figure 3. A clustered image map was generated using the Euclidean distance and the complete linkage clustering algorithm. On the figure, each entry of the matrix is colored according to its value; rows represent metabolites, columns represent subjects. Dendrograms are shown on the left side (for patients) and on top (for metabolites). Color bars on the bottom of the picture indicate: (**A**) whether the subjects belongs to the Wilson disease (WD) group or healthy control (HC) group; (**B**) sex distribution: male (M) or female (F); (**C**) nephropathy status: no, yes, HC or data not available (N.A.); (**D**) psychosomatic status: no, yes, HC or N.A.; (**E**) form of the disease: trembling, akinetic-rigid, extrapyramidal, beforeneurology, or HC.

We further annotated the enrichment of modules using patients' clinical characteristics. The color bars on the bottom of Figure 3 indicate the clinical and anthropometric annotation of the patients. In total, five bars are presented on the CIM, indicating:

(A)   HC/WD patients;

(B)   Sex;

(C)   Nephropathy status;

(D)   Psychosomatic status; and

(E)   Form of the disease.

However, all WD patients clustered in a group on the left side turned out to be females (Figure 3). We tested associations with the Cu serum concentration, the severity of motor system dysfunction according to the Shvab scale, age, and the debut age of subjects. There was no clear clustering according to the mentioned parameters (Figure S5A–D).

To test whether there were any differences in separate metabolites between selected groups, we conducted analysis of covariance (ANCOVA) to compare the adjusted means between groups taking into account the variability of the age and sex of patients. To identify which groups were different, pairwise comparisons of the adjusted means with the following Bonferroni multiple testing correction were applied. In the WD1 module patients, 10-HDoHE, 11-HETE, 12-HEPE, 12-HETE, 13-HDoHE, 15-HETE, 16-HDoHE, 5-HETE, and OEA were significantly different from HC (Figure S4). In the WD2 module patients, 10-HDoHE, 9-HODE, and AA were significantly different from HC. The mix module was not different from HCs (Figure S4).

2.6. Pathway Enrichment Analysis

Differences in separate metabolism branches are often a specific trait of biological processes [14–16]. This is why after independent analysis of the compounds, we took a step forward and investigated oxylipins as groups. Concentrations of compounds were summed according to their acid precursors (AA, DHA, EPA, ALA, DGLA, EA, EPA) (Figure 4A) or via the metabolic pathways they were derived from (cyclooxygenase (COX), cytochrome P450 monooxygenase (CYP), lipoxygenase (LOX), or non-enzymatic reactive oxygen species (ROS)) (Figure 4B). It should be mentioned that in both cases, only the derivatives were summed up; free acids were grouped into the "others" unit. The classification used was in accordance with [15]. Then, a similarity matrix was calculated, and a complete linkage algorithm was performed for the acid precursor matrix (Figure 4A). To simplify the analysis between acid precursor and enzyme pathways, the second CIM was plotted using the order of the corresponding row as in Figure 4A, and clustering was performed only in columns (Figure 4B).

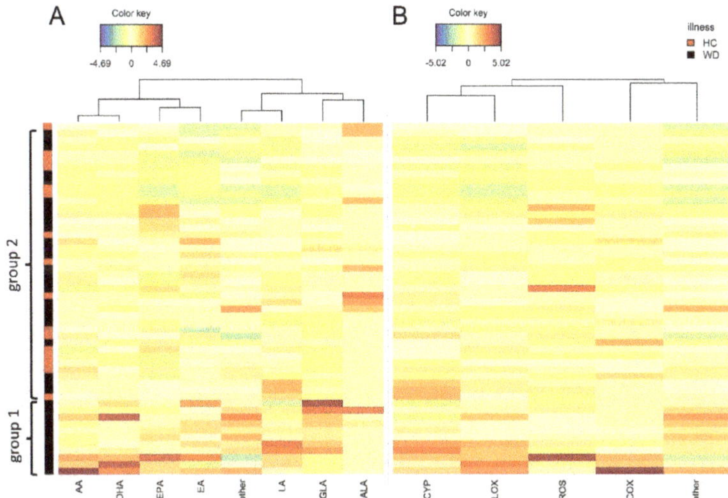

Figure 4. A clustered image map was performed using the Euclidean distance and complete linkage clustering algorithm. All polyunsaturated fatty acids (PUFA) derivatives were summed up according to their (**A**) initial substrate of biochemical pathways or (**B**) pathways' enzyme origin. In the figure, each entry of the matrix is colored according to its value, rows represent subjects, columns represent metabolites. Dendrograms are shown on the top (for metabolites). The color bar on the left side of the picture indicates whether a subject is WD (black) or a HC (red). Abbreviations: AA: arachidonic acid; DHA: docosahexaenoic acid, EPA: eicosapentaenoic acid; EA: AEA and OEA; LA: linoleic acid; DGLA: dihomo-γ-linolenic acid; ALA: α-linolenic acid; CYP: cytochrome P450 monooxygenase; LOX: lipoxygenase; ROS: reactive oxygen species; COX: cyclooxygenase.

The size of the cluster described in Section 2.5 increased. It was enriched with AA, DHA, EPA, LA, and DGLA metabolites, which indicates respective changes of their concentrations. It should be noted that the correlation between changes in the amount of metabolites of the CYP and LOX pathways was greater than that between their changes and changes in COX metabolites. However, this clustering was not explained by the clinical and demographic parameters mentioned in Section 2.4 and the reason for this standing apart is still unclear. Patients could be subdivided into two clusters independently of the grouping strategy (acids or enzymes), which is similar to the results presented in Figure 3. The bottom cluster of patients was associated with an overall upregulated content of oxylipins. On the other hand, the second cluster of patients did not show such an association.

After analysis of the grouped CIM, we determined a subgroup of patients as group 1. Patients from that group had significantly different concentrations of LOX metabolites (speaking about enzymatic pathways) and significant changes in the concentrations of AA, DGLA, and DHA derivatives and free fatty acids (Figure S5).

3. Discussion

Although WD is an autosomal recessive metabolic disorder, it possesses uncertain phenotype–genotype correlations and variability in its clinical manifestations. Understanding the biology of WD pathogenesis and improving diagnostic methods is an urgent problem posed before the science community [27]. Recently developed methods make possible quantitative measurement and analysis of a large number of markers, which, taken together, make up profiles. Oxylipin profiles are unique in that they reflect the activity level of a variety of biochemical processes in the organism and participate in the regulation of various signaling cascades. They possess a characteristic "fingerprint" when certain changes occur, and reflect dynamic characteristics of the organism, which is why interest in oxylipin profiles continues to grow.

Oxylipin profiles were investigated for the study of mechanisms, and as diagnostic markers for diseases, such as Alzheimer's disease [24], female breast cancer [26], alcohol-related liver disease [25], atherosclerotic diseases [28], and coronary artery disease [29]. Importantly, every disease was characterized by a special set of oxylipins: 9-HODE [26], 20-HETE [25], 8-HETE, LTB4, 9-HODE and 13-HODE [28], 9-HETE, and F(2)-isoprostanes [29]. Although t-test statistics of the oxylipin profiles of 39 (22 female and 17 male) WD patients and 16 (11 female and 5 male) donors allowed four substances (12-HHT, EPA, PGE2, and PGD2) to be revealed, which differ in WD vs HD, it is still not possible to suggest these substances as diagnostic markers, and further investigations are required. However, importantly, our data add new information concerning the biology of WD pathogenesis.

Indeed, analysis of data by VolcanoPlot showed that the patient vs. healthy groups differed significantly across three lipids. PLS-DA analysis revealed five more lipids that explained the difference in oxylipin profiles among WD and HC. Among them, two acids (EPA, OEA); two metabolites of LA, 9-HODE and 9-KODE, which can be attributed to LOX or non-enzymatic branches of metabolism; three metabolites of AA (12-HHT, PGD2, PGE2), attributed to the COX branch; and one AA metabolite from CYP branches of metabolism (14,15-DHET) were included. Although oxylipins possess multiple effects, and the same compound can be traced through various processes [14,15], the following processes can be characterized by the respective compounds: Oxidative stress (9-HODE, 9-KODE, OEA, EPA), inflammatory markers (9-HODE, 9-KODE, PGE2, 12-HHT, PGD2), and peroxisome proliferator-activated receptor (PPAR) agonists (9-HODE, 9-KODE, OEA, EPA, 14,15-DHET). The data allow us to make assumptions about the possible signaling pathways involved in this pathology.

The ability of free Cu ions to participate in the formation of reactive oxygen species (ROS) and induce cellular toxicity is known [30]. In the presence of reducing agents (e.g., the superoxide anion radical), $Cu2+$ can be reduced to $Cu+$, which catalyzes the formation of hydroxyl radicals from hydrogen peroxide via the Haber–Weiss and Fenton reaction [31]. Therefore, the role of oxidative stress in the pathogenesis of WD is currently under investigation, and peroxisome impairment is suggested to be involved in WD pathophysiology [9,30,31]. Oxidized LA metabolites (HODE/KODE) are traditionally classified as oxidative stress markers [32]. Note that 9-HODE was also marketed as the most upregulated oxylipin species in the plasma of breast cancer patients, which indicates the possibility of oxidative stress involvement in this disease [26]. An interesting finding of our work is an increase of OEA and EPA, substances that are known preventers of oxidative stress [33,34]. Although oxidative stress can be viewed as a common disruption in various pathologies, which may suggest similarities in the oxidized forms of lipids, the variations in the oxylipin profiles obtained in various diseases do not support this point of view [24–26,28,29]. Our data sheds light onto some appropriate compensatory mechanisms that may be involved in WD.

An increase in 12-HHT, PGE2, and PGD2 points to the involvement of inflammatory processes in the pathogenesis of WD. This is in accordance with the data obtained in the animal model of WD (the Long-Evans Cinnamon rats), which is characterized by an increase in COX expression, the main enzyme in the synthesis of these substances [35]. Interestingly, dietary omega-3 PUFAs suppress acute hepatitis, prolong the survival of rats, and seem to lead to a decrease in COX expression [35]. Our data, showing an increase in 12-HHT, PGE2, and PGD2, are in accordance with this observation, because it is known that these omega-6 derivatives are inflammatory markers, which may be decreased by supplementation with dietary omega-3 PUFAs [36].

Importantly, the elevation of some oxylipins may lead to the activation of PPARs [37]. Three subtypes of PPAR (PPARα, PPARβ, PPARγ) are active regulators at the lipid metabolism and inflammation crossroad [38]. Recent studies have shown that PPARα and PPARγ are associated with steatosis and impairment of the antioxidant system in the liver of WD patients [39]. Inconsistent with this finding, a transcriptome analysis of the liver in the mouse model of Wilson's disease under copper-transporting, P-type ATPase gene Atp7b knockout identified the PPAR signaling pathway as a high-copper-responsive target pathway [40].

It is noteworthy that while PPARγ increased, PPARα mRNA expression is decreased with increased severity of WD [41]. This may reveal the "PPAR triad" mechanism that was conjectured for other cells, which respond to an excess of various types of PPAR ligands [42]. There are few data concerning the mechanisms of different PPAR-type changes in the presence of their ligands' excess at the organism level. Our data suggest that the increased number of PPAR ligands in the blood of WD patients may be associated with some kind of regulatory compensatory mechanism associated with the PPARs system. Our data also single out this signaling pathway for further consideration as being involved in the clinical manifestations of WD. Indeed, among the investigated substances, PPARα agonists were determined by OEA [41], EPA [37], and 14,15-DHET [43]. 9-HODE is an endogenous activator and ligand of PPARγ [19]. In this context, the question remains regarding the role of PGD2 and PGE2. Besides action via specific G-protein-coupled receptors, these prostaglandins are converted in the course of inflammatory reactions into prostaglandins 15d-PGJ2 and PGA2, respectively. Compounds with anti-inflammatory properties are formed, which activate PPARα and PPARγ [18]. At present, the question remains open whether an increase in PPAR agonists in the blood plasma of patients with WD is a protective mechanism that weakens the severity of the clinical course of the disease or, conversely, aggravates the symptoms. We were not able to find data on the use of synthetic agonists of PPAR in WD models. It is likely that fibrates and thiazolidinediones may be used as potential therapeutic agents for WD. Further research is required to elucidate the molecular mechanisms by which PPAR agonists may exert their effects in WD pathogenesis.

Beside PPARs, we cannot exclude oxylipins' involvement in the regulation of the activity of other nuclear receptors, because it is known that lipid-related nuclear receptors change in WD patients or animal models [22,23]. Decreased binding of the nuclear receptors FXR, RXR, HNF4α, and LRH-1 to promoter response elements and decreased mRNA expression of nuclear receptor target genes [22] and dysregulation of LXR/RXR heterodimers [23] may also be compensated by increased lipid agonist concentrations. Our data are consistent with these results, but further research is required to understand the mechanisms.

An important finding of our work is the subdivision of the patients into groups relative to the oxylipin profiles. Patients from the group had significantly different concentrations of LOX metabolites and had significant changes in the concentration of AA, DGLA, and DHA derivatives and free fatty acids. It is not yet clear which parameter underlies this division, since we did not find any correlation with gender or age. The subdivision into groups can reflect the various states of the innate immunity system. Along with various cytokines, oxylipins are part of the innate immunity system, being proinflammatory substances, as well as mediators of resolution [13]. Innate immune traits are more affected by the environment [44]. Environmental factors, including diet, exercise, stress, and toxins, profoundly impact the phenotypes of diseases, and WD is among them [7]. It is worthwhile to assume that the observed separation of patients by the oxylipin profile reflects a phenotypic response to environmental factors and therefore the current state of the innate immune system. This aspect requires further investigation of the metabolites that we found to be characteristic of WD.

4. Materials and Methods

4.1. Reagents

The oxylipins standards were as follows: 6-keto PGF1α-d4 (cat.no. 315210), TXB2-d4 (cat.no. 319030), PGF2α-d4 (cat.no. 316010), PGE2-d4 (cat.no. 314010), PGD2-d4 (cat.no. 312010), leukotriene (LT) C4-d5 (cat.no. 10006198), LTB4-d4 (cat.no. 320110), 5(S)-HETE-d8 (cat.no. 334230), 12(S)-HETE-d8 (cat.no. 334570), 15(S)-HETE-d8 (cat.no. 334720), oleoyl ethanolamide-d4 (cat.no. 9000552), EPA-d5 (cat.no, 10005056), DHA-d5 (cat.no. 10005057), and AA-d8 (cat. No. 390010) (Cayman Chemical, Ann Arbor, MI, USA). An Oasis®PRIME HLB solid-phase lipid extraction cartridge (60 mg, 3 cc, cat.no. 186008056) was obtained from Waters, Eschborn, Germany.

4.2. Population and Study Design

This was an observational study with 55 recruited people: 39 patients with WD and 16 healthy controls. In total, 39 individuals with Wilson's disease admitted to the regular inpatient treatment in the Research Center of Neurology (Moscow, Russia) were recruited for the study. Inclusion criteria for the WD patients included the following clinical and laboratory signs of the disease: Debut of the illness in childhood, adolescence, or adulthood (most often up to 35 years old); combined brain and internal organ damage (liver cirrhosis, hepatolienal syndrome, portal hypertension, tubular nephritis, etc.); damage to the central nervous system in the form of extrapyramidal syndrome; and systemic discuprinosis with impaired copper-ligand metabolism.

WD exclusion criteria included the following: Disease manifestation after 35 years; autosomal dominant type of inheritance; the presence of anamnestic, clinical, or paraclinical signs of another disease that can cause similar symptoms; hallucinations not related to medication; the presence of dementia or signs of impaired cortical function (aphasia, apraxia, etc.); the slowing down of vertical saccades or vertical gaze paralysis; a positive history of inflammatory diseases; chronic diseases and metabolic disorders; treatment with nonsteroidal anti-inflammatory drugs (NSAIDs) or corticosteroids during the last month; and pregnancy or breast-feeding during the study visit.

In total, 16 healthy individuals not affected by neurodegenerative disorders as verified by clinical examination were included in the study. They were recruited among people undergoing periodic health examinations at the same center. The exclusion criteria for healthy controls were the same as for patients with WD.

The Ethics Committee of the Research Center of Neurology approved this study (protocol №4-4/19 15.05.19), and informed written consent was obtained from each patient and control according to the guidelines approved under this protocol (Article 20, Federal Law "Protection of Health Right of Citizens of Russian Federation" N323- FZ, 11.21.2011).

4.3. Clinical Evaluation

The criteria for inclusion in the group of patients with WD consisted of clinical and laboratory signs of the disease: Debut in childhood, adolescence, and adulthood (most often up to 35 years old); combined damage to the brain and internal organs (cirrhosis, hepatolienal syndrome, portal hypertension, tubular nephritis, etc.); central nervous system (CNS) damage in the form of extrapyramidal syndrome, including tremor, stiffness, dysarthria, dysphagia, cognitive impairment, and dysphoria; and extra-neural symptoms, including hepatosplenomegaly, hemorrhages, failure of levodopa treatment (prescribed in connection with Parkinson's syndrome). Criteria for laboratory diagnosis of WD: Kaiser–Fleischer corneal ring (when using a slit lamp); a decrease in the concentration of ceruloplasmin copper ligand protein in the blood serum; hypersecretion of copper with urine; increase in the concentration of free copper in blood serum; a decrease in the concentration of total copper in serum; decreased serum zinc concentration; increased copper concentration in liver biopsy specimens; DNA diagnostics (detection of mutations in the ATP7B gene); and high therapeutic effect when using copper eliminating chelates (D penicillamine, trientin) and zinc preparations. An additional diagnostic criterion for WD was the result of neuroimaging (computed tomography and magnetic resonance imaging (CT, MRI)), showing an atrophic process in the cerebral hemispheres, cerebellum, subcortical structures with a corresponding expansion of subarachnoid spaces, and the ventricular system, as well as foci in the area of lenticular nuclei, globus pallidus, and visual hillock [26,27].

4.4. Blood Sample Collection

Taking into account the reported serum oxylipin variety during daytime [45], all blood sample collection was conducted in the morning in the fasted state. The plasma was obtained immediately after blood sampling, aliquoted, and stored at −80 °C for further analysis.

4.5. UPLC-MS/MS Conditions and Sample Preparation

Samples were prepared for MS analysis by the solid-phase extraction (SPE) method using an Oasis®PRIME HLB cartridge (60 mg, 3 cc). For eicosanoid extraction, plasma (900 µL) was deproteinized with 1 mL of methanol, vortexed, and centrifuged at 12,000 rpm for 5 min at ambient temperature. The supernatant was diluted 1:6 with mQ water containing 0.1% formic acid for the next steps of SPE. Then, the sample was loaded, and the cartridge was washed with 2 mL of 15% methanol containing 0.1% formic acid, after which the lipids were sequentially eluted with 500 µL of anhydrous methanol and 500 µL of acetonitrile. The resulting samples were concentrated by evaporation of the solvent under a gentle stream of nitrogen and stored at −80 °C. For the identification of lipid mediators, the respective lipid extracts were analyzed using an 8040 series UPLC-MS/MS mass spectrometer (Shimadzu, Japan) in multiple reaction monitoring mode at a unit mass resolution for both the precursor and product ions [46]. The selected molecular ions were fragmentized in the gas phase by collision-induced dissociation and analyzed by tandem (MS/MS) mass spectrometry. The studied metabolites were identified and quantified according to the comparison of their multiple reaction monitoring parameters, retention times, and peak areas with the parameters obtained for deuterated internal standard compounds of the same classes (6-keto PGF1α-d4, TXB2-d4, PGF2α-d4, PGE2-d4, PGD2-d4, leukotriene (LT) C4-d5, LTB4-d4, 5(S)-HETE-d8, 12(S)-HETE-d8, 15(S)-HETE-d8, oleoyl ethanolamide-d4, EPA-d5, DHA-d5, AA-d8) (Table S1) using a commercial software method package Lipid Mediator Version 2 (Shimadzu, Tokyo, Japan) according to the manufacturer's instructions.

Prior to analysis, the plasma oxylipin detection method was validated according to food and drug administration (FDA) recommendations [47]. The stock ethanol solution containing 2 ng of each of the 15 deuterated oxylipin standards was prepared. Calibration samples (1.4, 1, 0.4, 0.2, and 0 ng/probe) were obtained by further dilution of the stock solution containing the same total volume of plasma. For each standard intraday, the interday reproducibility and relative standard deviation (RSD, %) were determined. The accuracy was measured with spiked standards for 3 concentration ranges:

0.2–0.8 ng/probe, 0.9–1.3 ng/probe, and 1.4–2 ng/probe. The limit of detection (LOD) was determined as the signal to noise ratio = 3, and the limit of quantification (LOQ) as the signal to noise ratio = 3. Signal to noise ratio was calculated as the standard error of the regression/slope. Results are presented in Table S2.

4.6. Experimental Data Analysis and Statistics

Comparison of the relative concentrations was performed using the two-sample two-sided t-test, followed by Bonferroni–Holm correction for multiple comparisons. $p < 0.05$ was considered as statistically significant.

Metabolomics data was analyzed using the mixOmics R package version 6.1.1 [48]. After data normalization on internal standards, peak area mean centering and unit variance scaling was applied. Class separation was analyzed by partial least square discriminant analysis (PLS-DA). The quality of the built model was estimated using leave-one-out cross validation. Each round of cross-validation included training on the bigger data subset and validation on the randomly selected sample. The model's predictive performance was estimated based on the validation results combined over rounds. This procedure was repeated for different numbers of components in the PLS-DA model. The overall error, balanced error rate, and AUC were obtained for each number of components (Figures S1 and S2, Table S5).

After building the PLS-DA model with 3 components, VIP scores for each investigated metabolite were calculated. A VIP score is a weighted sum of squares of the PLS loadings regarding the explained variation in each projection. A cutoff for VIP-scores was accepted as 1.5 according to the metabolomics standard initiative (level MSI = 1).

Analysis of covariance (ANCOVA) was used to compare the means of single metabolites between studied groups, taking into account sex and age. ANCOVA was performed using rstatix package for R. Pairwise comparisons of relative metabolites' concentrations was performed using function emmeans_test (rstatix package), also taking into consideration age and sex as covariates. Analysis was followed by Bonferroni–Holm correction for multiple comparisons. $p < 0.05$ was considered as statistically significant.

5. Conclusions

In conclusion, our findings reveal alterations of the plasma oxylipin profiles in Wilson's disease patients, and heterogeneity in patients relative to the oxylipin profiles. Eight lipids were found to vary between HC and WD: EPA, OEA, 9-HODE, 9-KODE, 12-HHT, PGD2, PGE2, and 14,15-DHET; among them, PGE2, PGD2, 12-HHT, and EPA changed significantly (based on the pairwise comparison of means adjusted for sex and age).

The biological significance of these compounds indicates the involvement of oxidative stress damage, inflammatory processes, and PPAR signaling pathways in this disease. The data reveal novel possible therapeutic targets and intervention strategies for treating WD.

Supplementary Materials: The following are available online at http://www.mdpi.com/2218-1989/10/6/222/s1, Figure S1: The Balanced Error Rate (BER) and overall error rate estimated via cross-validation for different component numbers, Figure S2: ROC (receiver operating characteristic) curve for chosen number of components, Figure S3: A clustered image map was generated using euclidean distance and complete linkage clustering algorithm, Figure S4: Relative concentrations of separate metabolites which changed significantly in WD1 and/or WD2 in comparison with HC, Figure S5: Relative concentrations of summed metabolites which changed significantly between group1 and group2, Table S1: UPLC-MS/MS parameters of the identified lipids, Table S2: Intraday reproducibility, Table S3: Interday reproducibility, Table S4: Accuracy, LOD, LOQ, Table S5: AUC (Area Under Curve) values and p-value estimated for different number of components, Table S6: Mean +/− standard deviation of relative concentrations for healthy controls (HC) and Wilson disease patients (WD), Table S7: Source acid and metabolic enzyme for the analyzed oxylipins.

Author Contributions: Experimental procedures, N.V.A., A.A.A., S.V.G., D.V.C., V.V.P. and A.V.L.; writing—original draft preparation, N.V.A., D.V.C. and M.G.S.; writing—review and editing, A.V.L., M.G.S., R.B.K and T.N.F.; clinical analysis, A.V.L., V.V.P., R.B.K. and T.N.F.; performed the MS analysis, D.V.C. and S.V.G.; data analysis, N.V.A.; supervision, T.N.F. and M.G.S. All authors have read and agreed to the published version of the manuscript.

Funding: The reported study was funded by RFBR according to the research project № 19-29-01243.

Acknowledgments: The publication has been prepared with the support of the "RUDN University Program 5-100" (MS/MS analysis).

Conflicts of Interest: The authors declare no competing interests.

Abbreviations

AA	Arachidonic acid
COX	Cyclooxygenase
CYP450	Cytochrome P450 monooxygenase
DHA	Docosahexaenoic acid
DiHOME	Dihydroxyoctadecamonoenoic acid
HC	Healthy control
HDoHE	Hydroxydocosahexaenoic acid
HETE	Hydroxyeicosatetraenoic acid
HODE	Hydroxyoctadecadienoic acid
LA	Linoleic acid
LOX	Lipoxygenase
PG	Prostaglandin
PUFAs	Polyunsaturated fatty acids
WD	Wilson disease
UPLC-MS/MS	Ultra-performance liquid chromatography-tandem mass spectrometry

References

1. Bull, P.C.; Thomas, G.R.; Rommens, J.M.; Forbes, J.R.; Cox, D.W. The Wilson disease gene is a putative copper transporting P-type ATPase similar to the menkes gene. *Nat. Genet.* **1993**, *5*, 327–337. [CrossRef]

2. Compston, A. Progressive lenticular degeneration: A familial nervous disease associated with cirrhosis of the liver, by S. A. Kinnier Wilson, (From the National Hospital, and the Laboratory of the National Hospital, Queen Square, London) Brain 1912: 34; 295–509. *Brain* **2009**, *132*, 1997–2001. [CrossRef] [PubMed]

3. Woimant, F.; Djebrani-Oussedik, N.; Poujois, A. New tools for Wilson's disease diagnosis: Exchangeable copper fraction. *Ann. Transl. Med.* **2019**, *7*, S70. [CrossRef] [PubMed]

4. Hermann, W. Classification and differential diagnosis of Wilson's disease. *Ann. Transl. Med.* **2019**, *7*, S63. [CrossRef] [PubMed]

5. Ala, A.; Walker, A.P.; Ashkan, K.; Dooley, J.S.; Schilsky, M.L. Wilson's disease. *Lancet* **2007**, *369*, 397–408. [CrossRef]

6. Huster, D. Wilson disease. *Best Pract. Res. Clin. Gastroenterol.* **2010**, *24*, 531–539. [CrossRef]

7. Kieffer, D.A.; Medici, V. Wilson disease: At the crossroads between genetics and epigenetics—A review of the evidence. *Liver Res.* **2017**, *1*, 121–130. [CrossRef]

8. Huster, D.; Khne, A.; Bhattacharjee, A.; Raines, L.; Jantsch, V.; Noe, J.; Schirrmeister, W.; Sommerer, I.; Sabri, O.; Berr, F.; et al. Diverse functional properties of wilson disease ATP7B variants. *Gastroenterology* **2012**, *142*, 947–956.e5. [CrossRef] [PubMed]

9. Gromadzka, G.; Kruszyńska, M.; Wierzbicka, D.; Litwin, T.; Dziezyc, K.; Wierzchowska-Ciok, A.; Chabik, G.; Członkowska, A. Gene variants encoding proteins involved in antioxidant defense system and the clinical expression of Wilson disease. *Liver Int.* **2015**, *35*, 215–222. [CrossRef]

10. Ferenci, P.; Stremmel, W.; Członkowska, A.; Szalay, F.; Viveiros, A.; Stättermayer, A.F.; Bruha, R.; Houwen, R.; Pop, T.L.; Stauber, R.; et al. Age and Sex but Not ATP7B Genotype Effectively Influence the Clinical Phenotype of Wilson Disease. *Hepatology* **2019**, *69*, 1464–1476. [CrossRef]

11. Mordaunt, C.E.; Kieffer, D.A.; Shibata, N.M.; Członkowska, A.; Litwin, T.; Weiss, K.H.; Zhu, Y.; Bowlus, C.L.; Sarkar, S.; Cooper, S.; et al. Epigenomic signatures in liver and blood of Wilson disease patients include hypermethylation of liver-specific enhancers. *Epiganetics Chromatin* **2019**, *12*, 10. [CrossRef] [PubMed]

12. Mazi, T.A.; Sarode, G.V.; Czlonkowska, A.; Litwin, T.; Kim, K.; Shibata, N.M.; Medici, V. Dysregulated choline, methionine, and aromatic amino acid metabolism in patients with wilson disease: Exploratory metabolomic profiling and implications for hepatic and neurologic phenotypes. *Int. J. Mol. Sci.* **2019**, *20*, 5937. [CrossRef] [PubMed]

13. Sarode, G.V.; Kim, K.; Kieffer, D.A.; Shibata, N.M.; Litwin, T.; Czlonkowska, A.; Medici, V. Metabolomics profiles of patients with Wilson disease reveal a distinct metabolic signature. *Metabolomics* **2019**, *15*, 43. [CrossRef] [PubMed]

14. Serhan, C.N. Pro-resolving lipid mediators are leads for resolution physiology. *Nature* **2014**, *510*, 92–101. [CrossRef]

15. Gabbs, M.; Leng, S.; Devassy, J.G.; Monirujjaman, M.; Aukema, H.M. Advances in Our Understanding of Oxylipins Derived from Dietary PUFAs. *Adv. Nutr.* **2015**, *6*, 513–540. [CrossRef]

16. Layé, S.; Nadjar, A.; Joffre, C.; Bazinet, R.P. Anti-Inflammatory Effects of Omega-3 Fatty Acids in the Brain: Physiological Mechanisms and Relevance to Pharmacology. *Pharmacol. Rev.* **2018**, *70*, 12–38. [CrossRef]

17. Chistyakov, D.V.; Astakhova, A.A.; Sergeeva, M.G. Resolution of inflammation and mood disorders. *Exp. Mol. Pathol.* **2018**, *105*, 190–201. [CrossRef]

18. Burstein, S.H. The chemistry, biology and pharmacology of the cyclopentenone prostaglandins. *Prostaglandins Other Lipid Mediat.* **2020**, *148*, 106408. [CrossRef]

19. Nagy, L.; Tontonoz, P.; Alvarez, J.G.A.; Chen, H.; Evans, R.M. Oxidized LDL regulates macrophage gene expression through ligand activation of PPARγ. *Cell* **1998**, *93*, 229–240. [CrossRef]

20. Moran, J.H.; Weise, R.; Schnellmann, R.G.; Freeman, J.P.; Grant, D.F. Cytotoxicity of linoleic acid diols to renal proximal tubular cells. *Toxicol. Appl. Pharmacol.* **1997**, *146*, 53–59. [CrossRef]

21. Huster, D.; Purnat, T.D.; Burkhead, J.L.; Ralle, M.; Fiehn, O.; Stuckert, F.; Olson, N.E.; Teupser, D.; Lutsenko, S. High copper selectively alters lipid metabolism and cell cycle machinery in the mouse model of Wilson disease. *J. Biol. Chem.* **2007**, *282*, 8343–8355. [CrossRef] [PubMed]

22. Wooton-Kee, C.R.; Jain, A.K.; Wagner, M.; Grusak, M.A.; Finegold, M.J.; Lutsenko, S.; Moore, D.D. Elevated copper impairs hepatic nuclear receptor function in Wilson's disease. *J. Clin. Investig.* **2015**, *125*, 3449–3460. [CrossRef] [PubMed]

23. Hamilton, J.P.; Koganti, L.; Muchenditsi, A.; Pendyala, V.S.; Huso, D.; Hankin, J.; Murphy, R.C.; Huster, D.; Merle, U.; Mangels, C.; et al. Activation of liver X receptor/retinoid X receptor pathway ameliorates liver disease in Atp7B−/− (Wilson disease) mice. *Hepatology* **2016**, *63*, 1828–1841. [CrossRef] [PubMed]

24. Morris, J.K.; Piccolo, B.D.; John, C.S.; Green, Z.D.; Thyfault, J.P.; Adams, S.H. Oxylipin profiling of alzheimer's disease in nondiabetic and type 2 diabetic elderly. *Metabolites* **2019**, *9*, 177. [CrossRef] [PubMed]

25. Gao, B.; Lang, S.; Duan, Y.; Wang, Y.; Shawcross, D.L.; Louvet, A.; Mathurin, P.; Ho, S.B.; Stärkel, P.; Schnabl, B. Serum and Fecal Oxylipins in Patients with Alcohol-Related Liver Disease. *Dig. Dis. Sci.* **2019**, *64*, 1878–1892. [CrossRef]

26. Chocholoušková, M.; Jirásko, R.; Vrána, D.; Gaték, J.; Melichar, B.; Holčapek, M. Reversed phase UHPLC/ESI-MS determination of oxylipins in human plasma: A case study of female breast cancer. *Anal. Bioanal. Chem.* **2019**, *411*, 1239–1251. [CrossRef]

27. Ryan, A.; Nevitt, S.J.; Tuohy, O.; Cook, P. Biomarkers for diagnosis of Wilson's disease. *Cochrane Database Syst. Rev.* **2019**, *11*, 1465–1858. [CrossRef]

28. Rago, B.; Fu, C. Development of a high-throughput ultra performance liquid chromatography-mass spectrometry assay to profile 18 eicosanoids as exploratory biomarkers for atherosclerotic diseases. *J. Chromatogr. B Anal. Technol. Biomed. Life Sci.* **2013**, *936*, 25–32. [CrossRef]

29. Shishehbor, M.H.; Zhang, R.; Medina, H.; Brennan, M.L.; Brennan, D.M.; Ellis, S.G.; Topol, E.J.; Hazen, S.L. Systemic elevations of free radical oxidation products of arachidonic acid are associated with angiographic evidence of coronary artery disease. *Free Radic. Biol. Med.* **2006**, *41*, 1678–1683. [CrossRef]

30. Ercal, N.; Gurer-Orhan, H.; Aykin-Burns, N. Toxic metals and oxidative stress part I: Mechanisms involved in metal-induced oxidative damage. *Curr. Top. Med. Chem.* **2001**, *1*, 529–539. [CrossRef]

31. Gaetke, L.M.; Chow, C.K. Copper toxicity, oxidative stress, and antioxidant nutrients. *Toxicology* **2003**, *189*, 147–163. [CrossRef]

32. Santoro, N.; Caprio, S.; Feldstein, A.E. Oxidized metabolites of linoleic acid as biomarkers of liver injury in nonalcoholic steatohepatitis. *Clin. Lipidol.* **2013**, *8*, 411–418. [CrossRef] [PubMed]

33. Nakamura, N.; Kumasaka, R.; Osawa, H.; Yamabe, H.; Shirato, K.I.; Fujita, T.; Murakami, R.I.; Shimada, M.; Nakamura, M.; Okumura, K.; et al. Effects of eicosapentaenoic acids on oxidative stress and plasma fatty acid composition in patients with lupus nephritis. *In Vivo (Brooklyn)* **2005**, *19*, 879–882.

34. Payahoo, L.; Khajebishak, Y.; Jafarabadi, M.A.; Ostadrahimi, A. Oleoylethanolamide supplementation reduces inflammation and oxidative stress in obese people: A clinical trial. *Adv. Pharm. Bull.* **2018**, *8*, 479–487. [CrossRef]

35. Du, C.; Fujii, Y.; Ito, M.; Harada, M.; Moriyama, E.; Shimada, R.; Ikemoto, A.; Okuyama, H. Dietary polyunsaturated fatty acids suppress acute hepatitis, alter gene expression and prolong survival of female Long-Evans Cinnamon rats, a model of Wilson disease. *J. Nutr. Biochem.* **2004**, *15*, 273–280. [CrossRef]

36. Calder, P.C. Polyunsaturated fatty acids, inflammation, and immunity. *Lipids* **2001**, *36*, 1007–1024. [CrossRef]

37. Grygiel-Górniak, B. Peroxisome proliferator-activated receptors and their ligands: Nutritional and clinical implications—A review. *Nutr. J.* **2014**, *13*, 17. [CrossRef]

38. Bensinger, S.J.; Tontonoz, P. Integration of metabolism and inflammation by lipid-activated nuclear receptors. *Nature* **2008**, *454*, 470–477. [CrossRef]

39. Nagasaka, H.; Miida, T.; Inui, A.; Inoue, I.; Tsukahara, H.; Komatsu, H.; Hiejima, E.; Fujisawa, T.; Yorifuji, T.; Hiranao, K.; et al. Fatty liver and anti-oxidant enzyme activities along with peroxisome proliferator-activated receptors γ and α expressions in the liver of Wilson's disease. *Mol. Genet. Metab.* **2012**, *107*, 542–547. [CrossRef]

40. He, K.; Chen, Z.; Ma, Y.; Pan, Y. Identification of high-copper-responsive target pathways in Atp7b knockout mouse liver by GSEA on microarray data sets. *Mamm. Genome* **2011**, *22*, 703–713. [CrossRef]

41. Fu, J.; Oveisi, F.; Gaetani, S.; Lin, E.; Piomelli, D. Oleoylethanolamide, an endogenous PPAR-α agonist, lowers body weight and hyperlipidemia in obese rats. *Neuropharmacology* **2005**, *48*, 1147–1153. [CrossRef] [PubMed]

42. Aleshin, S.; Strokin, M.; Sergeeva, M.; Reiser, G. Peroxisome proliferator-activated receptor (PPAR)β/δ, a possible nexus of PPARα- and PPARγ-dependent molecular pathways in neurodegenerative diseases: Review and novel hypotheses. *Neurochem. Int.* **2013**, *63*, 322–330. [CrossRef] [PubMed]

43. Fang, X.; Hu, S.; Xu, B.; Snyder, G.D.; Harmon, S.; Yao, J.; Liu, Y.; Sangras, B.; Falck, J.R.; Weintraub, N.L.; et al. 14,15-Dihydroxyeicosatrienoic acid activates peroxisome proliferator-activated receptor-α. *Am. J. Physiol. Hear. Circ. Physiol.* **2006**, *290*, H55–H63. [CrossRef] [PubMed]

44. Mangino, M.; Roederer, M.; Beddall, M.H.; Nestle, F.O.; Spector, T.D. Innate and adaptive immune traits are differentially affected by genetic and environmental factors. *Nat. Commun.* **2017**, *8*, 13850. [CrossRef] [PubMed]

45. Ostermann, A.I.; Greupner, T.; Kutzner, L.; Hartung, N.M.; Hahn, A.; Schuchardt, J.P.; Schebb, N.H. Intra-individual variance of the human plasma oxylipin pattern: Low inter-day variability in fasting blood samples: Versus high variability during the day. *Anal. Methods* **2018**, *10*, 4935–4944. [CrossRef]

46. Chistyakov, D.V.; Azbukina, N.V.; Astakhova, A.A.; Goriainov, S.V.; Chistyakov, V.V.; Tiulina, V.V.; Baksheeva, V.E.; Kotelin, V.I.; Fedoseeva, E.V.; Zamyatnin, A.A.; et al. Comparative lipidomic analysis of inflammatory mediators in the aqueous humor and tear fluid of humans and rabbits. *Metabolomics* **2020**, *16*, 27. [CrossRef]

47. FDA. Guidance for Industry Process Validation: General Principles and Practices. 2011. Available online: https://www.fda.gov/files/drugs/published/Process-Validation--General-Principles-and-Practices.pdf (accessed on 27 May 2020).

48. Rohart, F.; Gautier, B.; Singh, A.; Lê Cao, K.A. mixOmics: An R package for omics feature selection and multiple data integration. *PLoS Comput. Biol.* **2017**, *13*, e1005752. [CrossRef]

 *metabolites*

Article

Lipidomic Phenotyping Reveals Extensive Lipid Remodeling during Adipogenesis in Human Adipocytes

Florian Miehle [1,2], Gabriele Möller [1], Alexander Cecil [1], Jutta Lintelmann [1], Martin Wabitsch [3], Janina Tokarz [1], Jerzy Adamski [1,2,4,5] and Mark Haid [1,*]

[1] Research Unit Molecular Endocrinology and Metabolism, Helmholtz Zentrum München, German Research Center for Environmental Health GmbH, 85764 Neuherberg, Germany; florian.miehle@helmholtz-muenchen.de (F.M.); gabriele.moeller@helmholtz-muenchen.de (G.M.); alexander.cecil@helmholtz-muenchen.de (A.C.); lintelmann@helmholtz-muenchen.de (J.L.); janina.tokarz@helmholtz-muenchen.de (J.T.); adamski@helmholtz-muenchen.de (J.A.)

[2] German Center for Diabetes Research (DZD), 85764 Neuherberg, Germany

[3] Division of Pediatric Endocrinology and Diabetes, Department of Pediatrics and Adolescent Medicine, University Medical Center Ulm, 89075 Ulm, Germany; martin.wabitsch@uniklinik-ulm.de

[4] Lehrstuhl für Experimentelle Genetik, Technische Universität München, Freising-Weihenstephan, 85764 Neuherberg, Germany

[5] Department of Biochemistry, Yong Loo Lin School of Medicine, National University of Singapore, Singapore 117596, Singapore

* Correspondence: mark.haid@helmholtz-muenchen.de; Tel.: +49-893-187-3234

Received: 28 April 2020; Accepted: 23 May 2020; Published: 26 May 2020

Abstract: Differentiation of preadipocytes into mature adipocytes is a highly complex cellular process. At lipidome level, the adipogenesis remains poorly characterized. To investigate the lipidomic changes during human adipogenesis, we used the Lipidyzer™ assay, which quantified 743 lipid species from 11 classes. The undifferentiated human SGBS cell strain showed a heterogeneous lipid class composition with the most abundant classes, phosphatidylethanolamines (PE), phosphatidylcholines (PC), and sphingomyelins (SM). The differentiation process was accompanied by increased ceramide concentrations. After completion of differentiation around day 4, massive lipid remodeling occurred during maturation, characterized by substantial synthesis of diacylglycerols (DAG), lysophosphatidylethanolamines (LPE), PC, PE, SM, and triacylglycerols (TAG). Lipid species composition became more homogeneous during differentiation to highly concentrated saturated and monounsaturated long-chain fatty acids (LCFA), with the four most abundant being C16:0, C16:1, C18:0, and C18:1. Simultaneously, the amount of polyunsaturated and very long-chain fatty acids (VLCFA) markedly decreased. High negative correlation coefficients between PE and PC species containing VLCFA and TAG species as well as between ceramides and SM imply that PE, PC, and ceramides might have served as additional sources for TAG and SM synthesis, respectively. These results highlight the enormous remodeling at the lipid level over several lipid classes during adipogenesis.

Keywords: adipocytes; adipogenesis; differential mobility spectrometry (DMS); lipidomics; lipidyzer; mass spectrometry; metabolomics; phenotyping; Simpson-Golabi-Behmel syndrome (SGBS)

---

1. Introduction

Overweight and obesity have increased dramatically in recent decades and now affect hundreds of millions of people worldwide, reaching pandemic levels [1]. Obesity has an adverse impact on a range of physiological processes, thereby increasing the risk of developing diseases like type 2 diabetes [2], cardiovascular diseases [2,3], and some types of cancer [4]. Overweight and obesity are mostly characterized by an excess of white adipose tissue (WAT). Adipocytes, the main constituent of this tissue, control the energy balance by storing triacylglycerols in periods of energy excess and breaking down these lipids during energy deprivation. However, the physiological role of adipocytes is much more complex than simply acting in energy storage. These cells secrete numerous diverse lipids and proteins controlling and regulating various bodily functions like appetite, immunological and inflammatory responses, and blood pressure and thereby act as an endocrine organ [5,6].

The development of adipocytes from precursor cells is known as adipogenesis. Within this differentiation process, fibroblast-like preadipocytes differentiate into lipid-laden and insulin-responsive adipocytes. This highly complex process involves the concerted interaction of a cascade of transcription factors like peroxisome proliferator-activated receptor gamma (PPARγ) and CCAAT/enhancer-binding proteins (C/EBPs) as well as different metabolic pathways including the TCA cycle, fatty acid synthesis, glycolysis, and polyamine biosynthesis [7–9]. The differentiation process has been characterized predominantly in murine cells, using different omics approaches like transcriptomics [10,11], metabolomics [9,12], proteomics [13–15], as well as the combination of transcriptomics and metabolomics [16]. However, lipids and their precise composition changes were so far sparsely characterized due to a lack of appropriate "high-resolution" mass spectrometry (MS) methods and internal standards. Several years ago, Roberts and coworkers characterized the levels of free fatty acids as well as the total levels of fatty acids of triacylglycerols (TAG) and glycerophospholipids in differentiating murine 3T3-L1 cells [9]. Two other studies analyzed the levels of some phosphatidylcholines (PC) using the same cell model, but without resolving the lipid isobars in experimental setups that were focused on the analysis of polar analytes [16,17]. Liaw and coworkers compared differentiated 3T3-L1 cells with primary mouse ear-derived mesenchymal stem cells and brown BAT-C1 adipocytes with a global lipid profiling method [18]. Additionally, they investigated the differentiation of murine adipocytes. However, they did not longitudinally analyze the whole adipogenesis process but compared only undifferentiated with fully differentiated 3T3-L1 cells. Even less knowledge on lipids has been compiled in the past for human adipocytes. Collins and coworkers analyzed the levels of fatty acid compositions of TAG and phospholipids without further class separation in primary adipocytes of subcutaneous origin [19]. To the best of our knowledge, the adipogenic process in a human cell model was so far not characterized with high-resolution lipidomics approaches. Therefore, we studied the development of human preadipocytes into mature adipocytes on a lipidomics scale with the recently developed Lipidyzer™ (SCIEX, Darmstadt, Germany) method. We were able to simultaneously quantify 743 lipid species from 11 lipid classes. In combination with multivariate statistics, we could uncover correlations that suggest that extensive lipid remodeling occurs between several lipid classes during adipogenesis. These findings might contribute to the elucidation of new therapy strategies in obesity and other lipid metabolism affected disorders.

2. Results

2.1. Analytical Method Validation

In order to study the process of adipogenesis of human SGBS cells we decided to use the novel targeted Lipidyzer™ technology. Although originally developed and validated for the fast and automated analysis of human plasma samples [20], recently published studies also showed good performance of the Lipidyzer™ assay with other matrices than plasma [21–25]. With the aim to obtain a meaningful "fit for purpose method", we investigated the analytical performance regarding linearity and repeatability for the SGBS cells.

For linearity evaluation, we used nine different volumes of cell homogenates (10, 20, 40, 50, 60, 80, 100, 200, and 300 μL) of undifferentiated and differentiated (day 15) SGBS cells and determined mean concentrations for the single lipid classes. For most classes, the coefficients of determination (R^2) of the linear regressions were higher than 0.9 for both sample types, the differentiated and the undifferentiated cells, meaning we had good linearity of the method (Figure S1, Table S1). However, two classes, the dihydroceramides (DCER) and free fatty acids (FFA), revealed insufficient linearity for both sample types. The measurements of lactosylceramides (LCER) showed high linearity in the undifferentiated ($R^2 = 0.9734$) but low linearity in the differentiated cells ($R^2 = 0.0479$).

For repeatability evaluation, we calculated the relative standard deviations (CV) of the QC pooled samples. With the exception of DCER, all other lipid classes revealed CV values < 15% (Table S2). We also analyzed the background signals of the lipid classes by defining a lower cut-off value for the Lipidyzer™ method at a value of 1.5× the value measured in the blank. In total, 12 lipid classes were present in QC samples in quantities higher than 1.5× the blank values (Table S2). Only FFA showed up in lower apparent amounts because their measured concentrations were already high in blank samples. In conclusion, the Lipidyzer™ assay showed good linearity and repeatability for most of investigated lipids in SGBS cells.

Finally, we decided to exclude the FFA and DCER data from the data set due to their insufficient reliability in the validation testing. However, we left the LCER in the data set as we think that the drop of concentration between undifferentiated and differentiated cells is a noticeable result of our study.

2.2. Cellular Lipid Composition Undergoes Remodeling During Adipogenesis to Mainly TAG

Alterations in the lipid content of some lipid classes (e.g., TAG, phospholipids) in murine cells undergoing the process of adipogenesis have long been known [9,16,18,19]. However, concentration changes of many lipid species from multiple lipid classes at the different stages of human adipogenesis have not been investigated so far. We used the Lipidyzer™ technology to follow the differentiation process of the human SGBS cell strain by quantifying the lipids of samples at days 0, 4, 8, 12, 16, and 20 of adipogenesis.

To track the successful cell differentiation of preadipocytes into lipid-laden adipocytes, we monitored the cellular process by microscopy (Figure S2) and analyzed relative mRNA levels of the main adipogenic transcription factors *PPARG* and *CEBPA* (Figure S3). Microscopic analysis showed enormous lipid storage in droplets starting between day 4 and 8. The analyses of the relative mRNA expression levels showed strong upregulation of *PPARG* (40.7 ± 9.5 fold change at day 12 compared to day 0) and *CEBPA* (52.0 + 8.8 fold change at day 12 compared to day 0). These data demonstrate successful differentiation of SGBS cells into adipocytes.

We were able to simultaneously quantify 743 lipid species of 11 different lipid classes with the accurate identification of lipid isobars. To investigate putative differences between the lipid concentration levels at different time points, we conducted partial least squares-discriminant analysis (PLS-DA, Figure 1) and principal component analysis (PCA, Figure S4). PLS-DA shows a clear separation of the different time points of adipogenesis using the first two principal components with 68.5% and 12.7% of explained variance (Figure 1). While component 1 was sufficient to separate the early phase of SGBS differentiation (days 0, 4, and 8), the second component was necessary for separation of the later stages of differentiation (days 12, 16, and 20). Component 1 was mostly influenced by TAG species, whereas their influence on component 2 was lower (Table S3).

Next, we were interested in the lipid class compositions at the six time points of differentiation (Figure 2A). In preadipocytes, that is, at the start of differentiation (day 0), a very heterogeneous lipid composition was observed with the most dominant classes being the phosphatidylethanolamines (PE; 32.1% ± 0.9%), phosphatidylcholines (PC; 26.3% ± 0.4%), sphingomyelins (SM; 19.7% ± 0.6%), and TAG (10.3% ± 0.5%). While the relative proportions of PE, PC, and SM declined tremendously during the ongoing cell differentiation, the fraction of TAG increased from initially 10.3 ± 0.5% to finally 96.9 ± 0.4% at day 20. The relative fractions of all other lipid classes decreased strongly during the differentiation process to fraction sizes of finally 1.2% and lower. To conclude, the different stages of cell differentiation could be clearly distinguished based on the relative lipid compositions (Figure 2A) as well as by PLS-DA (Figure 1). Interestingly, the relative lipid class compositions did not reveal strong changes after day 8 of differentiation.

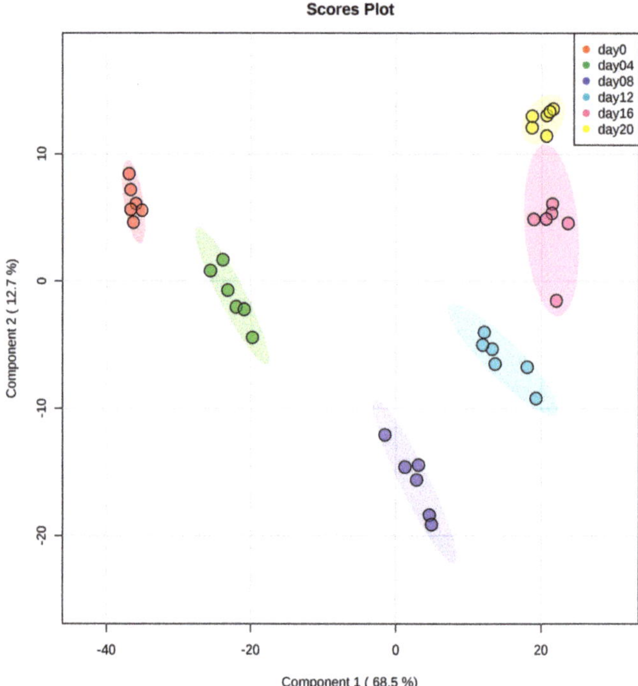

Figure 1. Partial least squares-discriminant analysis (PLS-DA) score plot showing very clear clustering of lipid species regarding the different time points of adipogenesis. While component 1 (68.5% variance) was sufficient to separate the early phase of differentiation (days 0, 4, and 8), the second component was necessary for separation of the later stages of differentiation (days 12, 16, and 20, 12.7% variance). The color code for the data points of the different days of adipogenesis is shown in the box inside the figure. Illustrated are also the 95% confidence intervals for each group. Cross-validation and permutation results confirmed the model to be predictive and not overfitted (R2: 0.97, Q2: 0.97; $p < 5 \times 10^{-4}$ (0/2000 permutation numbers), test statistics selected by separation distance (B/W)). Each group consisted of six samples.

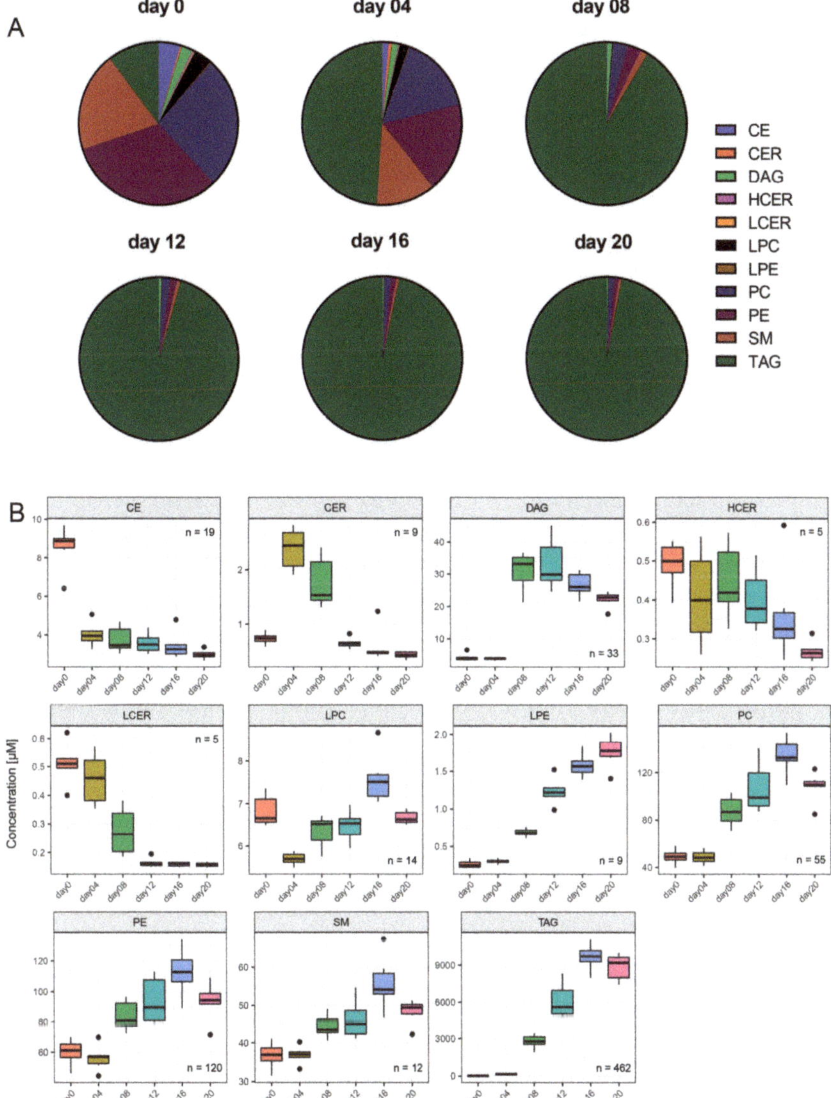

Figure 2. Changes in lipid class compositions and concentrations in specific lipid classes in the course of adipogenesis showed enormous lipid remodeling during the ongoing differentiation process. (**A**): Relative lipid class compositions at different days of adipogenesis in molarity %. Prior to differentiation at day 0, a very heterogeneous class distribution could be observed. This composition became more uniform during adipogenesis due to the predominance of triacylglycerols (TAG). The relative fraction for TAG increased from 10.3 ± 0.5 molarity % at day 0 to 96.9 ± 0.4 molarity % at day 20. (**B**): The concentration profiles of different analyzed lipid classes showed individual time courses during ongoing adipogenesis. Cholesteryl esters (CE), HCER, and LCER concentrations decreased strongly during adipogenesis, whereas LPE, PC, PE, and TAG concentrations strongly increased. Ceramide (CER) concentrations increased from day 0, peaking at day 4, and then decreased below the starting concentration levels. DAG species had their concentration maximum from days 8 to 12. LPC fluctuated during the whole differentiation process.

The time courses of the lipid species concentrations revealed an interesting alternative perspective on adipogenesis (Figure 2 B). Overall, we observed significant changes over time in the concentrations of 725 out of 743 lipids, that is, for 97.7% of all quantified lipid species. When looking at the lipid classes, only three of them, namely, CE ($p = 4.52 \times 10^{-4}$, Table S4), HCER ($p = 5.23 \times 10^{-3}$), and LCER ($p = 2.80 \times 10^{-5}$), showed continuous and significant decreases in concentration levels from day 0 to day 20 (Figure 2B). CER concentrations increased significantly from day 0 to day 4 of differentiation ($p = 1.51 \times 10^{-5}$) and subsequently decreased significantly below the limit of detection (LOD). In contrast, the TAG concentration levels increased continuously ($p = 4.01 \times 10^{-6}$). The TAG concentration levels were the highest among all classes and levels started from 19.2 ± 2.0 µM on day 0 to 8874.3 ± 1072.1 µM on day 20. LPE, PC, and PE concentrations also increased strongly during adipogenesis, whereas SM concentrations showed only a mild increase during differentiation ($p = 5.04 \times 10^{-5}$). In contrast, DAG concentrations increased from day 4 to day 8 and remained at a high level ($p = 3.40 \times 10^{-5}$). LPC concentrations fluctuated around the starting value during differentiation ($p = 1.50 \times 10^{-4}$).

2.3. The Most Abundant Fatty Acids in Differentiated Human SGBS Cells Are C16:0, C16:1, C18:0, and C18:1

As we were interested in the concentration changes of the single fatty acid (FA) species during SGBS adipogenesis, we subsequently focused on a detailed analysis of the fatty acids bound to the lipid backbone. Figure 3 illustrates the time courses of the concentrations for the single FA side chains summarized over all lipid classes. With the exception of C20:4, all LCFA as well as the medium-chain FA (MCFA) lauric acid (C12:0) showed strong increases in concentration during adipogenesis. Among this group of FA, C18:1, C16:0 (palmitic acid), C16:1, and C18:0 (stearic acid)—in descending order—were the most abundant. In contrast to the MCFA and nearly all LCFA, most of the VLCFA decreased during adipogenesis. FA C22:0, C22:1, and C22:2 showed fluctuating concentration profiles.

A detailed analysis of the time courses of the concentrations for the single FA side chains separately for each lipid class allowed an even more detailed glimpse into adipogenesis (Figure 4). Illustrations on an enlarged scale can be found in Figure S5A–F. We were able to identify individual FA concentration changes over time of differentiation, which were strongly dependent on the lipid class.

The FA species of the three classes of ceramides, namely CER, HCER, and LCER, showed similar behavior in that the concentrations of the LCFA and VLCFA side chains decreased during adipogenesis. In contrast, the LCFA in DAG, LPE, PC, PE, and TAG substantially increased. In particular, the concentrations of C18:1 and C16:0 increased strongly during adipogenesis; for example, TAG-C16:0 increased from 2.5 µM at day 0 to 1515.1 µM at day 20. In contrast, VLCFA were present at only very low levels in the classes of DAG, LPE, PC, PE, and TAG.

The FA composition of the SM differed strongly from the compositions of the other classes in that the SM comprised the highest absolute amounts of VLCFA. Especially C22:0 (behenic acid) and C24:0 (lignoceric acid) were present in SM at considerable concentrations. Their concentrations were increased up to a factor of 320 at day 20 compared with the other classes. The fatty acid concentrations of the LPC did not change markedly during adipogenesis. In contrast, the LCFA of the CE decreased to half the maximal concentrations and the VLCFA even more during the differentiation process.

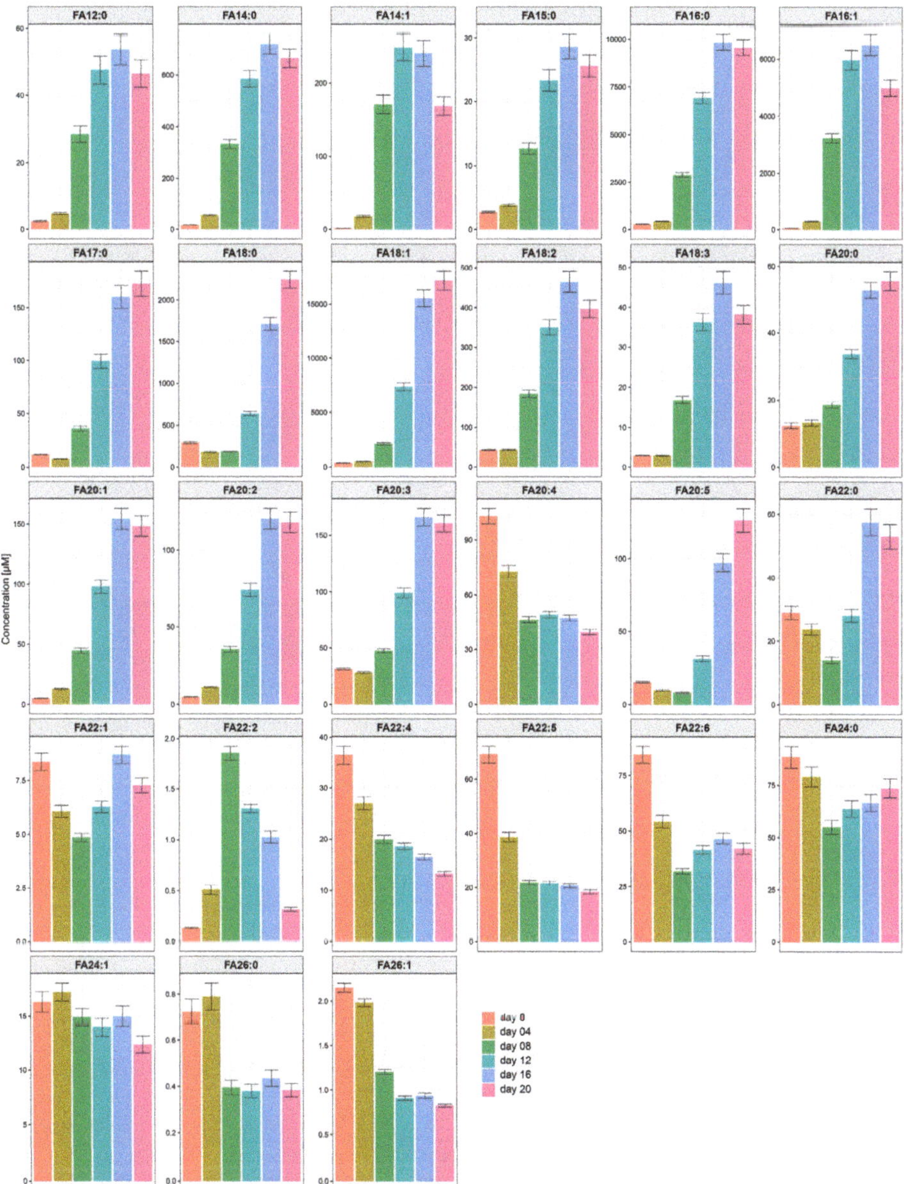

Figure 3. Time courses for the concentrations of the single FA side chains bound in the lipids summarized over all classes. Analysis of summarized lipids' side chain concentrations over all classes revealed four dominant fatty acids, namely FA 18:1, FA 16:0, FA 16:1, and FA 18:0. Their concentrations increased strongly during adipogenesis together with the other LCFA (with the exception of FA 20:4) and the MCFA 12:0. In contrast, the very VLCFA mostly decreased during adipogenesis.

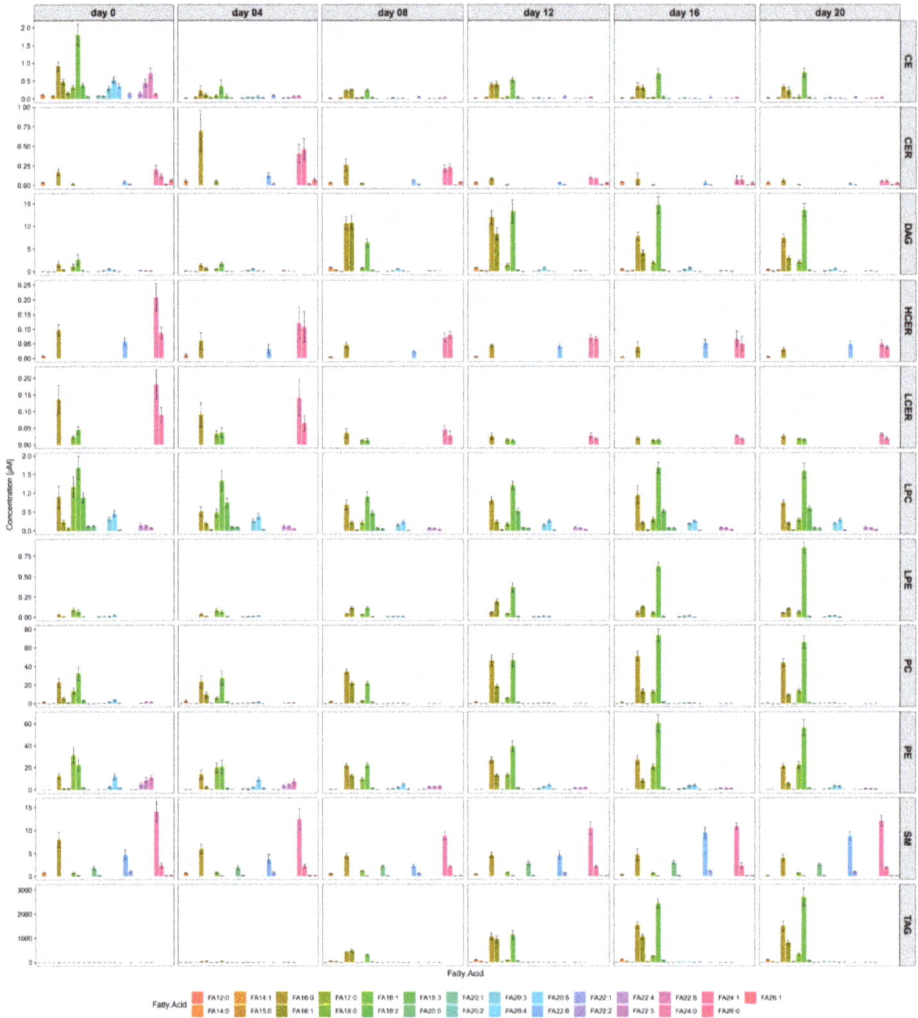

Figure 4. Fatty acid concentrations and compositions changed markedly during adipogenesis in all 11 lipid classes. At the start of differentiation, the lipids had a very heterogeneous side chain distribution with high concentration levels of LCFA and VLCFA. The concentrations of the VLCFA decreased during adipogenesis in all classes, with the exception of the class SM. The concentration courses of the LCFA were more complex because their levels increased markedly during adipogenesis within the classes DAG, LPE, PC, PE, and TAG, but decreased in the class CE. The FA concentration course of the SM differed strongly from the other classes because the VLCFA remained at high levels during adipogenesis.

2.4. Correlations between Concentration Profiles of Lipid Species from Different Lipid Classes Reveal Extensive Lipid Remodeling during Adipogenesis

The opposing trends in concentrations of some lipid classes raised the questionwhether they might be the result of an underlying regulatory network. To reveal possible associations between the lipid species, we computed pairwise Spearman's rank correlations of lipid species concentration trajectories (Figure 5). The lipids could be assigned to six clusters (Figure 5A) and for each of the clusters the average concentration changes over time are displayed in Figure 5B. Species of different lipid classes were found to be distributed between the different clusters (Figure 6A and Figure S6).

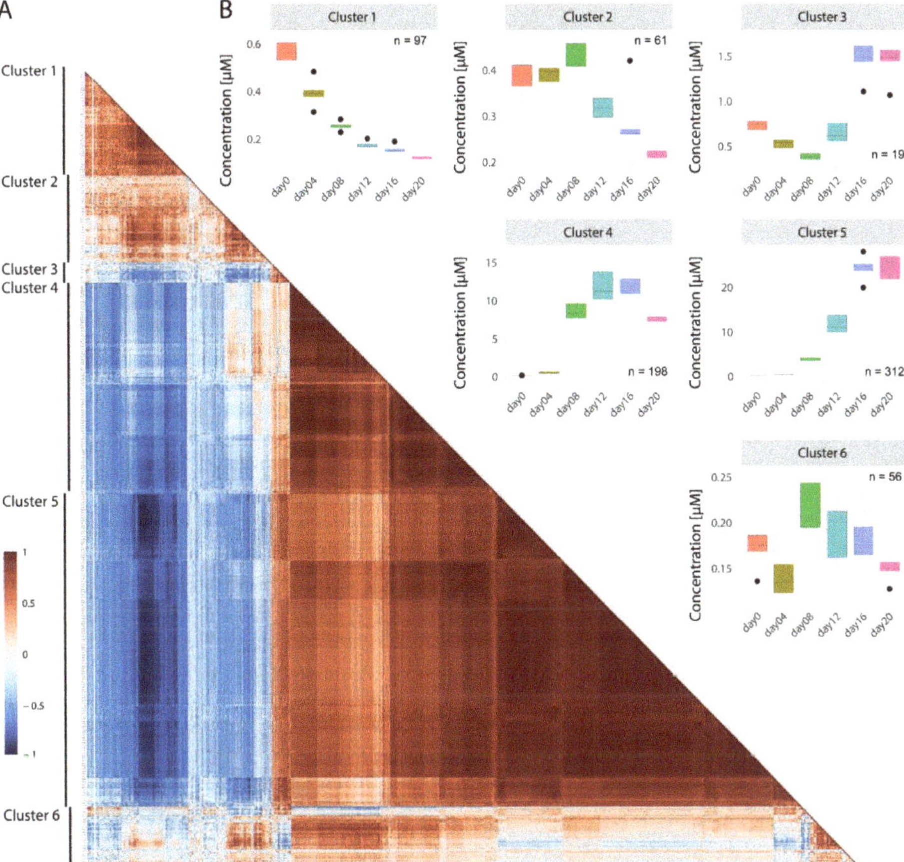

Figure 5. Spearman's rank correlation analysis of lipid concentration trajectories during adipogenesis showed strong clustering and correlation of the lipid species. Panel (**A**) illustrates the matrix of the analysis where each square indicates the Spearman's rank correlation coefficient. Positive correlations between the variables are shown in red, while negative correlations are shown in blue. Correlation matrix enables the assignment of six lipid clusters. Species were clustered using Ward's clustering algorithm. Panel (**B**) shows the changes of the average lipid concentration in the different clusters with time.

Cluster 1 (n = 97 lipid species) consisted of lipid species with continuously decreasing concentration profiles (Figure 5B). In cluster 1, lipid species with the highest average FA side chain lengths (expressed as average total number of C-atoms) as well as number of double bonds (DB) were found, independent of the lipid class (Figure 6B–D). The cluster was dominated by PE (37.6%) and PC (20.8%) species. However, with the exception of SM, lipids from all other lipid classes were also clustered here (Figure 6A). Interestingly, 80% of all LCER species of the dataset could be found in this cluster (Figure S6A). In addition, more than 74.3% of the lipid species within this cluster had at least one FA side chain with at least 20 C-atoms and a high degree of desaturation (Table S5). Therefore, this cluster can be considered as the "PUFA cluster".

Cluster 2 (n = 61) comprised species that exhibited decreasing concentrations starting at day 8. This cluster contained lipids of all classes except those of LPE. The species had approximately the third highest total number of C-atoms and DB (Figure 6B–D). Remarkably, cluster 2 contained a relatively high number of sphingolipids: 56.6% of all CER, 40.0% of all HCER, and 41.7% of all SM species of the dataset (Figure S6B) could be found in this cluster. Therefore, it can be characterized as the "sphingolipid and PUFA cluster".

Cluster 3 species (n = 19) showed a fluctuating concentration course with a decrease in concentrations until day 8, followed by an increase to above the starting values. Four lipid classes were part of this cluster: TAG (42.1%), DAG (10.5%), PC (15.8%), and PE (31.6%). Remarkably, 15 out of 19 lipids (79%) contained at least one FA with 18 carbon atoms.

Lipids of clusters 4 and 5 (n = 198 and 312, respectively) were generally characterized by a strong concentration increase during adipogenesis. However, cluster 4 species increased substantially only until days 12–16 and decreased thereafter to half their maximum concentrations, while cluster 5 species reached a plateau at day 16. In cluster 4, TAG was the most abundant lipid class (75.8%), followed by PE (13.1%), and DAG (6.1%), while CE and all ceramide classes (CER, HCER, and LCER) were completely absent. The species of this cluster had the lowest number of C-atoms and DB within the dataset (Figure 6B–C). In cluster 5, the relative amount of TAG was even higher (88%) than in cluster 4 (Figure 6A). Owing to the high number of TAG species, clusters 4 and 5 also showed the highest lipid concentration values among all clusters (Figure 2B). Therefore, clusters 4 and 5 can be considered as "TAG clusters."

The lipids in cluster 6 (n = 56) exhibited a fluctuating averaged concentration profile throughout the investigated time period. Lipid species of all classes except LCER were represented in this cluster, which was in general rather heterogeneous in terms of composition (Figure 6A).

The correlations between the clusters might reveal new insights into lipid remodeling. In general, we observed positive correlations between clusters 1 and 2 as well as between clusters 4 and 5. Remarkably, lipids from clusters 1 and 2 were strongly negatively correlated with lipids in clusters 4 and 5, which might be an indicator of lipid remodeling. Specifically, many TAG species from cluster 5 had strong negative correlations with more than a dozen PE species of cluster 1 (between −0.7 and −0.96). These PE species mostly contained polyunsaturated and VLCFA, whereas the TAG species were carrying at least one LCFA (Table S5).

Moreover, some sphingomyelins, especially SM 20:0 (cluster 4), SM 22:0 (cluster 5), and SM 24:0 (cluster 5), had high negative Spearman's correlation coefficients (down to −0.84) with species from classes CER, HCER, and LCER (all in clusters 1 and 2, Table S5). These results point to regulatory interactions between the lipid species over several lipid classes.

In addition, we also found high positive correlation coefficients (mostly between 0.85 and 0.98) between DAG and TAG species. Those DAG and TAG with very high co-correlations mostly contained one or two of the most abundant fatty acids (C16:0, 16:1, 18:0, and 18:1) as side chains. Additionally, these TAG were characterized by total C-atom numbers between 42 and 54, which further implies that LCFA were the main constituents.

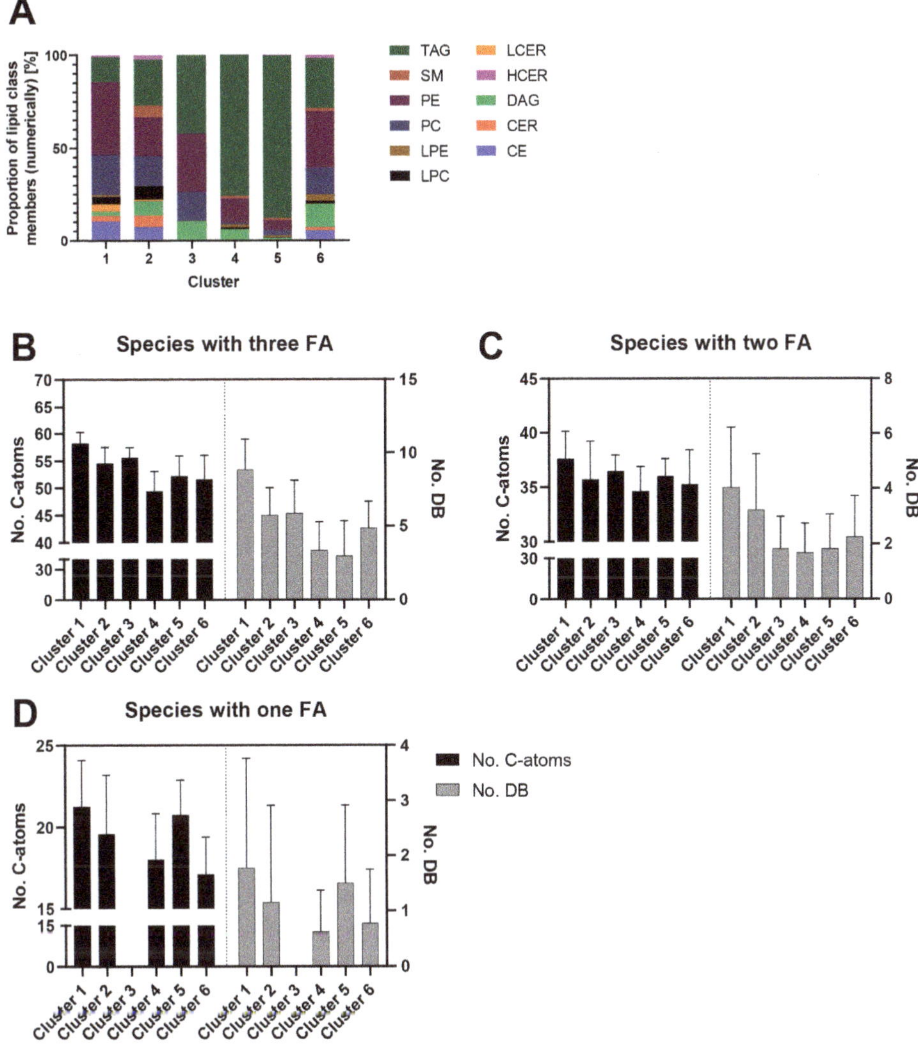

Figure 6. Clusters differed widely in their lipid class and lipid species composition. Panel (**A**) highlights the cluster compositions by showing the relative numerical proportions of the lipid class members. Panels (**B–D**) show the analysis of the FA side chain lengths and the number of DB based on the number of side chains of the lipid classes. Panel (**B**) shows the analysis of the class with three bound FA side chains, panel C that of lipid classes with two bound FA side chains, and panel D that of classes with only one FA side chain. Lipid species with patterns of decreasing concentration throughout adipogenesis had in general the highest numbers of DB and chain lengths, independent of their number of side chains.

3. Discussion

In the present study we characterized the different stages of human adipogenesis by using the Lipidyzer™ method [26] which enabled the simultaneous quantification of 743 lipid species of 11 different lipid classes in differentiating human SGBS cells. This global lipid analysis enabled us to identify correlations between lipid species over several classes and opened the possibility to generate hypotheses on lipid remodeling during adipogenesis.

Prior to the adipogenesis characterization, we analyzed the method performance in terms of linearity, repeatability, and background signals to be able to apply this novel methodology for accurate lipid analysis to SGBS cell culture samples. We used representative samples for undifferentiated and differentiated SGBS cells and validated the Lipidyzer™ method according to recently published studies that already showed good performance of the Lipidyzer™ method [21–25].

The lipid classes CE, CER, DAG, HCER, LCER, LPC, LPE, PC, PE, SM, and TAG could be analyzed with good linearities as indicated by coefficients of determination above 0.9 in at least one SGBS sample type. Furthermore, these lipid classes could be measured with high precision (CVs < 12%) which was determined by repeated analyses of QC samples from pooled SGBS homogenates. However, DCER and FFA had to be discarded from the data set, because the observed concentration levels were mostly in the range of the blank samples and therefore led to poor linearity and low precision. The detected high background signals of FFA in blank samples indicate high contaminations of these species that can be often found in glassware, pipette tips, and even in high-grade organic solvents [27]. We conclude that the Lipidyzer™ method was suited to reliably quantify 743 lipids from 11 lipid classes for the analysis of SGBS cell samples.

We are the first to use this novel method for the determination of lipid levels in differentiating human adipocytes. Additionally, we are also the first who characterized the human adipogenesis with this broad coverage of different lipid classes using only one method. Liaw and coworkers investigated the endpoints of adipogenesis (i.e., pre-adipocytes vs. fully differentiated adipocytes, day 12) in murine adipocytes of a large amount of lipids using an LC-MS/MS^{ALL} shotgun lipidomics approach [18]. They identified a shift from highly unsaturated VLCFA bound to the backbones of TAG, SM, cardiolipins, and ether-linked monoalkyldiacylglycerols in preadipocytes to more saturated LCFA in differentiated 3T3-L1 adipocytes. We could confirm these findings for VLCFA-carrying SM since we identified a decrease in their concentration levels until day 16. However, we measured an increase from day 16 to 20 almost reaching the concentration levels of day 0. The authors investigated adipogenesis at day 12, possibly missing the increase in the late phase of differentiation we observed. On the other hand, and different to the study of Liaw et al., we observed members of further lipid classes, namely CE, CER, DAG, HCER, LCER, LPC, PC, and PE species, having strongly decreased amounts of highly unsaturated VLCFA in fully differentiated adipocytes when compared to preadipocytes.

We followed the adipogenesis process over 20 days by analyzing samples from six time points (days 0, 4, 8, 12, 16, and 20). Successful cell differentiation of preadipocytes into lipid-laden adipocytes was confirmed on two different levels. First, we could observe the expected morphological changes during cell differentiation by microscopy. Second, the upregulation of *PPARG* and *CEBPA* expression showed strong transcriptional activation of cell differentiation. The very tight clustering of the data for samples from the same time point in the PCA and PLS-DA plots demonstrated generally high quality of the data obtained by the targeted lipidomics technique. Furthermore, the shift from differentiating (days 0–4) to maturating (days 4–12) SGBS cells became clearly visible by the change from PC1 to PC2 as the main contributor for cluster separation in the score plots. The observed shift coincides with findings from Halama and coworkers, who characterized murine adipogenesis of 3T3-L1 cells using a combined metabolomics and transcriptomics approach [16]. This shift during adipogenesis can be explained by the change from differentiation to maturation medium at day 4.

During ongoing adipogenesis we observed on the one hand strongly decreasing levels of CE, CER, HCER, and LCER, and on the other hand, substantially increasing levels of nearly all investigated glycerophospholipid classes, namely LPE, PC, PE, SM, as well as TAG. However, the temporal concentration courses of the individual lipid classes followed a particular pattern that was dependent on the developmental phases of the cells.

During the differentiation phase, we observed increasing concentrations of CER until day 4. The ceramides are known to be involved in signaling activity in cell cycle arrest and the inhibition of cell proliferation during early adipogenesis [28]. Both processes are required for the induction of cell differentiation of preadipocytes into adipocytes [29]. Thus, the time courses of CER concentrations give rise to the hypothesis that at least some of these compounds were involved during cell differentiation signaling. After day 4, when cell differentiation was completed and maturation started, we observed decreasing levels of CER together with HCER and LCER accompanied by a simultaneous increase of SM. This could be explained by the remodeling of all ceramides into SM. This is further supported by the fact that we observed high negative Spearman's correlation coefficients between several SM species and CER, HCER, and LCER from the sphingolipid and PUFA cluster (cluster 2).

The differentiation phase was also characterized by decreasing concentration levels of CE. The CE operate as transport intermediates of cholesterol, which is an important component of the cell membrane [30,31]. Decreasing CE concentration levels indicate the release of cholesterol and its insertion into the cell membranes [32]. Incorporation of cholesterol decreases the flexibility of plasma membranes and thereby enables the morphological changes of the membranes that are important for differentiation and maturation. Furthermore, the cholesterol might also have been incorporated into triglyceride lipid droplet surfaces, serving as an intracellular free cholesterol reservoir [31]. These hypotheses are supported by negative correlations down to −0.78 between CE in the PUFA clusters 1 and 2 and TAG in the TAG clusters 4 and 5.

The second phase of adipogenesis, the maturation phase, was characterized by a strong increase of TAG from micromolar to millimolar levels. It is not surprising that TAG became the dominant lipid constituent of the adipocytes, as this reflects the function of adipocytes as sites for the storage and supply of fatty acids. The biosynthesis of TAG from DAG is corroborated by high positive correlations between these two lipid classes [33,34]. We observed a biphasic pattern of DAG and TAG concentration courses during adipogenesis. After a lag phase until day 4, considerable production of TAG via DAG reached a maximum at day 16 and then slowly declined. We conclude that the massive synthesis of TAG and their precursors started only after full differentiation from preadipocytes to adipocytes. As Collins and coworkers have shown by using ^{13}C-labeled substrates, the massive generation of TAG is presumably based on de novo lipogenesis from glucose provided in the cell culture medium [19]. However, our correlation analysis revealed strong negative correlations between several PE species containing VLCFA and TAG. This might be an indicator of a possible contribution of PE containing VLCFA as an additional source for TAG synthesis during adipogenesis. The PE species might have been catabolized by phospholipase C to DAG, re-esterified to TAG, and incorporated into lipid droplets [35,36]. The increase of intracellular lipid depots is accompanied by expansion of the cell surface and volume, which requires larger amounts of the major membrane lipid classes like PC, PE, SM, and cholesterol [37–39]. Indeed, we observed simultaneous increases of PC, PE, and SM after day 4. Furthermore, LPC and LPE, which can be regarded as metabolic intermediates of PC and PE, also increased during adipogenesis [40].

Surprisingly, we observed the odd chain fatty acids C15:0 and C17:0 in CE, DAG, LPC, PC, PE, and TAG at non-negligible concentrations. Both fatty acids showed strongly increasing concentration time courses during adipogenesis. Roberts et al. also quantified increased odd chain fatty acid levels during cell differentiation of murine 3T3-L1 adipocytes [9]. We speculate that the increasing levels during adipogenesis might be explained by sequential peroxisomal fatty acid α-oxidation, which has been shown to occur in differentiating adipocytes [41].

Furthermore, some of the time courses of certain lipids might be explained by influences of the composition of the culture medium. To investigate a possible contribution of FA from the culture medium, we measured the FBS-containing medium for the cultivation of preadipocytes as well as the differentiation medium with the Lipidyzer™ method. Lipids with very long-chain PUFAs were highly concentrated in the FBS-supplemented medium compared with the levels in the differentiation medium lacking FBS (Figure S7). Therefore, we hypothesize that the measured concentration profiles of these lipid species during adipogenesis could have been artificially influenced by the cultivation with FBS-containing medium before the start of differentiation. As a result of this, the decreasing concentration levels of the VLCFA might also be explained by a lack of supply of these FA in the FBS-free differentiation and maturation media during adipogenesis.

It is important to keep in mind that cell culture experiments reflect artificial conditions. First, in vitro experiments often require high concentrations of growth factors, hormones, or several stimulation factors, which do not reflect physiological in vivo conditions. For instance, the differentiation of SGBS and other pre-adipocyte cells requires the corticosteroid dexamethasone and the insulin sensitizer rosiglitazone, which might strongly influence the lipidome. Indeed, Jeucken and Breuwens recently showed that rosiglitazone has an effect on the lipidome of HeLa cells [42]. In addition, the PPARγ agonist rosiglitazone as well as the endogenous myokine irisin can induce browning of white adipocytes [43]. Some recently published manuscripts showed this propensity of the SGBS cells towards a beige phenotype [43–46]. The used protocol for the SGBS cells requires an initial four-day stimulation with rosiglitazone for the induction of differentiation. This short time period might have an influence on the lipidome. However, undifferentiated SGBS cells behave very similar to human primary preadipocytes and the fully differentiated cells cannot be morphological distinguished from human primary adipocytes [47]. Moreover, one study compared SGBS cells, derived from subcutaneous adipose tissue of a male infant, with primary subcutaneous adipocytes from obese female patients [44]. The different confounders, obesity and sex, might also have a significant influence on the differentiation capacity and therefore also on the comparison of the two cell models. In addition, SGBS cells carry an FTO risk allele and the cells do not have a Simpson-Golabi-Behmel syndrome typical mutation in the glypican-3-gene (GPC3) what the name of the cell strain would suggest [47]. Nevertheless, future efforts will be necessary to confirm our findings in human primary subcutaneous adipocytes. Second, the lipid synthesis in adipose tissue *in vivo* is based not only on de novo synthesis from mostly glucose but also on circulating fatty acids in the bloodstream [48]. To sum up, cell culture experiments are indeed helpful to shed light on several cellular processes separately. However, owing to their simplicity and artificial nutrition, they cannot represent physiological in vivo conditions.

It also has to be mentioned that the Lipidyzer™ technology has some limitations. First, it was developed for the analysis of human plasma, which of course has a different lipid composition from (pre)adipocytes. The internal standard concentrations were therefore optimized for human plasma and may not match the actual situation in cell culture samples. Another issue concerns quantification. Although the Lipidyzer™ uses up to 10 internal standards (IS) per lipid class, there is no IS for each individual analyte. In addition, the method does not use a calibration curve for absolute quantification. Therefore, the measured absolute concentrations should be interpreted carefully. Furthermore, the Lipidyzer™ method is capable to determine the lipid species at the fatty acyl/alkyl level, except for the class of TAG. In case of TAG, the method cannot distinguish the *sn-1*, *sn-2*, and *sn-3* positions of the glycerol backbone and can as well not determine the exact positions of double bonds in the side chains. Besides, as the Lipidyzer™ method is a commercial assay with specialized and standardized software it is not possible to include other lipids into the method. This unfortunately limits the availability of further interesting lipid species such as for example signaling phospholipids.

Despite these limitations, we think that targeted analytical methods using multiple internal standards per lipid class—like the Lipidyzer™ technology—should be the methods of choice for the quantitative analysis of longitudinal samples with strongly different analyte concentrations and matrix conditions between sampling points. It has recently been shown by Chamberlain et al. that "due to the presence of matrix effects in untargeted, non-quantitative metabolomics, the signal intensity of any single analyte cannot be directly compared to the signal intensity of that same analyte (or any other analyte) between any two different matrices" [49]. This is of particular importance for ESI-MS-based lipid analytics, because matrix effects can vary considerably between lipid classes and even lipid species of the same class can respond differently to matrix effects depending on acyl chain length and degree of unsaturation [50,51]. Chamberlain et al. further concluded that "due to differences in ionization efficiency, the signal intensity of any single analyte cannot be directly compared to the signal intensity of any other analyte, even in the same matrix." Thus, any kind of correlation or network analysis would be hampered with non-targeted or shotgun approaches. The application of IS can avoid or at least reduce the negative impact of matrix effects on the results. Non-targeted metabolomics or shotgun lipidomics approaches not including IS are more susceptible to matrix influences and should be therefore interpreted very carefully.

4. Materials and Methods

4.1. Cell Culture, Cell Harvesting and Homogenization

The Simpson Golabi Behmel syndrome (SGBS) preadipocyte cell strain was provided by Martin Wabitsch. The cells were cultivated and differentiated for 20 days, as described previously [52]. In brief, 50,000 preadipocytes per well were seeded in six-well plates in DMEM/F-12 medium (Thermo Fisher Scientific, Waltham, MA, USA), supplemented with 10% FBS (Biochrom, Berlin, Germany), 3.3 mM biotin (Merck, Darmstadt, Germany), and 1.7 mM pantothenate (Merck, Darmstadt, Germany) and grown at 37 °C and 5% CO_2 in a humidified atmosphere. Cell differentiation was initiated when cells reached about 90% confluence. At that point, the medium was exchanged for serum-free medium supplemented with 10 µg/mL transferrin (Merck, Darmstadt, Germany), 0.2 nM triiodothyronine (T3; Merck, Darmstadt, Germany), 250 nM hydrocortisone (Merck, Darmstadt, Germany), 20 nM human insulin (Merck, Darmstadt, Germany), 25 nM dexamethasone (Merck, Darmstadt, Germany), 250 µM 3-isobutyl-1-methylxanthine (IBMX; Merck, Darmstadt, Germany), and 2 µM rosiglitazone (Biomol, Hamburg, Germany). After 4 days and then every fourth day, thereafter, the medium was replaced with serum-free medium containing 10 µg/mL transferrin, 0.2 nM T3, 250 nM hydrocortisone, and 20 nM human insulin (maturation medium). Cell morphology was monitored by microscopy.

Cell samples for quantitative real-time PCR (qRT-PCR) analyses were taken in quadruplicates (biological replicates) at each of 5 time points beginning with the start of differentiation (representing day 0), then on day 2, 4, 8, and 12 of adipogenesis. Cell samples were processed as described below. Cell samples for Lipidyzer™ analyses were taken at six time points beginning with the start of differentiation (representing day 0) and then on every fourth day of differentiation until day 20. At each time point, six samples (biological replicates) were collected. Harvesting, homogenization of cells, and normalization of measured metabolite concentrations to the cell number were performed as recently reported [53], with minor modifications. In brief, after one washing step with 6 mL of warm PBS per well of a six-well plate, the cells were scraped off the wells in 500 µL of extraction solvent of ice-cold 80% methanol per well using rubber-tipped cell scrapers (Sarstedt, Nümbrecht, Germany). Harvested cell-solvent suspensions of four wells were pooled into pre-cooled 2 mL microtubes (Sarstedt, Nümbrecht, Germany) containing 400 mg of glass beads (Bertin, Frankfurt, Germany). The samples were stored at −80 °C until further use. Homogenization of cells was performed immediately before analyses at 4–10 °C twice for 25 s at 5500 rpm using a Precellys24 (PeqLab, Erlangen, Germany). The resulting homogenates were used for lipidomics measurement by FIA-(DMS)-MS/MS as well as for DNA quantification (DNA content indirectly reflects the cell number of the sample and was determined for normalization purposes) [53].

4.2. RNA Isolation and Quantitative Real-Time PCR (qRT-PCR)

Total RNA of four independent biological replicates per group was extracted from cells using miRNeasy mini kit (Qiagen, Hilden, Germany) according to the manufacturer's protocol. Synthesis of cDNA was performed by using the RevertAid First Strand cDNA Synthesis Kit (ThermoFisher Scientific, Dreieich, Germany) according to the manufacturer's instructions. Total RNA was reverse transcribed using an anchored oligo(dT)$_{18}$ primer (5′-TTTTTTTTTTTTTTTTTTVN-3′) in a final concentration of 0.5 µM for priming cDNA synthesis. For qRT-PCR, primers were designed using Primer3 to span at least one exon-intron boundary to avoid falsified amplification results [54]. Primers were synthesized by Metabion (Planegg, Germany) and sequences were as follows: PPARG_for (5′-GACCACTCCCACTCCTTTGA-3′), PPARG_rev (5′-GAGATGCAGGCTCCACTTTG-3′), CEBPA_for

(5′-AACAGCTGAGCCGCGAACTG-3′), CEBPA_rev (5′-CGGAATCTCCTAGTCCTGGCT-3′), TBP_for (5′-CAGCCGTTCAGCAGTCAA-3′), TBP_rev (5′-CTGCGGTACAATCCCAGAAC-3′). The amplification was performed on a QuantStudio Flex 7 Real-time PCR system (ThermoFisher Scientific, Dreieich, Germany) in triplicates using Power SYBR Green PCR Mastermix with ROX as passive reference (ThermoFisher Scientific, Dreieich, Germany) as follows: Denaturation at 95 °C for 10 min, 39 amplification and quantification cycles with 95 °C for 15 s and 60 °C for 1 min, and finally a melting curve program (95 °C for 15 s, followed by 60–95 °C with a heating rate of 0.1 °C/s) and continuous fluorescence measurement. The cycle threshold (CT) values were determined using the QuantStudio Flex 7 Real-time PCR system software. Relative gene expression was calculated using the comparative $2^{-\Delta\Delta CT}$ method [55]. Amplification efficiencies were determined based on the slope of the calibration curve consisting of five different cDNA concentrations each measured in triplicates and efficiencies were as follows: *PPARG* (106.3%), *CEBPA* (96.3%), and *TBP* (88.7%). The fold-change values for gene expression were normalized by the respective efficiencies using a published procedure [55]. Relative gene expression data for *PPARG* and *CEBPA* were subsequently normalized to the reference gene of tata-box binding protein (TBP; in pre-experiments tested to be suited) and the gene expression of samples at day 0 of adipogenesis.

4.3. Hoechst Assay for DNA Quantification

For DNA quantification, the fluorochrome Hoechst 33342 (ThermoFisher Scientific, Waltham, MA, USA) was diluted in PBS to a final concentration of 20 µg/mL. A total of 80 µL of this solution was pipetted into each of the wells of a black 96-well plate (F96; Nunc, Thermo Fisher, Schwerte, Germany). Then, 20 µL of vortexed cell homogenates or plain solvent (80% MeOH; blanks) was added to the Hoechst solution and mixed by pipetting. The plate was incubated in the dark for 30 min at room temperature. Fluorescence signals were read using a GloMax Multi Detection System (Promega, Mannheim, Germany), equipped with a UV filter ($\lambda_{ex.}$ = 365 nm; $\lambda_{em.}$ = 410–460 nm, Promega, Mannheim, Germany) [53].

4.4. Lipid Extraction and Targeted Lipidomics Analysis

The Lipidyzer™ method (SCIEX, Darmstadt, Germany) was used to analyze the cellular lipidome. It detects lipids with fatty acid side chains of medium-chain (MCFA; C12), long-chain (LCFA; C13–C21), and very long-chain (VLCFA; C22–C26) lengths from 13 classes of lipids including cholesterol esters (CE), ceramides (CER), dihydroceramides (DCER), diacylglycerols (DAG), free fatty acids (FFA), hexosylceramides (HCER), lactosylceramides (LCER), lysophosphatidylcholines (LPC), lysophosphatidylethanolamines (LPE), phosphatidylcholines (PC), phosphatidylethanolamines (PE), sphingomyelins (SM), and triacylglycerols (TAG). The method allows the identification of lipid species at the fatty acyl/alkyl level (exception: TAG) [56]. The Lipidyzer™ method determines the sum of the numbers of C-atoms and double bonds (DB) for one fatty acid side chain as well as the sum of the C-atoms and DB of all three side chains. The notation rules from Liebisch and coworkers only know the case that either no fatty acid is known (e.g., TAG 52:2) or all three (e.g., TAG 16:0_18:1_18:1) [56]. Therefore, the nomenclature for TAG species in our study was adopted to these recommendations. The internal standard (IS) mixture (Avanti Polar Lipids, Inc., AL, USA) was prepared in accordance to the Lipidyzer™ manual. For QC samples, 250 µL of pooled cell homogenates were used, consisting in equal parts of undifferentiated, differentiating (day 8 of differentiation), and maturely differentiated cells (day 16). Three reference plasma samples (SCIEX, Darmstadt, Germany) of 100 µL in volume were spiked each with 50 µL of the QC spike mixture (SCIEX, Darmstadt, Germany) to investigate inter-run and inter-project effects. Lipids were extracted by two-phase separation using methyl *tert*-butyl ether (MTBE), methanol, and water [57]. Briefly, 250 µL of cell homogenates for the main experiments or 10–300 µL for method evaluation experiments, QC samples, or QC spiked plasma samples were transferred to 1.5 mL safe-lock reaction tubes (Eppendorf, Hamburg, Germany). For each time point, we took in total six biological independent cell samples. Next, 160 µL of MeOH and 900 μL of MTBE were added to each tube and incubated for 30 min at 900 rpm and room temperature in a shaker. For phase separation, 500 μL of H_2O was added to each tube, the mixtures were vortexed, and the tubes were centrifuged at 15,000× *g* for 4 min at RT. The upper organic phases were transferred into glass vials. The extraction step was repeated once and organic phases were combined. Organic solvents were evaporated to complete dryness under a stream of gaseous nitrogen and residuals were reconstituted with 250 μL of sample running buffer (10 mM ammonium acetate in dichloromethane:methanol (50:50 *v/v*)). Samples were then analyzed with the Lipidyzer™ method, consisting of a Sciex 5500 MS/MS QTRAP system (SCIEX, Darmstadt, Germany) equipped with a SelexION ion source for differential mobility spectrometry (DMS), in accordance with the manufacturer's instructions [24]. A sample volume of 50 μL was injected with a Shimadzu Nexera X2 liquid chromatography system (SCIEX, Darmstadt, Germany) at an isocratic flow rate of 7 μL/min. Data were acquired automatically with the Lipidyzer™ Workflow Manager software (version 1.0, SCIEX, Darmstadt, Germany). The obtained concentration values in nmol/g supplied by the software were converted to μmol/L with the assumption that 1 mL of cell culture sample was equal to 1 mL of plasma which is equal to 1 g [58]. Converted concentration values were normalized by Hoechst assay results.

4.5. Data Analysis

To trace the process of adipogenesis, the concentrations of the single lipid species, summed concentrations of lipid classes, concentrations of fatty acids, and percentage compositions were analyzed. Lipid species were completely excluded from the data set if concentration values were missing (NA) in more than 33.3% of the samples within a time point. Missing values were replaced by the respective minimal lipid species concentration measured, divided by $\sqrt{2}$ and multiplied by a randomly chosen factor between 0.75 and 1.25.

Statistical analyses and graphical illustrations of the lipidomic data were performed using the software MetaboAnalyst 4.0 [59], GraphPad Prism 8.1.1, and R 3.5.1 [60].

Univariate statistical analyses were performed using the Mann–Whitney U test and Kruskal–Wallis test with Dunn's post-hoc test. Spearman's rank correlation analysis was used to test correlations between lipid species. Prior to PCA and other multivariate statistical analyses, lipid concentrations were log-normalized and auto-scaled (mean-centered and divided by the standard deviation of each variable) to achieve a normal distribution of the data set. Averaged concentrations are shown with standard deviations.

5. Conclusions

We used the Lipidyzer™ technology to study the cellular lipidome during the development of preadipocytes into maturating and finally mature adipocytes in a human cell culture model. The switch from differentiating preadipocytes to maturating adipocytes became clearly visible at the lipidome level. The differentiation process was accompanied by increased concentrations of ceramides that are known to be involved in cell differentiation signaling. While these ceramide species decreased after completion of differentiation around day 4, massive lipid remodeling occurred during maturation of the adipocytes. This maturating phase was characterized by the substantial synthesis of DAG and TAG species. We furthermore observed increases of membrane lipids like PC, PE, and SM as well as their biosynthetic precursors. Moreover, we could also show that the compositions of the lipid species itself became more homogeneous during differentiation to highly concentrated saturated/monounsaturated LCFA with the four most abundant fatty acids being C16:0, C16:1, C18:0, and C18:1. Interestingly, VLCFA constantly decreased in almost all investigated lipid classes during the maturation process. High negative correlation coefficients between membrane lipids containing VLCFA and TAG species imply that these lipids might have served as additional sources for TAG synthesis. However, further studies are necessary to shed light on other lipid classes, lipid synthesis, degradation, and remodeling pathways during adipogenesis. For instance, fluxome-based approaches could be helpful to follow especially polyunsaturated and VLCFA within the preadipocytes. Moreover, a multi-omics approach might be helpful to detect connections between different pathways within the lipidome as well as the connection with pathways of more polar metabolites. In addition, it would also be interesting to compare the lipidomes of SGBS cells and freshly isolated human adipocytes since the metabolism in these cells might better reflect the original situation in humans. We could also show that the cultivation of cells with FBS-containing medium might influence the metabolism of the cells from several days up to weeks later. We recommend analyzing the medium used for cell culture lipidomics/metabolomics studies to prevent misinterpretation of the data.

Supplementary Materials: The following are available online at http://www.mdpi.com/2218-1989/10/6/217/s1, Figure S1: Analytical evaluation of the Lipidyzer™ method for undifferentiated and differentiated cells revealed strong linearity for most of the analyzed lipid classes; Figure S2: Representative microscopic images of SGBS cells showed successful differentiation based on morphological changes and a strong increase in number and size of lipid droplets; Figure S3: Transcripts of the main adipogenic transcription factors PPARγ (*PPARG*) and C/EBPα (*CEBPA*) were highly upregulated during adipogenesis; Figure S4: Principal component analysis (PCA) score plot showing very clear clustering of lipid species regarding the different time points of adipogenesis; Figure S5: Fatty acid concentrations and compositions changed markedly during adipogenesis in all 11 lipid classes; Figure S6: Clusters from the spearman's rank correlation analysis consisted of distinct lipid class compositions; Figure S7: Polyunsaturated and very long-chain fatty acids were highly concentrated in FBS-containing medium used for cultivation before differentiation start; Table S1: Linearity testing of method; Table S2: Repeatability and background signal testing; Table S3: Top 30 variable importance in projection (VIP) of PLS-DA for component 1 and 2, respectively; Table S4: Significances of Kruskal Wallis test; Table S5: Spearman's rank correlation coefficients.

Author Contributions: Conceptualization, J.A. and M.H.; methodology, F.M. and J.T.; software, F.M., A.C., and M.H.; validation, F.M. and J.L.; formal analysis, F.M. and M.H.; investigation, F.M.; resources, J.A.; data curation, F.M. and A.C.; writing—original draft preparation, F.M.; writing—review and editing, F.M., G.M., J.T., J.L., M.W., J.A. and M.H.; visualization, F.M.; supervision, M.H.; project administration, J.A., J.T., and G.M.; funding acquisition, J.A. All authors have read and agreed to the published version of the manuscript.

Funding: Part of this study was supported by EIT Health (eithealth.eu). EIT Health is supported by EIT, a body of the European Union.

Acknowledgments: The authors are grateful to Gabriele Zieglmeier for her technical assistance with the cell cultivation. Furthermore, we thank Maximilian Schmidtke for his support in RT-qPCR experiments.

Conflicts of Interest: The authors declare that they have no conflicts of interest with the contents of this article.

References

1. Chooi, Y.C.; Ding, C.; Magkos, F. The epidemiology of obesity. *Metabolism* **2019**, *92*, 6–10. [CrossRef] [PubMed]

2. Singh, G.M.; Danaei, G.; Farzadfar, F.; Stevens, G.A.; Woodward, M.; Wormser, D.; Kaptoge, S.; Whitlock, G.; Qiao, Q.; Lewington, S.; et al. The age-specific quantitative effects of metabolic risk factors on cardiovascular diseases and diabetes: A pooled analysis. *PLoS ONE* **2013**, *8*, e651740. [CrossRef] [PubMed]

3. Czernichow, S.; Kengne, A.P.; Stamatakis, E.; Hamer, M.; Batty, G.D. Body mass index, waist circumference and waist-hip ratio: Which is the better discriminator of cardiovascular disease mortality risk? Evidence from an individual-participant meta-analysis of 82 864 participants from nine cohort studies. *Obes. Rev.* **2011**, *12*, 680–687. [CrossRef] [PubMed]

4. Ackerman, S.E.; Blackburn, O.A.; Marchildon, F.; Cohen, P. Insights into the Link Between Obesity and Cancer. *Curr. Obes. Rep.* **2017**, *6*, 195–203. [CrossRef] [PubMed]

5. Ali, A.T.; Hochfeld, W.E.; Myburgh, R.; Pepper, M.S. Adipocyte and adipogenesis. *Eur. J. Cell Biol.* **2013**, *92*, 229–236. [CrossRef] [PubMed]

6. Lefterova, M.I.; Lazar, M.A. New developments in adipogenesis. *Trends Endocrinol. Metab.* **2009**, *20*, 107–114. [CrossRef]

7. Tontonoz, P.; Spiegelman, B.M. Fat and Beyond: The Diverse Biology of PPARγ. *Annu. Rev. Biochem.* **2008**, *77*, 289–312. [CrossRef]

8. Otto, T.C.; Lane, M.D. Adipose Development: From Stem Cell to Adipocyte. *Crit. Rev. Biochem. Mol. Biol.* **2005**, *40*, 229–242. [CrossRef]

9. Roberts, L.D.; Virtue, S.; Vidal-Puig, A.; Nicholls, A.W.; Griffin, J.L. Metabolic phenotyping of a model of adipocyte differentiation. *Physiol. Genom.* **2009**, *39*, 109–119. [CrossRef]

10. Guo, X.; Liao, K. Analysis of gene expression profile during 3T3-L1 preadipocyte differentiation. *Gene* **2000**, *251*, 45–53. [CrossRef]

11. Ross, S.E.; Erickson, R.L.; Gerin, I.; DeRose, P.M.; Bajnok, L.; Longo, K.A.; Misek, D.E.; Kuick, R.; Hanash, S.M.; Atkins, K.B.; et al. Microarray analyses during adipogenesis: Understanding the effects of Wnt signaling on adipogenesis and the roles of liver X receptor alpha in adipocyte metabolism. *Mol. Cell. Biol.* **2002**, *22*, 5989–5999. [CrossRef] [PubMed]

12. Green, C.R.; Wallace, M.; Divakaruni, A.S.; Phillips, S.A.; Murphy, A.N.; Ciaraldi, T.P.; Metallo, C.M. Branched-chain amino acid catabolism fuels adipocyte differentiation and lipogenesis. *Nat. Chem. Biol.* **2016**, *12*, 15–21. [CrossRef] [PubMed]

13. Molina, H.; Yang, Y.; Ruch, T.; Kim, J.-W.; Mortensen, P.; Otto, T.; Nalli, A.; Tang, Q.-Q.; Lane, M.D.; Chaerkady, R.; et al. Temporal profiling of the adipocyte proteome during differentiation using a five-plex SILAC based strategy. *J. Proteome Res.* **2009**, *8*, 48–58. [CrossRef] [PubMed]

14. Welsh, G.I.; Griffiths, M.R.; Webster, K.J.; Page, M.J.; Tavaré, J.M. Proteome analysis of adipogenesis. *Proteom. Clin. Appl.* **2004**, *4*, 1042–1051. [CrossRef]

15. Ye, F.; Zhang, H.; Yang, Y.-X.; Hu, H.-D.; Sze, S.K.; Meng, W.; Qian, J.; Ren, H.; Yang, B.-L.; Luo, M.-Y.; et al. Comparative proteome analysis of 3T3-L1 adipocyte differentiation using iTRAQ-coupled 2D LC-MS/MS. *J. Cell. Biochem.* **2011**, *112*, 3002–3014. [CrossRef]

16. Halama, A.; Horsch, M.; Kastenmuller, G.; Moller, G.; Kumar, P.; Prehn, C.; Laumen, H.; Hauner, H.; Hrabe de Angelis, M.; Beckers, J.; et al. Metabolic switch during adipogenesis: From branched chain amino acid catabolism to lipid synthesis. *Arch. Biochem. Biophys.* **2016**, *589*, 93–107. [CrossRef]

17. Kirkwood, J.S.; Miranda, C.L.; Bobe, G.; Maier, C.S.; Stevens, J.F. 18O-Tracer Metabolomics Reveals Protein Turnover and CDP-Choline Cycle Activity in Differentiating 3T3-L1 Pre-Adipocytes. *PLoS ONE* **2016**, *11*, e0157118. [CrossRef]

18. Liaw, L.; Prudovsky, I.; Koza, R.A.; Anunciado-Koza, R.V.; Siviski, M.E.; Lindner, V.; Friesel, R.E.; Rosen, C.J.; Baker, P.R.; Simons, B.; et al. Lipid Profiling of In Vitro Cell Models of Adipogenic Differentiation: Relationships With Mouse Adipose Tissues. *J. Cell. Biochem.* **2016**, *117*, 2182–2193. [CrossRef]

19. Collins, J.M.; Neville, M.J.; Pinnick, K.E.; Hodson, L.; Ruyter, B.; van Dijk, T.H.; Reijngoud, D.-J.; Fielding, M.D.; Frayn, K.N. De novo lipogenesis in the differentiating human adipocyte can provide all fatty acids necessary for maturation. *J. Lipid Res.* **2011**, *52*, 1683–1692. [CrossRef]

20. Ubhi, B.K. Direct Infusion-Tandem Mass Spectrometry (DI-MS/MS) Analysis of Complex Lipids in Human Plasma and Serum Using the Lipidyzer Platform. *Methods Mol. Biol.* **2018**, *1730*, 227–236. [CrossRef]

21. Cao, Z.; Schmitt, T.C.; Varma, V.; Sloper, D.; Beger, R.D.; Sun, J. Evaluation of the Performance of Lipidyzer Platform and Its Application in the Lipidomics Analysis in Mouse Heart and Liver. *J. Proteome Res.* **2019**. [CrossRef] [PubMed]

22. Contrepois, K.; Mahmoudi, S.; Ubhi, B.K.; Papsdorf, K.; Hornburg, D.; Brunet, A.; Snyder, M. Cross-Platform Comparison of Untargeted and Targeted Lipidomics Approaches on Aging Mouse Plasma. *Sci. Rep.* **2018**, *8*, 17747. [CrossRef] [PubMed]

23. Lepropre, S.; Kautbally, S.; Octave, M.; Ginion, A.; Onselaer, M.-B.; Steinberg, G.R.; Kemp, B.E.; Hego, A.; Wéra, O.; Brouns, S.; et al. AMPK-ACC signaling modulates platelet phospholipids and potentiates thrombus formation. *Blood* **2018**, *132*, 1180. [CrossRef] [PubMed]

24. Franko, A.; Merkel, D.; Kovarova, M.; Hoene, M.; Jaghutriz, B.A.; Heni, M.; Konigsrainer, A.; Papan, C.; Lehr, S.; Haring, H.U.; et al. Dissociation of Fatty Liver and Insulin Resistance in I148M PNPLA3 Carriers: Differences in Diacylglycerol (DAG) FA18:1 Lipid Species as a Possible Explanation. *Nutrients* **2018**, *10*, 1314. [CrossRef]

25. Alarcon-Barrera, J.C.; von Hegedus, J.H.; Brouwers, H.; Steenvoorden, E.; Ioan-Facsinay, A.; Mayboroda, O.A.; Ondo-Mendez, A.; Giera, M. Lipid metabolism of leukocytes in the unstimulated and activated states. *Anal. Bioanal. Chem.* **2020**, *412*, 2353–2363. [CrossRef] [PubMed]

26. Lintonen, T.P.; Baker, P.R.; Suoniemi, M.; Ubhi, B.K.; Koistinen, K.M.; Duchoslav, E.; Campbell, J.L.; Ekroos, K. Differential mobility spectrometry-driven shotgun lipidomics. *Anal. Chem.* **2014**, *86*, 9662–9669. [CrossRef]

27. Yao, C.-H.; Liu, G.-Y.; Yang, K.; Gross, R.W.; Patti, G.J. Inaccurate quantitation of palmitate in metabolomics and isotope tracer studies due to plastics. *Metabolomics* **2016**, *12*, 143. [CrossRef]

28. Bourbon, N.A.; Yun, J.; Kester, M. Ceramide Directly Activates Protein Kinase C ζ to Regulate a Stress-activated Protein Kinase Signaling Complex. *J. Biol. Chem.* **2000**, *275*, 35617–35623. [CrossRef]

29. Reichert, M.; Eick, D. Analysis of cell cycle arrest in adipocyte differentiation. *Oncogene* **1999**, *18*, 459–466. [CrossRef]

30. Ikonen, E. Cellular cholesterol trafficking and compartmentalization. *Nat. Rev. Mol. Cell Biol.* **2008**, *9*, 125–138. [CrossRef]

31. Prattes, S.; Horl, G.; Hammer, A.; Blaschitz, A.; Graier, W.F.; Sattler, W.; Zechner, R.; Steyrer, E. Intracellular distribution and mobilization of unesterified cholesterol in adipocytes: Triglyceride droplets are surrounded by cholesterol-rich ER-like surface layer structures. *J. Cell Sci.* **2000**, *113*, 2977. [PubMed]

32. Wang, Y.; Castoreno, A.B.; Stockinger, W.; Nohturfft, A. Modulation of endosomal cholesteryl ester metabolism by membrane cholesterol. *J. Biol. Chem.* **2005**, *280*, 11876–11886. [CrossRef] [PubMed]

33. Weiss, S.B.; Kennedy, E.P.; Kiyasu, J.Y. The Enzymatic Synthesis of Triglycerides. *J. Biol. Chem.* **1960**, *235*, 40–44. [CrossRef] [PubMed]

34. Coleman, R.A.; Mashek, D.G. Mammalian triacylglycerol metabolism: Synthesis, lipolysis, and signaling. *Chem. Rev.* **2011**, *111*, 6359–6386. [CrossRef] [PubMed]

35. Fukami, K. Structure, Regulation, and Function of Phospholipase C Isozymes. *J. Biochem.* **2002**, *131*, 293–299. [CrossRef]

36. Fong, W.-K.; Sánchez-Ferrer, A.; Rappolt, M.; Boyd, B.J.; Mezzenga, R. Structural Transformation in Vesicles upon Hydrolysis of Phosphatidylethanolamine and Phosphatidylcholine with Phospholipase C. *Langmuir* **2019**, *35*, 14949–14958. [CrossRef]

37. Casares, D.; Escriba, P.V.; Rossello, C.A. Membrane Lipid Composition: Effect on Membrane and Organelle Structure, Function and Compartmentalization and Therapeutic Avenues. *Int. J. Mol. Sci.* **2019**, *20*, 2167. [CrossRef]

38. Haczeyni, F.; Bell-Anderson, K.S.; Farrell, G.C. Causes and mechanisms of adipocyte enlargement and adipose expansion. *Obes. Rev.* **2018**, *19*, 406–420. [CrossRef]

39. Kasturi, R.; Wakil, S.J. Increased synthesis and accumulation of phospholipids during differentiation of 3T3-L1 cells into adipocytes. *J. Biol. Chem.* **1983**, *258*, 3559–3564.

40. Gauster, M.; Rechberger, G.; Sovic, A.; Horl, G.; Steyrer, E.; Sattler, W.; Frank, S. Endothelial lipase releases saturated and unsaturated fatty acids of high density lipoprotein phosphatidylcholine. *J. Lipid Res.* **2005**, *46*, 1517–1525. [CrossRef]

41. Su, X.; Han, X.; Yang, J.; Mancuso, D.J.; Chen, J.; Bickel, P.E.; Gross, R.W. Sequential Ordered Fatty Acid α Oxidation and $\Delta 9$ Desaturation Are Major Determinants of Lipid Storage and Utilization in Differentiating Adipocytes. *Biochemistry* **2004**, *43*, 5033–5044. [CrossRef] [PubMed]

42. Jeucken, A.; Brouwers, J.F. High-Throughput Screening of Lipidomic Adaptations in Cultured Cells. *Biomolecules* **2019**, *9*, 42. [CrossRef] [PubMed]

43. Klusóczki, Á.; Veréb, Z.; Vámos, A.; Fischer-Posovszky, P.; Wabitsch, M.; Bacso, Z.; Fésüs, L.; Kristóf, E. Differentiating SGBS adipocytes respond to PPARγ stimulation, irisin and BMP7 by functional browning and beige characteristics. *Sci. Rep.* **2019**, *9*, 5823. [CrossRef]

44. Yeo, C.R.; Agrawal, M.; Hoon, S.; Shabbir, A.; Shrivastava, M.K.; Huang, S.; Khoo, C.M.; Chhay, V.; Yassin, M.S.; Tai, E.S.; et al. SGBS cells as a model of human adipocyte browning: A comprehensive comparative study with primary human white subcutaneous adipocytes. *Sci. Rep.* **2017**, *7*, 4031. [CrossRef]

45. Tews, D.; Fischer-Posovszky, P.; Fromme, T.; Klingenspor, M.; Fischer, J.; Rüther, U.; Marienfeld, R.; Barth, T.F.; Möller, P.; Debatin, K.M.; et al. FTO Deficiency Induces UCP-1 Expression and Mitochondrial Uncoupling in Adipocytes. *Endocrinology* **2013**, *154*, 3141–3151. [CrossRef] [PubMed]

46. Tews, D.; Fromme, T.; Keuper, M.; Hofmann, S.M.; Debatin, K.M.; Klingenspor, M.; Wabitsch, M.; Fischer-Posovszky, P. Teneurin-2 (TENM2) deficiency induces UCP1 expression in differentiating human fat cells. *Mol. Cell. Endocrinol.* **2017**, *443*, 106–113. [CrossRef]

47. Wabitsch, M.; Brenner, R.E.; Melzner, I.; Braun, M.; Möller, P.; Heinze, E.; Debatin, K.M.; Hauner, H. Characterization of a human preadipocyte cell strain with high capacity for adipose differentiation. *Int. J. Obes.* **2001**, *25*, 8–15. [CrossRef]

48. Shadid, S.; Koutsari, C.; Jensen, M.D. Direct Free Fatty Acid Uptake Into Human Adipocytes In Vivo Relation to Body Fat Distribution. *Diabetes* **2007**, *56*, 1369–1375. [CrossRef]

49. Chamberlain, C.A.; Rubio, V.Y.; Garrett, T.J. Impact of matrix effects and ionization efficiency in non-quantitative untargeted metabolomics. *Metabolomics* **2019**, *15*, 135. [CrossRef]

50. Triebl, A.; Trotzmuller, M.; Hartler, J.; Stojakovic, T.; Kofeler, H.C. Lipidomics by ultrahigh performance liquid chromatography-high resolution mass spectrometry and its application to complex biological samples. *J. Chromatogr. B Analyt. Technol. Biomed. Life Sci.* **2017**, *1053*, 72–80. [CrossRef]

51. Koivusalo, M.; Haimi, P.; Heikinheimo, L.; Kostiainen, R.; Somerharju, P. Quantitative determination of phospholipid compositions by ESI-MS: Effects of acyl chain length, unsaturation, and lipid concentration on instrument response. *J. Lipid Res.* **2001**, *42*, 663–672. [PubMed]

52. Fischer-Posovszky, P.; Newell, F.S.; Wabitsch, M.; Tornqvist, H.E. Human SGBS cells—A unique tool for studies of human fat cell biology. *Obes. Facts* **2008**, *1*, 184–189. [CrossRef] [PubMed]

53. Muschet, C.; Moller, G.; Prehn, C.; de Angelis, M.H.; Adamski, J.; Tokarz, J. Removing the bottlenecks of cell culture metabolomics: Fast normalization procedure, correlation of metabolites to cell number, and impact of the cell harvesting method. *Metabolomics* **2016**, *12*, 151. [CrossRef] [PubMed]

54. Untergasser, A.; Cutcutache, I.; Koressaar, T.; Ye, J.; Faircloth, B.C.; Remm, M.; Rozen, S.G. Primer3—New capabilities and interfaces. *Nucleic Acids Res.* **2012**, *40*, e115. [CrossRef]

55. Pfaffl, M.W. A new mathematical model for relative quantification in real-time RT-PCR. *Nucleic Acids Res.* **2001**, *29*, e45. [CrossRef]

56. Liebisch, G.; Vizcaíno, J.A.; Köfeler, H.; Trötzmüller, M.; Griffiths, W.J.; Schmitz, G.; Spener, F.; Wakelam, M.J.O. Shorthand notation for lipid structures derived from mass spectrometry. *J. Lipid Res.* **2013**, *54*, 1523–1530. [CrossRef]

57. Matyash, V.; Liebisch, G.; Kurzchalia, T.V.; Shevchenko, A.; Schwudke, D. Lipid extraction by methyl-tert-butyl ether for high-throughput lipidomics. *J. Lipid Res.* **2008**, *49*, 1137–1146. [CrossRef]

58. Vitello, D.J.; Ripper, R.M.; Fettiplace, M.R.; Weinberg, G.L.; Vitello, J.M. Blood Density Is Nearly Equal to Water Density: A Validation Study of the Gravimetric Method of Measuring Intraoperative Blood Loss. *J. Vet. Med.* **2015**, *2015*, 152730. [CrossRef]

59. Chong, J.; Soufan, O.; Li, C.; Caraus, I.; Li, S.; Bourque, G.; Wishart, D.S.; Xia, J. MetaboAnalyst 4.0: Towards more transparent and integrative metabolomics analysis. *Nucleic Acids Res.* **2018**, *46*, W486–W494. [CrossRef]

60. R Core Team. R: A language and environment for statistical computing. R Foundation for Statistical Computing, Vienna, Austria. Available online: https://www.R-project.org/ (accessed on 24 April 2020).

 *metabolites*

Article

Increased Expression of the Leptin Gene in Adipose Tissue of Patients with Chronic Kidney Disease–The Possible Role of an Abnormal Serum Fatty Acid Profile

Justyna Korczyńska [1], Aleksandra Czumaj [1], Michał Chmielewski [2], Maciej Śledziński [3], Adriana Mika [4] and Tomasz Śledziński [1,*]

[1] Department of Pharmaceutical Biochemistry, Faculty of Pharmacy, Medical University of Gdansk, 80-211 Gdansk, Poland; justyna.korczynska@gumed.edu.pl (J.K.); aleksandra.czumaj@gumed.edu.pl (A.C.)

[2] Department of Nephrology, Transplantology and Internal Medicine, Faculty of Medicine, Medical University of Gdansk, 80-211 Gdansk, Poland; chmiel@gumed.edu.pl

[3] Department of General, Endocrine and Transplant Surgery, Faculty of Medicine, Medical University of Gdansk, 80-214 Gdansk, Poland; msledz@gumed.edu.pl

[4] Department of Environmental Analytics, Faculty of Chemistry, University of Gdansk, 80-309 Gdansk, Poland; adrianamika@tlen.pl

* Correspondence: tomasz.sledzinski@gumed.edu.pl

Received: 9 February 2020; Accepted: 6 March 2020; Published: 8 March 2020

Abstract: Chronic kidney disease (CKD) is associated with an increased level of leptin and an abnormal fatty acid (FA) profile in the serum. However, there are no data on the associations between them, and the reason for increased serum levels in patients with CKD is not well elucidated. Recently, we found that a CKD-related abnormal FA profile caused significant changes in the expression of genes involved in lipid metabolism in hepatocytes. The aim of this study was to examine whether leptin gene expression in subcutaneous adipose tissue (SAT) of patients with CKD may contribute to increased serum levels of this adipokine and whether the abnormal serum FA profile observed in CKD patients has an impact on leptin gene expression in adipocytes. The FA profile was measured in serum samples from patients with CKD and controls by GC–MS. The relative mRNA levels of leptin were measured in SAT by Real-Time PCR. Moreover, the effect of the CKD-related abnormal FA profile on leptin gene expression was studied in in vitro cultured 3T3-L1 adipocytes. Patients with CKD had higher concentrations of serum leptin than controls and higher expression level of the leptin gene in SAT. They also had increased serum monounsaturated FAs and decreased polyunsaturated FAs. The incubation of adipocytes with FAs isolated from CKD patients resulted in an increase of the levels of leptin mRNA. Increased leptin gene expression in SAT may contribute to elevated concentrations of these adipokine in patients with CKD. CKD-related alterations of the FA profile may contribute to elevated serum leptin concentrations in patients with CKD by increasing the gene expression of this adipokine in SAT.

Keywords: chronic kidney disease; leptin; fatty acids; adipose tissue; adipocytes

---

1. Introduction

Chronic kidney disease (CKD), with its prevalence exceeding 5% of the general population, constitutes an important clinical issue [1]. Irrespective of the underlying disease, CKD increases the cardiovascular burden of the patients several times compared to people with preserved kidney function [2]. Potential mechanisms for increased cardiovascular risk in CKD include alterations in the lipid profile and serum adipokine levels [3–5]. Changes in adipokine levels have been discussed in the context of CKD progression and the risk of comorbidities [6–9]. Leptin is one of the hormones secreted by adipose tissue and has been recognized as a signaling molecule that regulates energy homeostasis [10,11]. This protein also plays an important role in immune regulation and inflammation, both closely associated with oxidative stress and endothelial dysfunction, which may affect the risk of cardiovascular disorders [12–14]. Previous studies showed increased concentrations of leptin in the serum of patients with CKD [7,15]; however, the mechanism of this change is not well elucidated.

CKD is also associated with significant lipid disorders. The most frequently described changes focus on cholesterol and triacylglycerols. Many studies, including our previous research, have also shown abnormal serum fatty acid (FA) profiles in patients with CKD [16–18]. Dyslipidemia observed in patients with CKD is one of the well-known risk factors of cardiovascular disease (CVD) or diabetes mellitus [19–21]. This is especially important because cardiovascular disease is the leading cause of death in this group of patients [19]. Changes in the lipid profile in patients with CKD may also actively participate in the deterioration of renal function, thus contributing to the worsening of the disease [20,22]. However, all of the consequences of these alterations are still not fully understood.

Leptin and FAs are critical factors for the crosstalk between adipose tissue and other metabolically important organs, including the kidneys. Although elevated leptin levels and alterations in the fatty acid profile are factors that can be involved in the pathogenesis and complications of CKD, no previous study has investigated the relationship between them. It is still unclear whether changes in leptin levels in CKD are caused by a reduced glomerular filtration rate, increased production in adipose tissue, or both. Our previous study showed that an altered FA profile in patients with CKD significantly changed hepatocyte metabolism [16]. Thus, in the present study, we aimed to examine the associations of leptin serum levels and its gene expression in adipose tissue. Moreover, we studied the effect of an abnormal serum FA profile in patients with CKD on the expression of the leptin gene in adipocytes.

2. Results

2.1. Leptin Levels in Serum and mRNA Levels in Subcutaneous Adipose Tissue of Study Subjects

The mean serum leptin concentrations among patients with CKD (29.24 ± 16,5 ng/mL) were significantly elevated, at almost three times higher than the mean value for healthy controls (11.96 ± 5.3 ng/mL) (Figure 1a). When we analyzed serum concentrations of leptin separately in women and men, we found that both male and female CKD patients had almost three times higher serum leptin concentrations than controls; however, the leptin concentrations in women (both in CKD patients and controls) were about two times higher than in men (in CKD women 37.7 ± 15 vs. 15.8 ± 9.9 in healthy women, $p < 0.01$; in CKD men 19.6 ± 10 vs. 7.54 ± 5.2 in healthy man $p < 0.05$). The sex-related differences in serum leptin concentrations are in agreement with results of other researchers [23]. The relative mRNA level of the leptin gene in subcutaneous adipose tissue of patients with CKD was approximately three times higher than that in controls (Figure 1b).

2.2. Serum Fatty Acid Profile of Study Subjects

The profile of FA in the serum of control subjects and patients with CKD is shown in Table 1. We observed several alterations in the FA profile, including various individual FAs and main groups of FAs (Table 1). The total saturated FA (SFA) and monounsaturated FA (MUFA) contents in the serum were significantly higher in the CKD group than those in the control group. At the same time, patients with CKD had lower levels of total n-3 polyunsaturated FAs (n-3 PUFAs) and n-6 polyunsaturated FAs (n-6 PUFAs) in the serum. This section may be divided by subheadings. It should provide a concise and precise description of the experimental results, their interpretation, and the experimental conclusions that can be drawn.

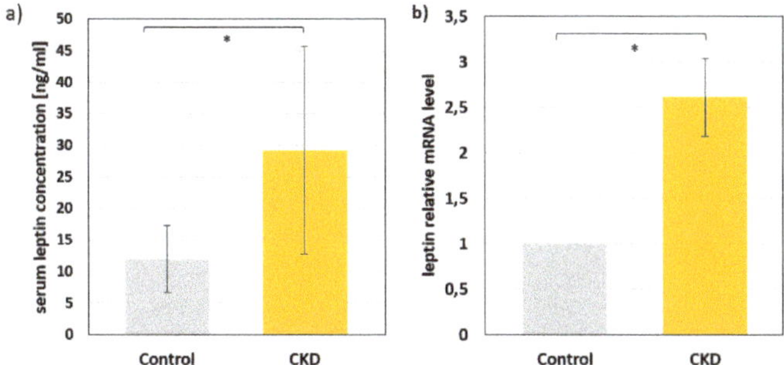

Figure 1. The serum concentrations of leptin (**a**) and leptin mRNA levels (**b**) in subcutaneous fat tissue of patients with chronic kidney disease (CKD) and control subjects. Data are shown as the mean ± SD, * $p < 0.05$. Serum leptin concentrations were assayed in 46 CKD patients and 57 healthy subjects. Leptin mRNA levels were assayed in adipose tissue obtained from 22 patients with CKD and 11 healthy subjects.

Table 1. The percent content of the main classes of fatty acids in the serum of patients with chronic kidney disease (CKD) and the control group. The data are presented as fatty acid proportions (%). Values are the mean ± SD.

FA	CONTROL	CKD
14:0	1.16 ± 0.31	1.11 ± 0.50
16:0	22.9 ± 1.62	24.1 ± 2.10 *
18:0	6.96 ± 0.72	6.80 ± 0.98
OTHER SFAs	1.22 ± 0.09	1.36 ± 0.09
TOTAL SFAs	32.3 ± 1.83	33.4 ± 3.02 *
14:1	0.07 ± 0.02	0.05 ± 0.03
16:1	2.81 ± 0.85	2.95 ± 0.79
18:1	25.7 ± 3.15	28.9 ± 3.63 *
OTHER MUFAs	0.49 ± 0.08	0.65 ± 0.11 *
TOTAL MUFAs	29.1 ± 1.08	32.6 ± 1.22 *
18:3 n-3	0.31 ± 0.11	0.20 ± 0.09 *
20:5 n-3	0.94 ± 0.60	0.61 ± 0.26 *
22:6 n-3	1.03 ± 0.43	0.83 ± 0.38 *
OTHER N-3 PUFAs	0.37 ± 0.1	0.34 ± 0.13
TOTAL N-3 PUFAs	2.66 ± 1.04	1.98 ± 0.71 *
18:2 n-6	26.1 ± 3.59	22.9 ± 4.79 *
20:4 n-6	5.31 ± 1.14	4.53 ± 1.31 *
OTHER N-6 PUFAs	1.42 ± 0.34	1.10 ± 0.32
TOTAL N-6 PUFAs	32.8 ± 3.82	28.6 ± 5.39 *

* Statistically significant compared to controls at $p < 0.05$. SFAs—saturated fatty acids; MUFAs— monounsaturated fatty acids; PUFAs—polyunsaturated fatty acids.

2.3. The Effect of the CKD-Related Abnormal Fatty Acid Profile on the Expression of Leptin in In Vitro Cultured Adipocytes

To examine whether reported alterations in the proportion of particular serum FA groups in CKD had an impact on adipose leptin gene expression, we treated 3T3-L1 adipocytes with a selected representative SFA (palmitic acid 16:0, PA), MUFA (oleic acid 18:1, OA), n-3 PUFA (docosahexaenoic acid 22:6 n-3, DHA), and n-6 PUFA (arachidonic acid 20:4 n-6, AA). All FAs were used at three different concentrations (25, 50, and 100 µM). After 48 h of incubation, PA and OA, FAs that are elevated in the serum of patients with CKD, increased the expression of the leptin gene, whereas DHA and AA–FAs, which are decreased in the serum of patients with CKD, decreased the expression of the leptin gene (Figure 2). Almost all these changes were statistically significant. All observed effects were dose-dependent.

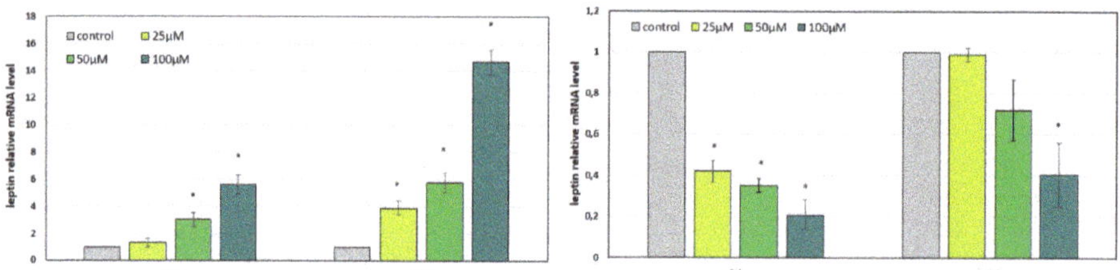

Figure 2. The relative mRNA level of leptin in 3T3-L1 adipocytes cultured for 48 h with various concentrations of palmitic acid 16:0 (PA), oleic acid 18:0 (OA), arachidonic acid 20:4 n-6 (AA), and docosahexaenoic acid 22:6 n-3 (DHA) or without FA supplementation (control). * Significantly different compared to the control ($p < 0.05$). Data are presented as the mean ± SD. All experiments were run in three independent attempts.

We are aware that, in patients with CKD, alterations in the proportion of all particular serum FA groups occur at the same time. To examine the combined effect of all FA disorders, we decided to use a complete set of FAs isolated from CKD patients and healthy controls serum samples. Incubation of adipocytes with a set of FAs from patients with CKD resulted in a significantly increased mRNA level of the leptin gene in comparison to the mRNA level from adipocytes incubated with FA-mix from healthy subjects (Figure 3). We did not find any structural changes in the cells after treatment with a set of FAs isolated from patients with CKD or from healthy subjects (Figure 4).

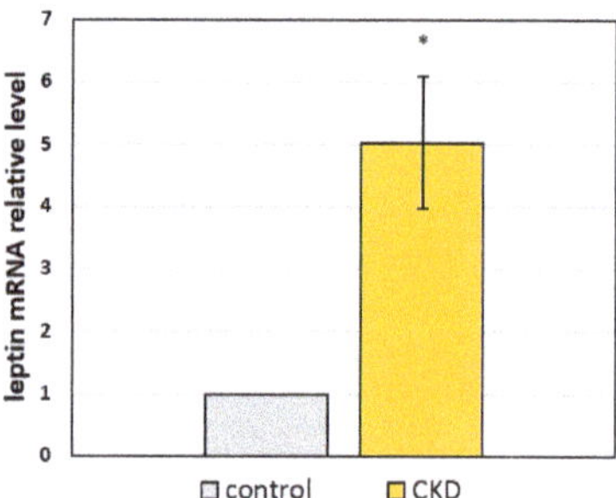

Figure 3. The relative mRNA level of leptin in 3T3-L1 adipocytes cultured for 48 h with a set of FAs extracted from the serum of control subjects (control) or patients with stage 5 CKD (CKD). * $p < 0.05$ compared to the control. Data are presented as the means ± SD. All experiments were run in three independent attempts.

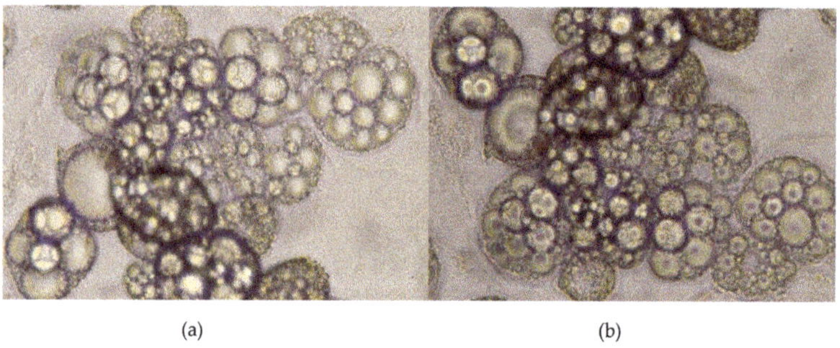

(a)   (b)

Figure 4. Representative sample of 3t3 adipocytes before (**a**) and after (**b**) treatment by FA isolated from serum of CKD patients.

3. Discussion

Our results demonstrated that, compared to healthy volunteers, patients with CKD had higher serum leptin levels. Concerning leptin, this result is in line with most clinical studies of leptin alteration in kidney diseases in adults and children [15,24–27]. Interestingly, a few studies have shown contradictory results [28–30]. However, in these cases, patients with CKD have experienced significant weight loss during dialysis. In our study, there was no difference in BMI between the CKD and control groups. In patients with CKD, an increased level of leptin may predict poor prognosis. A number of studies have indicated that leptin is involved in CKD progression and CKD complications [14,31]. Leptin plays a role in various types of kidney cells (e.g., glomerular mesangial cells, glomerular endothelial cells, and podocytes), and high leptin levels can lead to renal injury symptoms. Leptin inhibits the expression of nephrin, podocin, podoplanin, and podocalyxin (the podocyte-associated molecules necessary for the proper functioning of the renal filtration barrier) and promotes the production of reactive oxygen species (a known factor involved in the pathogenesis of CKD) [14,31]. The late stages of CKD are related to protein–energy wasting (reduced body protein and fat mass, and usually reduced protein and energy intake), and since leptin inhibits appetite, it may contribute to a further deterioration of nutritional status or even to malnutrition [32].

We show that increased serum levels of leptin observed in patients with CKD are accompanied by increased expression of the leptin gene in adipose tissue. Previous studies suggested that decreased kidney function (and as a consequence, decreased renal clearance of adipokines) may contribute to the elevated circulating levels of leptin in patients with CKD [33]. However, our study showed increased SAT mRNA levels and serum protein levels of leptin, which suggests that its production in SAT may contribute to its serum concentrations. To date, the literature data on this subject are very poor. Nordfors et al. reported elevated leptin gene expression only in adipose tissue in patients with chronic renal failure with inflammation compared to patients with chronic renal failure with no inflammation and no changes in leptin expression between patients with chronic renal failure and healthy controls [33]. Nonetheless, they examined a small number of patients (15 patients with chronic renal failure, including two with inflammation) and used in situ hybridization histochemistry for gene expression quantification. In turn, Witasp et al. reported downregulation of the leptin gene in abdominal subcutaneous adipose tissue of patients with advanced CKD with 'uremic–metabolic syndrome' [34]. Since there are no data on how the expression of the leptin gene is regulated in the adipose tissue of patients with CKD, we tried to find a molecular mechanism for the increased expression of leptin in the SAT of our patients.

The consequences of changes in the FA profile are usually considered in the context of dietary intake. Our recent study showed that CKD-related alterations of the FA profile influence hepatocyte metabolism. Thus, in the present study, we examined the hypothesis that CKD-related FA alteration can affect leptin gene expression in adipocytes. We found specific effects of FAs from various groups. However, since each FA can have a different effect on adipose tissue, alterations in the FA profile during CKD should be considered together as a whole. Thus, we made an effort to address this issue by using total FAs extracted from patients and control serum. To the authors' knowledge, this is a unique research approach. Our data demonstrated, for the first time, that the altered serum FA profile observed in patients with CKD increased adipocyte expression of leptin. Thus, it may also be responsible for the increased expression of this adipokine in SAT and its elevated circulating levels.

The study has limitations that ought to be mentioned. Our data are derived from patients at one point of chronic progressive disease. Further studies should consider the cross-sectional selection of patients. Another limitation is that our study cohort was not very numerous and there are relatively large deviations in some parameters; however, the differences were statistically significant. Thus, our study can serve as a proof-of-concept report. The expression of leptin gene has been assayed only at the mRNA levels. In our in vitro experiment, we used only one cell line as an adipocyte model. Adipose tissue is a highly heterogeneous organ with cell- and depot-specific functions [35–37]. However, we chose 3T3-L1 differentiated adipocytes because they are a well-characterized and widely used cell line in metabolic, molecular, and endocrine studies, including studies on leptin expression levels [38–41].

In conclusion, our study showed that increased leptin gene expression in SAT may contribute to elevated concentrations of this adipokine in patients with CKD. Moreover, alteration of the serum FA profile observed in the course of CKD might contribute to CKD-related elevated serum leptin levels, through induction of its gene expression in adipocytes.

4. Materials and Methods

4.1. Study Subjects

Forty-six patients with stage 5 CKD (27 males and 19 females; pre-dialysis and dialyzed, recruited from the Outpatient Unit of the Department of Nephrology, Transplantology and Internal Medicine in the Medical University of Gdansk) and 57 healthy subjects or metabolically healthy subjects (31 males and 26 females) who underwent hernia surgeries, matched for age and weight, with no known kidney disease, were included in this study. Among the abovementioned participants, subcutaneous adipose tissue, which is available during kidney transplantation, was taken from 22 patients with CKD and from 11 metabolically healthy subjects who underwent hernia surgeries at the Department of Surgery, Medical University of Gdansk. After overnight fasting, blood samples were taken from all the participants of the study. The study was performed in agreement with the principles of the Declaration of Helsinki of the World Medical Association. Experimental protocols received approval from the Local Bioethics Committee at the Medical University of Gdansk (protocol numbers: NKBBN/664/2013-2014, NKBBN/614-276/2014). The general characteristics of the study subjects and selected laboratory parameters are presented in Table 2. Patients with CKD had higher creatinine, blood urea nitrogen, and triacylglycerols, in comparison with control individuals.

Table 2. Selected biochemical and anthropometric characteristics of the study subjects.

Parameter	CONTROL	CKD
AGE (years)	47 ± 14.9	51 ± 13.0
BMI (kg/m^2)	26.0 ± 3.8	25.9 ± 4.8
CREATININE (mg/dL)	0.9 ± 0.2	6.15 ± 2.5 *
BUN (mg/dL)	15.1 ± 3.6	44.8 ± 25.0 *
ALBUMIN (g/L)	39.5 ± 3.9	37.5 ± 4.6
CRP (mg/dL)	2.1 ± 2.5	4.9 ± 6.1 *
TG (mg/dL)	115.6 ± 59.6	150.2 ± 73.7 *
TC (mg/dL)	195.9 ± 45.5	200.6 ± 48.0
GLUCOSE (mg/dL)	96.2 ± 20.2	102.7 ± 27.8
INSULIN (mU/mL)	9.4 ± 5.4	9.6 ± 6.2
HOMA-IR	2.35 ± 2.0	2.8 ± 2.7

* Statistically significant compared to controls at $p < 0.05$. Values are the mean ± SD. BMI—body mass index; BUN—blood urea nitrogen; CRP—C-reactive protein; HOMA-IR—homeostatic model assessment of insulin resistance; TG—triacylglycerols TC—total cholesterol.

4.2. Materials and Reagents

From Avantor Performance Materials Poland (Gliwice, Poland) methanol, chloroform, dichloromethane, n-hexane (all HPLC-grade), hydrochloric acid, and potassium hydroxide were acquired. DMEM, glucose, bovine calf serum, glutamine, penicillin/streptomycin solution, fetal bovine serum, dexamethasone, 3-isobutyl-1-methylxanthine, insulin, Oil Red O solution, palmitic acid, oleic acid, docosahexaenoic acid, arachidonic acid phosphate-buffered saline, FAME mix, Nuclease-Free Water, boron trifluoride–methanol solution, and 19-methyleicosanoic acid were obtained from Sigma-Aldrich (St. Louis, MO, USA). Eppendorf laboratory consumables were used for experiments (Hamburg, Germany).

4.3. Serum Leptin Assay

For the detection of serum leptin concentrations, the Leptin Human ELISA Clinical Range kit (BioVendor, Brno, Czech Republic) was used according to the manufacturer's instructions. In brief, samples from the control group and samples from the CKD group were incubated in microplate wells coated with polyclonal anti-human leptin antibodies. Bound leptin was detected by horseradish peroxidase-conjugated polyclonal anti-human leptin antibody. Tetramethylbenzidine was used as a substrate for peroxidase, and color intensity was determined by measuring the absorbance at 450 nm.

4.4. Serum Fatty Acid Profile Analysis

Total lipids were extracted from the serum of patients with CKD and healthy subjects, using the method described by Folch et al., with a mixture of chloroform:methanol (2:1, *v/v*) [42]. Then, the lipid extracts were dried by evaporation under a stream of nitrogen and alkaline hydrolyzed with 0.5 M KOH in methanol, at 90 °C, for 3 h. Next, the mixture was acidified with 6 M HCl, and 1 mL of water was added. Fatty acids were extracted three times, with 1 mL of n-hexane, and evaporated under a stream of nitrogen. To obtain fatty acid methyl esters (FAMEs), 10% boron trifluoride–methanol solution was added to each sample, which was then heated at 55 °C, for 90 min. After 1.5 h, 1 mL of water was added to the mixture, and FAMEs were extracted three times with 1 mL of n-hexane and dried under nitrogen stream. Fatty acid profiles were analyzed by gas chromatography–mass spectrometry (GC–MS), using a QP-2010SE apparatus (Shimadzu, Kyoto, Japan), as described previously [16]. In brief, a 30 m 0.25 mm i.d. ZB-5MSi capillary column was used (film thickness 0.25 μm). Temperature of the column was set between 60 and 300 °C (4 °C/min). Helium was used as a carrier gas at the column head pressure of 100 kPa, and FAME ionization was carried out with 70 eV electron energy. Full-scan mode was applied, with mass scan range *m/z* 45–700. Then, 19-methyleicosanoic acid was used as an internal standard. FAMEs were identified by comparison with reference standards (37 FAME Mix, Sigma-Aldrich, St. Louis, MO, USA) and NIST2011 reference library.

4.5. Adipocyte Culture, Differentiation, and Treatment

The 3T3-L1 cell line was obtained from American Type Culture Collection (Manassas, VA, USA). Pre-adipocytes were cultured in expansion medium (Dulbecco's modified Eagle's medium (DMEM) with 4.5 mg/mL of glucose supplemented with 10% bovine calf serum, 4 mM glutamine, 100 IU/mL of penicillin, and 100 IU/mL of streptomycin), at 37 °C, in a 5% CO_2 incubator. Cells were seeded at approximately 3000 cells per cm^2. Two days after reaching confluence, cells were differentiated by replacing the expansion medium with high-glucose DMEM containing 10% fetal bovine serum (FBS), 1.0 μM dexamethasone, 0.5 mM 3-isobutyl-1-methylxanthine (IBMX), and 1.0 μg/mL of insulin. After 48 h, the differentiation medium was replaced by DMEM with 10% FBS, 10 μg/mL of insulin, 4 mM glutamine, 100 IU/mL of penicillin, and 100 IU/mL of streptomycin. Media were replaced every other day. After 10 days, the adipocytes were fully differentiated (confirmed by Oil red O staining). The 3T3-L1 fully differentiated adipocytes were used for experiments. Cells were supplemented for 48 h, with one selected fatty acid: PA, OA, DHA, and AA or with the full FA set isolated from pooled serum of 10 randomly selected patients with stage 5 CKD, or 10 healthy controls. From pooled patient or control serum samples, total lipids were extracted and hydrolyzed, as previously described in Czumaj et al. (2019) [16]. A mixture of fatty acids isolated from the patients and controls sera was added to the cell culture at exactly the same concentration as in the study participants' sera. The FA set isolated from the serum of patients with CKD was characterized by a higher content of MUFAs and SFAs and a lower content of n-3 and n-6 PUFAs than FAs isolated from serum of control subjects [16].

4.6. Gene Expression Analyses

Total RNA isolation from in vitro cultured adipocytes and adipose tissue depots was carried out, using an RNeasy Lipid Tissue Mini Kit (Qiagen, Hilden, Germany) according to the manufacturer's instructions. The RNA concentration and integrity were assessed by an Experion automated electrophoresis system (Bio-Rad, Hercules, CA, USA). One microgram of RNA was reverse transcribed, using the RevertAid First Strand cDNA Synthesis Kit (Thermo Fisher Scientific, Waltham, MA, USA). Quantitative real-time PCR was carried out in a CFX Connect Real-Time System (Bio-Rad), using the SensiFAST SYBR No-ROX Kit (Bioline Meridian Bioscience, Cincinnati, OH, USA). The comparative Ct method ($\Delta\Delta$Ct) was used for relative quantification of gene expression. The β-actin gene was used for normalization.

4.7. Statistics

Data are presented as the mean ± SD. Statistical analyses were performed by using two-tailed the Student's *t*-test for two-group comparisons (including biochemical and anthropometric characteristics of the study subjects presented in Table 2) or analysis of variance (ANOVA), followed by post hoc correction (Bonferroni) for multi-group comparisons. The threshold of statistical significance was defined as $p < 0.05$. During all analyses, every sample was run in duplicate. The cell culture experiment was run in three independent attempts. All statistical analyses were performed by using STATISTICA 12 (TIBCO Software Inc., Palo Alto, CA, USA).

Author Contributions: Conceptualization, A.C., J.K., and T.Ś.; methodology, M.C., M.Ś., and A.M.; formal analysis, A.M.; investigation, J.K., A.C., and A.M.; resources, M.C. and M.Ś.; writing—original draft preparation, J.K., A.C., and A.M.; writing—review and editing, A.C. and T.Ś.; visualization, J.K.; supervision, T.Ś.; funding acquisition, A.C. All authors have read and agreed to the published version of the manuscript.

Funding: This study was supported by a grant from the National Science Centre of Poland (grant no. NCN 2013/11/B/NZ5/00118) and grants from the Medical University of Gdansk (grants no. ST40, ST4). The publication of the article was supported by the project POWR.03.02.00-00-I026/17-00, co-financed by the European Union through the European Social Fund under the Operational Programme Knowledge Education Development 2014–2020.

Acknowledgments: We are indebted to Julian Swierczynski for criticism and discussion of the manuscript.

Conflicts of Interest: The authors declare no conflict of interest. The funders had no role in the design of the study; in the collection, analyses, or interpretation of data; in the writing of the manuscript; or in the decision to publish the results.

Abbreviations

AA	Arachidonic acid
BMI	Body mass index
BUN	Blood urea nitrogen
CKD	Chronic kidney disease
CRP	C-reactive protein
CVD	Cardiovascular disease
DHA	Docosahexaenoic acid
DMEM	Dulbecco's modified Eagle's medium
FA	Fatty acid
FAME	Fatty acid methyl esters
FBS	Fetal bovine serum
GC-MS	Gas chromatography-mass spectrometry
HOMA-IR	Homeostatic model assessment of insulin resistance
IBMX	3-isobutyl-1-methylxanthine
MUFA	Monounsaturated fatty acid
OA	Oleic acid
PA	Palmitic acid
PUFA	Polyunsaturated fatty acid
SAT	Subcutaneous adipose tissue
SFA	Saturated fatty acid
TC	Total cholesterol
TG	Triacylglycerols

References

1. Zdrojewski, Ł.; Zdrojewski, T.; Rutkowski, M.; Bandosz, P.; Król, E.; Wyrzykowski, B.; Rutkowski, B. Prevalence of chronic kidney disease in a representative sample of the Polish population: Results of the NATPOL 2011 survey. *Nephrol. Dial. Transplant.* **2016**, *31*, 433–439. [CrossRef] [PubMed]
2. Foley, R.N.; Parfrey, P.S.; Sarnak, M.J. Clinical epidemiology of cardiovascular disease in chronic renal disease. *Am. J. Kidney Dis.* **1998**, *32*, S112–S119. [CrossRef] [PubMed]
3. Chang, Y.; Zhang, Y.; Youl Hyun, Y.; Ryu, S.; Choi, Y.; Cho, J.; Kwon, M.-J.; Lee, K.-B.; Kim, H.; Jung, H.-S.; et al. Metabolically healthy obesity and development of chronic kidney disease: A cohort study. *Ann. Intern. Med.* **2016**, *164*, 305–312. [CrossRef] [PubMed]
4. Spoto, B.; Pisano, A.; Zoccali, C. Insulin resistance in chronic kidney disease: A systematic review. *Am. J. Physiol. Physiol.* **2016**, *311*, F1087–F1108. [CrossRef]
5. Kovesdy, C.P.; Furth, S.L.; Zoccali, C. Obesity and Kidney Disease: Hidden Consequences of the Epidemic. *Am. J. Hypertens.* **2017**, *30*, 328–336. [CrossRef]
6. Yu, J.Z.; Kalantar-Zadeh, K.; Rhee, C.M. Adiponectin and Leptin in Kidney Disease Patients. In *Endocrine Disorders in Kidney Disease*; Springer International Publishing: Cham, Switzerland, 2019; pp. 277–290.
7. Marchelek-Mysliwiec, M.; Wisniewska, M.; Nowosiad-Magda, M.; Safranow, K.; Kwiatkowska, E.; Banach, B.; Dołegowska, B.; Dołegowska, K.; Stepniewska, J.; Domanski, L.; et al. Association Between Plasma Concentration of Klotho Protein, Osteocalcin, Leptin, Adiponectin, and Bone Mineral Density in Patients with Chronic Kidney Disease. *Horm. Metab. Res.* **2018**, *50*, 816–821. [CrossRef]
8. Canpolat, N.; Sever, L.; Agbas, A.; Tasdemir, M.; Oruc, C.; Ekmekci, O.B.; Caliskan, S. Leptin and ghrelin in chronic kidney disease: Their associations with protein-energy wasting. *Pediatr. Nephrol.* **2018**, *33*, 2113–2122. [CrossRef]
9. Noor, S.; Alam, F.; Fatima, S.S.; Khan, M.; Rehman, R. Role of Leptin and dyslipidemia in chronic kidney disease. *Pak. J. Pharm. Sci.* **2018**, *31*, 893–897.
10. Abella, V.; Scotece, M.; Conde, J.; Pino, J.; Gonzalez-Gay, M.A.; Gómez-Reino, J.J.; Mera, A.; Lago, F.; Gómez, R.; Gualillo, O. Leptin in the interplay of inflammation, metabolism and immune system disorders. *Nat. Rev. Rheumatol.* **2017**, *13*, 100–109. [CrossRef]
11. Friedman, J. The long road to leptin. *J. Clin. Invest.* **2016**, *126*, 4727–4734. [CrossRef]
12. Akchurin, O.M.; Kaskel, F. Update on inflammation in chronic kidney disease. *Blood Purif.* **2015**, *39*, 84–92. [CrossRef] [PubMed]
13. Ding, N.; Liu, B.; Song, J.; Bao, S.; Zhen, J.; Lv, Z.; Wang, R. Leptin promotes endothelial dysfunction in chronic kidney disease through AKT/GSK3β and β-catenin signals. *Biochem. Biophys. Res. Commun.* **2016**, *480*, 544–551. [CrossRef] [PubMed]
14. Mao, S.; Fang, L.; Liu, F.; Jiang, S.; Wu, L.; Zhang, J. Leptin and chronic kidney diseases. *J. Recept. Signal Transduct.* **2018**, *38*, 89–94. [CrossRef] [PubMed]
15. Lim, C.C.; Teo, B.W.; Tai, E.S.; Lim, S.C.; Chan, C.M.; Sethi, S.; Wong, T.Y.; Sabanayagam, C. Elevated serum leptin, adiponectin and leptin to adiponectin ratio is associated with chronic kidney disease in Asian adults. *PLoS ONE* **2015**, *10*, e0122009. [CrossRef] [PubMed]
16. Czumaj, A.; Śledziński, T.; Carrero, J.-J.; Stepnowski, P.; Sikorska-Wisniewska, M.; Chmielewski, M.; Mika, A. Alterations of fatty acid profile may contribute to dyslipidemia in chronic kidney disease by influencing hepatocyte metabolism. *Int. J. Mol. Sci.* **2019**, *20*, 2470. [CrossRef] [PubMed]
17. Szczuko, M.; Kaczkan, M.; Drozd, A.; Maciejewska, D.; Palma, J.; Owczarzak, A.; Marczuk, N.; Rutkowski, P.; Małgorzewicz, S. Comparison of fatty acid profiles in a group of female patients with Chronic Kidney Diseases (CKD) and Metabolic Syndrome (MetS)–similar trends of changes, different pathophysiology. *Int. J. Mol. Sci.* **2019**, *20*, 1719. [CrossRef] [PubMed]
18. Madsen, T.; Christensen, J.H.; Svensson, M.; Witt, P.M.; Toft, E.; Schmidt, E.B. Marine n-3 polyunsaturated fatty acids in patients with end-stage renal failure and in subjects without kidney disease: A comparative study. *J. Ren. Nutr.* **2011**, *21*, 169–175. [CrossRef]
19. Hager, M.R.; Narla, A.D.; Tannock, L.R. Dyslipidemia in patients with chronic kidney disease. *Rev. Endocr. Metab. Disord.* **2017**, *18*, 29–40. [CrossRef]
20. Mikolasevic, I.; Žutelija, M.; Mavrinac, V.; Orlic, L. Dyslipidemia in patients with chronic kidney disease: Etiology and management. *Int. J. Nephrol. Renovasc. Dis.* **2017**, *10*, 35–45. [CrossRef]

21. Moradi, H.; Vaziri, N.D. Molecular mechanisms of disorders of lipid metabolism in chronic kidney disease Publication Date. *Front. Biosci.* **2018**, *23*, 146–161.

22. Tsimihodimos, V.; Dounousi, E.; Siamopoulos, K.C. Dyslipidemia in chronic kidney disease: An approach to pathogenesis and treatment. *Am. J. Nephrol.* **2008**, *28*, 958–973. [CrossRef] [PubMed]

23. Hellström, L.; Wahrenberg, H.; Hruska, K.; Reynisdottir, S.; Arner, P. Mechanisms behind gender differences in circulating leptin levels. *J. Intern. Med.* **2000**, *247*, 457–462. [CrossRef] [PubMed]

24. Heimburger, O.; Lonnqvist, F.; Danielsson, A.; Nordenstrom, J.; Stenvinkel, P. Serum immunoreactive leptin concentration and its relation to the body fat content in chronic renal failure. *J. Am. Soc. Nephrol.* **1997**, *8*, 1423–1430. [PubMed]

25. Merabet, E.; Dagogo-Jack, S.; Coyne, D.W.; Klein, S.; Santiago, J.V.; Hmiel, S.P.; Landt, M. Increased plasma leptin concentration in end-stage renal disease. *J. Clin. Endocrinol. Metab.* **1997**, *82*, 847–850. [CrossRef] [PubMed]

26. Jiang, Y.; Zhang, J.; Yuan, Y.; Zha, X.; Xing, C.; Shen, C.; Shen, Z.; Qin, C.; Zeng, M.; Yang, G.; et al. Association of increased serum leptin with ameliorated anemia and malnutrition in stage 5 chronic kidney disease patients after parathyroidectomy. *Sci. Rep.* **2016**, *6*, 27918. [CrossRef] [PubMed]

27. Daschner, M.; Tönshoff, B.; Blum, W.F.; Englaro, P.; Wingen, A.M.; Schaefer, F.; Wühl, E.; Rascher, W.; Mehls, O. Inappropriate elevation of serum leptin levels in children with chronic renal failure. European Study Group for Nutritional Treatment of Chronic Renal Failure in Childhood. *J. Am. Soc. Nephrol.* **1998**, *9*, 1074–1079.

28. Odamaki, M.; Furuya, R.; Yoneyama, T.; Nishikino, M.; Hibi, I.; Miyaji, K.; Kumagai, H. Association of the serum leptin concentration with weight loss in chronic hemodialysis patients. *Am. J. Kidney Dis.* **1999**, *33*, 361–368. [CrossRef]

29. Diez, J.J.; Iglesias, P.; Fernandez-Reyes, M.J.; Aguilera, A.; Bajo, M.A.; Alvarez-Fidalgo, P.; Codoceo, R.; Selgas, R. Serum concentrations of leptin, adiponectin and resistin, and their relationship with cardiovascular disease in patients with end-stage renal disease. *Clin. Endocrinol.* **2005**, *62*, 242–249. [CrossRef]

30. Ines, M.; Silva, B.; Vale, B.S.; Lemos, C.C.S.; Torres, M.R.S.G.; Bregman, R. Body adiposity index assess body fat with high accuracy in nondialyzed chronic kidney disease patients. *Obesity* **2013**, *21*, 546–552.

31. Frühbeck, G.; Catalán, V.; Rodríguez, A.; Ramírez, B.; Becerril, S.; Salvador, J.; Portincasa, P.; Colina, I.; Gómez-Ambrosi, J. Involvement of the leptin-adiponectin axis in inflammation and oxidative stress in the metabolic syndrome. *Sci. Rep.* **2017**, *7*, 1–8. [CrossRef]

32. Yadav, V.K.; Oury, F.; Suda, N.; Liu, Z.W.; Gao, X.B.; Confavreux, C.; Klemenhagen, K.C.; Tanaka, K.F.; Gingrich, J.A.; Guo, X.E.; et al. A serotonin-dependent mechanism explains the leptin regulation of bone mass, appetite, and energy expenditure. *Cell* **2009**, *138*, 976–989. [CrossRef] [PubMed]

33. Nordfors, L.; Lönnqvist, F.; Heimbürger, O.; Danielsson, A.; Schalling, M.; Stenvinkel, P. Low leptin gene expression and hyperleptinemia in chronic renal failure. *Kidney Int.* **1998**, *54*, 1267–1275. [CrossRef] [PubMed]

34. Witasp, A.; Carrero, J.J.; Heimbürger, O.; Lindholm, B.; Hammarqvist, F.; Stenvinkel, P.; Nordfors, L. Increased expression of pro-inflammatory genes in abdominal subcutaneous fat in advanced chronic kidney disease patients. *J. Intern. Med.* **2011**, *269*, 410–419. [CrossRef] [PubMed]

35. Kwok, K.H.M.; Lam, K.S.L.; Xu, A. Heterogeneity of white adipose tissue: Molecular basis and clinical implications. *Exp. Mol. Med.* **2016**, *48*, e215. [CrossRef] [PubMed]

36. Lee, Y.-H.; Kim, S.-N.; Kwon, H.-J.; Granneman, J.G. Metabolic heterogeneity of activated beige/brite adipocytes in inguinal adipose tissue. *Sci. Rep.* **2017**, *7*, 39794. [CrossRef]

37. Lynes, M.D.; Tseng, Y.-H. Deciphering adipose tissue heterogeneity. *Ann. N. Y. Acad. Sci.* **2018**, *1411*, 5–20. [CrossRef]

38. Kowalska, K.; Olejnik, A.; Rychlik, J.; Grajek, W. Cranberries (Oxycoccus quadripetalus) inhibit lipid metabolism and modulate leptin and adiponectin secretion in 3T3-L1 adipocytes. *Food Chem.* **2015**, *185*, 383–388. [CrossRef]

39. Kuroda, M.; Tominaga, A.; Nakagawa, K.; Nishiguchi, M.; Sebe, M.; Miyatake, Y.; Kitamura, T.; Tsutsumi, R.; Harada, N.; Nakaya, Y.; et al. DNA methylation suppresses leptin gene in 3T3-L1 adipocytes. *PLoS ONE* **2016**, *11*, e0160532. [CrossRef]

40. Tsubai, T.; Noda, Y.; Ito, K.; Nakao, M.; Seino, Y.; Oiso, Y.; Hamada, Y. Insulin elevates leptin secretion and mRNA levels via cyclic AMP in 3T3-L1 adipocytes deprived of glucose. *Heliyon* **2016**, *2*, e00194. [CrossRef]

41. Jenks, M.Z.; Fairfield, H.E.; Johnson, E.C.; Morrison, R.F.; Muday, G.K. Sex steroid hormones regulate leptin transcript accumulation and protein secretion in 3T3-L1 cells. *Sci. Rep.* **2017**, *7*, 8232. [CrossRef]
42. Folch, J.; Lees, M.; SLOANE STANLEY, G.H. A simple method for the isolation and purification of total lipides from animal tissues. *J. Biol. Chem.* **1957**, *226*, 497–509. [PubMed]

 *metabolites*

Review

Lipidomics Issues on Human Positive ssRNA Virus Infection: An Update

David Balgoma [1], Luis Gil-de-Gómez [2] and Olimpio Montero [3,*]

[1] Analytical Pharmaceutical Chemistry, Department of Medicinal Chemistry, Uppsala University, Husarg. 3, 75123 Uppsala, Sweden; david.balgoma@ilk.uu.se

[2] Center of Childhood Cancer Center, Children's Hospital of Philadelphia, Colket Translational Research Center, 3501 Civic Center Blvd, Philadelphia, PA 19104, USA; gildegomel@email.chop.edu

[3] Spanish National Research Council (CSIC), Boecillo's Technological Park Bureau, Av. Francisco Vallés 8, 47151 Boecillo, Spain

* Correspondence: olimpio.montero@dicyl.csic.es; Tel.: +34-983548209

Received: 16 July 2020; Accepted: 27 August 2020; Published: 31 August 2020

Abstract: The pathogenic mechanisms underlying the Biology and Biochemistry of viral infections are known to depend on the lipid metabolism of infected cells. From a lipidomics viewpoint, there are a variety of mechanisms involving virus infection that encompass virus entry, the disturbance of host cell lipid metabolism, and the role played by diverse lipids in regard to the infection effectiveness. All these aspects have currently been tackled separately as independent issues and focused on the function of proteins. Here, we review the role of cholesterol and other lipids in ssRNA+ infection.

Keywords: lipidomics; ssRNA+ virus; membrane fusion; lipid metabolism; cholesterol; sphingolipids; phosphatidylinositol; SARS-CoV

---

1. Introduction

The ongoing COVID-19 pandemic is developing (July 2020) worldwide with devastating global consequences, both for social organization and healthcare systems. COVID-19 illness is brought about by infection with the severe acute respiratory syndrome coronavirus SARS-CoV-2 [1,2], which is an enveloped positive single-stranded RNA virus (ssRNA+) [3]. The most abundant studies related to human diseases induced by ssRNA-positive viruses refer to *Picornaviridae*, *Coronaviridae*, and *Flaviviridae* [4].

This impact in a short time span has brought the Biology and Biochemistry of viral infection mechanisms to reach momentum. The infection mechanisms have been described for diverse unrelated viral families [5], with the majority of them being DNA viruses. Within *Picornaviridae*, *Coronaviridae*, and *Flaviviridae*, Rhino and Poliovirus (Picornaviridae), SARS-CoV, Middle East Respiratory Syndrome Coronavirus (MERS-CoV), Hepatitis C virus (HCV), West Nile virus (WNV) and Dengue virus (DENV) fall within the viruses whose life cycle biology is better known. Nonetheless, knowledge regarding virus entry mechanisms and other related features of the virus life cycle has been gained from the research on the influenza virus from the *Orthomysoviridae* family and the human immunodeficiency virus from the *Retroviridae* family. Consequently, these and other unrelated viruses will be also considered in this review from the point of view of the different aspects that affect the lipidomics of the viral infection.

All ssRNA+ viruses initially infect mammal cells through the interaction of virus proteins with any given host cell protein. Further fusion of the virus and host cell membranes is required for the viral genetic material to get into the cell. Once inside the cell, the genomic and subgenomic viral RNAs are translated into the virus proteins; these then lead the virus replication, which is a process that involves modulation of the host cell lipid metabolism [3,5,6]. Consequently, along with other features, current lipid studies about the aforementioned virus infection focus their research on membrane fusion and modulation of the lipid metabolism of the host cell. These two processes are considered separated disciplines of the infection.

The fight against the virus infection encompasses primarily the inhibition of the binding of the viral spike protein to the host cell's receptor protein. Consequently, most of the current research focuses on the role played by viral proteins but the lipid environment, where the proteins carry out their function and regulation, is considered secondarily [7]. Nevertheless, improving the knowledge on how the lipids are involved in the mechanisms of infection may provide clues to develop treatments and better counteract the virus-induced pathology [3]. To fill this gap, here, we review the main aspects regarding the lipidome regulation of the viral infection mechanism by ssRNA+ viruses.

2. Virus Entry: Lipid Rafts and Membrane Domains

2.1. Membrane Mechanical Properties Required for Virus Infection

The initial step in virus infection is the binding of any viral structural glycoprotein to a receptor of the host cell. The spike protein accounts for such function in coronaviruses (CoVs) and other enveloped viruses. After the virus is attached to the host cell protein, the process of membrane fusion starts to get the viral genome into the host cell. This process implies viral envelope and host cell membrane fusion, for which an energetically cost-effective barrier must be overcome. For example, in coronavirus, membrane fusion is driven by the fusion peptide (class I), which is localized within the spike protein (S protein) and becomes active after cleavage of the S protein at specific sites by host proteases or pH-dependent mechanisms [4,6,8]. A different mechanism of attachment and endocytosis drives the virus entry in the case of HCV. This mechanism is more complex than that of coronaviruses and involves interaction of the virus envelope E1 and E2 proteins (class II fusion loop) with several host cell proteins [9–11]. However, a membrane fusion-driven pore is also required in HCV to deliver the viral genetic material into the host cell cytoplasm.

Two main mechanisms of membrane fusion have been described: viral endocytosis by host cell membrane (endocytic pathway), and both viral and host cell plasma membrane fusion (non-endocytic pathway). After docking of the virus to the attachment factor or the receptor on the host cell surface, the virus may internalize its genomic material or the entire particle [12–14]. The non-endocytic pathway encompasses the direct delivery of the genetic material through a pore formed in the cell membrane by the induction of viral proteins at neutral pH. This pathway is typical of non-enveloped viruses. The endocytic pathway is more complex and harnesses the host cell endocytosis machinery for the virus internalization. Three main ways have been described in the endocytic pathway, namely: the clathrin-mediated endocytosis (CME), the caveolae-mediated endocytosis (CavME), and the macropycnocytosis. The best-known endocytic mechanism is the clathrin-mediated endocytosis. The CME is used by small to intermediate-sized viruses. This mechanism uses vesicles coated by the protein clathrin, which forms a polyhedral lattice that surrounds the cell membrane-derived vesicle where the virus is internalized into the cell cytoplasm through the early endosomes. Clathrin coating is coordinated by the adaptor protein (AP-2) and other adaptors; it is less commonly AP-independent. The protein dynamin is involved in regulating the clathrin-coated vesicle (CCV) formation as well as its scission from the membrane. Some viruses proceed to membrane fusion at this stage for releasing their genome into the cytoplasm. The early endosomes have a pH of about 6.0 to 6.5; therefore, it is considered that membrane fusion is not strictly pH-dependent. Other viruses need a lower pH for the membrane fusion to be effective; thus, it is considered pH-dependent. A further step leading to endosome maturation to become late endosomes with a pH of about 5 has to proceed before the membrane fusion takes place and the genetic material is delivered to the cytoplasm. Sequential acidification of the virus proteins from the early to late endosomes has also been suggested through the self-organized endosomal network. Maturation of the early endosomes to late endosomes and trafficking between them is controlled by the Rab proteins, which are members of the Ras superfamily of small G proteins. Subsets of

Rab proteins differ between the early and late endosomes, and the Rab subset change is accompanied for by formation of the phosphoinositide PI(3,5)P$_2$ from the precursor PI(3)P. Regarding lipid composition, early endosome membrane lipids are primarily composed of unsaturated and short alkyl chains, whereas long and saturated alkyl chains, such as in gangliosides, are predominant in the membrane lipids of late endosmes. Membrane fusion in some viruses requires a further step in which late endosomes are fused with lysosomes, this step giving rise to the late endosome/lysosome pathway. Cholesterol depletion driven by its synthesis inhibition or extracting agents as methyl-β-cyclodextran (MβCD) is used to assess whether the virus entry takes place through the caveolae/raft endocytosis. This pathway in less known and encompasses the formation of initial endocytic vesicles enriched in cholesterol from lipid-rafts, with complex signaling routes that involve the activity of tyrosine kinases and phosphatases. Thereafter, the cargo is transported to the endoplasmic reticulum (ER) through early and late endosomes. Most of the viruses using this endocytic pathway have different gangliosides as receptors, mainly GM1, which has a high concentration in caveolae. Polyomavirus, which are non-enveloped DNA viruses that replicate in the nucleus, use preferently this endocytic pathway, but picornaviruses and the coronavirus HCoV-229E have also been reported to internalize through the Caveolae-mediated endocytosis [15]. Macropinocytosis is a phagocytic-like mechanism of virus entry that is currently utilized by the cell to internalized fluids; it is dependent on actin and implies the actin cytoskeleton rearrangement to enable internalization of the virus particle [14]. Macropinocytic vacuoles (macropinosomes) are formed after membrane ruffles fold to reach at its end the membrane again, and the vacuole is closed through self-membrane fusion. These vacuoles containing the viral particle may traffick afterwards through the early and late endosome network. Macropinocytosis is common to large-sized viruses. However, recent work [16] has shown that Ebola virus (EBOV) may use a macropinocytosis-like process to entry the host cell in a clathrin, caveolae, and dynamin-independent manner, but dependent of actin and a lipid raft. Conversely, this virus may use as well an endocytic pathway that is dependent on clathrin, caveolae, and dynamin. Which endocytic route is used by this virus depends on the host cell type. Description of the current methodologies used to study the entry route by viruses can be found in reference [14].

Some viruses may use different entry mechanisms, this feature being likely dependent upon the membrane lipid composition of the host cell they infect as well as the particular cell surface factor attachment used. CME is the entry route currently used by HCV, HIV-1, EBOV, rotaviruses, and some coronaviruses, even though other routes can also be used as for EBOV (see above). A reaction between clathrin and actin seems to be necessary for the effective entry of these viruses. Regulation by microtubules of the CME has been reported for flaviviruses. DENV, WNV, and Semliki Forest Virus (SFV, Alphavirus family, *Togaviridae*) have been found to depend on early endosomes (Rab5 protein marker) for entry but not late endosomes (Rab7 protein marker), which means that they do not have strict low pH requirements or depend on different acidification mechanisms for membrane fusion. Conversely, influenza avian virus (IAV) needs both early and late endosomes to entry, thus reflecting low pH dependence for membrane fusion. Marburg virus (MARV) may use for internalization a CME through the endo/lysosomal pathway. Coronaviruses differ in their internalization mechanism among strains. Thus, while HCoV-229E is known to use the Cav-ME route, SARS-CoVs use an endocytic pathway that is clathrin- and caveolae-independent but receptor and pH-sensitive, with lipid rafts playing an essential role [17]. This endocytic mechanism implies internalization of the receptor protein angiotensin-convering enzyme 2 (ACE2) along with the spike protein into the early endosomes, but the receptor is afterward recycled to the membrane via lysosomes. Nonetheless, previous studies showed that SARS-CoV could enter through a pH-independent direct membrane fusion as it could infect cells that do not express ACE2, such as enterocytes and hepatocytes [18]. Recent research on the virus SARS-CoV-2 points to pH-independent direct cell and viral membrane fusion, which is a process that is driven by the subunit S2 of the spike protein after cleavage by the cellular serine protease TMPRSS2 [19]. On the contrary, the infectious bronchitis virus (IBV), a gamma-coronavirus, was reported to use the CME pathway to entry, with vesicle scission being mediated by GTPase dynamin 1, and a dependence on low pH and lipid raft localization of the receptor. Tracking of the virus trip inside the cell was followed by using diverse inhibitors, cholesterol sequestering agents, and virus particles labeled with fluorescent markers. Membrane fusion takes place at the late endosome/lysosome step of the endocytic pathway, with deep rearrangement of the host cell cytoskeleton being induced by the endosomal viral cargo [15]. Accordingly, viruses may sequester on their own profit the diverse endocytic pathways that are currently used by the host cell, but variability of the proteins and even the general mechanisms may also exist as a consequence of virus specifity.

Membrane fusion has been described to proceed through the catalytic action of three different types of fusion peptides or fusion loops of class I, II, or III. These proteins afford the free energy necessary to overcome through conformational changes the kinetic barrier due to repulsive hydration strength. Most of the knowledge on the viral and host membrane fusion has been gained from the influenza virus and its type I fusion peptide hemagglutinin. A detailed description of the three fusion peptide-guided mechanisms involved in membrane fusion has been previously reviewed in [20–22]. Bringing the viral and the host membranes closer enough (c.a. 20 Å) for inducing the membrane fusion is a process that entails membrane curvature and changes in the lipid bilayer phase. They are driven by the insertion of a hydrophobic region of the fusion peptide, which requires dehydration of the inter-membrane space. Nonetheless, from experiments with no-protein fusogens, such as polyethilen glycol, it seems that membrane curvature stabilization is not a key player in membrane pore opening. The calculated displacement of lipids in the outer leaflet of the host membrane accounts for no more than 10% of the membrane area (about 3500 Å^2), which does not represent a substantial energetic demand [21]. This energetic burden has been demonstrated to be afforded by the cooperation of three fusion peptides in influenza virus membrane fusion [23], whereas two adjacent trimers of the fusion protein are required in West Nile virus [24]. This result points to the fact that the viral membrane curvature may not actually impose a constraint for proceeding to the hemifusion step and the formation of a steep curvature stalk, where the outer leafleats are merged. By the mesurement of electron density profiles through X-Ray reflectivity in stalks formed from bilayers in a lamellar state with different lipid compositions, Aeffner et al. [25] determined that the inter-bilayer separation should attain 9.0 ± 0.5 Å in order to facilitate dehydration and promote stalk formation. These authors also found that increasing the relative proportion of nonbilayer-forming, cone-shaped lipids, such as glycerophosphoethanolamine or cholesterol, favored the stalk formation by reducing the hydration energy barrier and, possibly, by contributing with their intrinsic negative curvature. As well, the energy required for dehydration was, in this study, found to decrease with the length of the acyl chains of the glycerophospholipids. However, the hemifusion stalk stage was not detected by Gui et al. [26] using fluorescence and electron microscopy. The results of this study show that such a stage might be an unstable intermediate that is quickly resolved toward the postfusion stage. Contrarily, localized point-like contacts were abundantly visualized in this study, where the dimples formed in the target membrane, about 5 nm wide, were drawn toward the virus surface. They were able to detect up to well-resolved four types of virus–target membrane contacts at pH 5.5 and 5.25 using liposomes of dioleylglycerophosphocholine, DOPC, with 20% cholesterol. At the lowest pH, a tight contact of the two membranes through an extended length of about 100 nm (catalogued by the authors as type III) was the predominant interaction, whose abundance was increased by about 3-fold in cholesterol-containing liposomes in comparison to only DOPC liposomes.

Using synthetic peptides that resemble the fusion peptide hemagglutinin and electron spin resonance (ESR), Ge and Freed [27] found that the most relevant effect of the synthetic fusion peptides was the induction of highly ordered membrane domains, which came motivated by virtue of electrostatic interactions between the peptide and negatively charged phospholipid headgroups. A similar effect was reported for two putative fusion peptides enclosed in the spike glycoprotein of SARS-CoV-1. It was found in this study that the inner water content in the lipid bilayer was dropped by the insertion of the fusion peptide as a consequence of increased lipid packing, but only in membranes containing negatively charged lipids, whereas the water content was only slightly altered in zwitterionic dipalmitoylglycerophosphocholine (DPPC) liposomes [28]. Additionally, the fusion peptides created opposing curvature stresses in the highly bended membranes containing nonbilayer-forming phospholipids. However, previous studies had pointed out that interaction with the lipid headgroups is not an essential factor in reaching the membrane hemifusion state [21,29]. In SARS-CoV, the possibility of existing two fusion peptides that act in coordination has been suggested [7]; one of the peptides would promote the dehydration process, while the other one would act in modifying/disturbing the lipid organization within the target membrane [26,28,30]. Hence, the catalytic role of the fusion peptide(s) is likely to tackle three properties of the target membrane in the virus entry machinery: (i) dehydration of the intermembrane space for the fusing membranes coming into the required proximity, (ii) to promote negative curvature to form the hemifusion stalk, and (iii) to alter the lipid packing density, which will be generated in the highly curved local dimples of the stalk [22,28]. The effectiveness of these three processes is likely to depend upon the membrane lipid composition. Further research is devoted to this issue, and new clues are expected to come from electron and fluorescence microscopy [31].

2.2. Raft Lipids Related to Virus Entry

Since the dominant phospholipid in the outer leaflet of most membranes is the bilayer-forming, positive charged diacylglycerophosphosphocholine (PC), the idea was raised that the viral docking to the receptor on the target cell and, consequently, the membrane fusion were likely to take place at specific microdomains with particular lipid composition, the so-called lipid rafts [27,32–36]. A special characteristic of the lipid rafts is the high content of cholesterol [37–39]. Even though a high content of sphingolipids and gangliosides is also a defining characteristic of lipid rafts (Figure 1), direct in vivo visualization still remains unresolved [39].

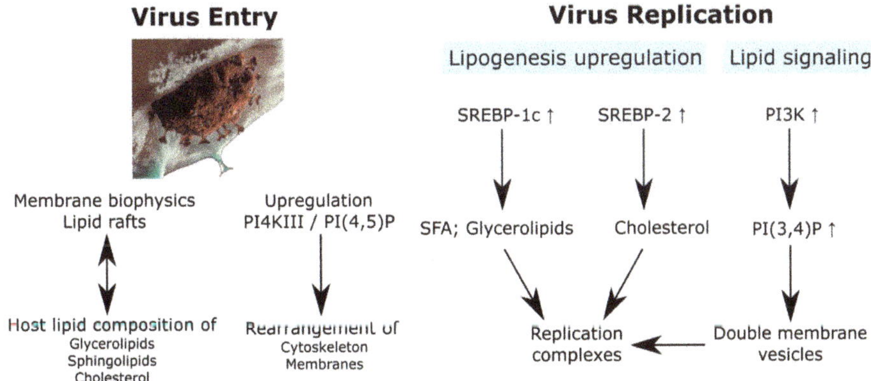

Figure 1. Relationship between the virus entry and replication with the lipidome. SREBP, sterol regulatory element binding protein; SFA, saturated fatty acid. SARS-CoV-2 artwork was modified from a work from *We Are Covert*, who allows anyone to use it for any purpose including unrestricted redistribution, commercial use, and modification.

An unexplored possibility is that rafts do not have a permanent localized existence, but they arise under the induction of certain proteins such as the hydrophobic insert of the viral fusion peptide or the fusion loop. This fact might be also responsible for bringing negatively charged lipids from the inner leaflet of the bilayer to its outer leaflet by flip-flop mechanisms. This hypothesis would explain the promotion of virus entry by the interaction of the fusion peptide with the negatively charged phospholipid headgroups [25,27] as well as the kinetics of the membrane fusion [25]. A number of studies have shown that the hemifusion step and pore widening are sped up after increasing the relative concentration of cholesterol in the bilayer composition, whereas either the depletion of cholesterol in the cell culture medium or the inhibition of cholesterol synthesis by statins was able to halt the viral infection at the virus entry step [26–28,40,41]. The effect of cholesterol on promoting membrane merging has also been observed for Bis-(monoacylglycero)-phosphate (BMP) [26]. This particular phospholipid was shown to be strictly necessary for Dengue virus (DENV) entry even at low endosomal pH [42]. As pointed out above, the exact role played by cholesterol is not known in detail, but its intrinsic negative curvature seems to be an essential characteristic in promoting the stalk formation during virus entry. However, a recent study shows that the cholesterol action is likely to involve a direct influence on the oligomeric state of the fusion peptide after insertion into the host cell membrane, as well as on the effects of the fusion peptide on the membrane reorganization and dynamics [43]. In another recent study, a new lipid-label-free methodology was used to measure the kinetics of influenza virus infection [44]. According to the results of this study, cholesterol is able to augment the efficiency of membrane fusion in a receptor binding-independent manner. Nevertheless, the rate of membrane fusion was not altered. These results led the authors to conclude that the positive effect of cholesterol in membrane lipid mixing is related to its capability to induce negative curvature. Since membrane mixing was achieved in this latter study without binding of the spike protein of the influenza virus to the host cell receptor, the catalytic effect of the fusion peptide might proceed in an independent way in this virus. Cleavage of the spike protein in SARS-CoV-1 does not seem to be also necessary for the fusion peptide to become fusogenic, but rearrangement of disulfide bridges in the S1 peptide after receptor binding are likely involved in the conformational changes driving the fusion mechanism [43,45]. Contrary to these latter results, which point to the fact that membrane fusion is independent of viral protein attachment to its receptor, Guo et al. reported lipid raft-dependent viral protein binding with the suppression of viral infection if the lipid rafts were disrupted with cholesterol drug-induced depletion; lipid rafts, as recognized by the caveolin-1 marker, were the membrane domain where structural proteins of the infectious bronchitis virus (IBV) co-localized but the nonstructural proteins did not [35]. The question regarding whether the lipid-raft domains may serve as platforms to concentrate the proteins required for viral entry and, even though some evidence exists, to activate signaling pathways inside the host cell still remains unsolved.

Sphingomyelins (SMs) are also common lipids found in lipid rafts, which contribute to make these membrane microdomains detergent-resistant [34]. The structure of a representative of this lipid class is illustrated in Figure 2. The ganglioside GM1, a sphingolipid, is used as a marker of lipid rafts [34]. Sphingolipids (SLs) promote to an extent higher than Chol the liquid-ordered phase in the outer leaflet of the membrane bilayer because of the long saturated acyl chains they currently contain (the R group in Figure 2 may extend to a length of up to 26 C), in addition to their capability to form intermolecular hydrogen bonds [46]. A relevant function of the lipid rafts has been suggested to be the connection between the events outside the cell with the pathways inside the cell, thus acting as 'signaling platforms'. With the aim of this function to be properly accomplished, the lipid rafts would act as concentrators of specific transmembrane proteins, mainly receptors, whose compatibility with the membrane phase would determine their selectivity. Thus, SLs would account for a role in connecting the outer leaflet with the inner leaflet through their long saturated acyl chains. Regarding virus entry, research has been primarily focused toward the role played by cholesterol, but a number of studies have also enlightened the SM influence on this early step of viral infection. The displacement of cholesterol by SMs and the other way round has been demonstrated, with the bilayer liquid-ordered phase being preferentially determined by the interaction between SM and cholesterol. This interaction would be controlled to a certain extent by the intracellular actin meshwork, which would also be responsible for the compartmentalization of the membrane into lipid-specific domains [47]. Furthermore, the actin role is possibly extended to the routing of the viral genomic material toward the replication place inside the host cell. The hydrolysis of SM by sphingomyelinases to render the corresponding ceramide in specific membrane domains is proposed to regulate the dynamics of cholesterol in the cell membrane, the effect of such regulation being the progressive disassembly of cholesterol from the liquid-ordered phase and its displacement. Since the interaction of ceramides with cholesterol has been suggested to be an apoptotic regulator, it can be expected that viral proteins would act in recruiting cholesterol to displace the ceramide and to avoid the programmed cell death. This fact is added to the other characteristics conferred by cholesterol to the membrane mechanical properties discussed above. To study the influence of ceramide on membrane fusion during Semliki Forest Virus (SFV, Alphavirus family, *Togaviridae*) infection, ceramide analogs have been used [48]. According to this experiment, in which cholesterol-containing PC plus PE liposomes were used, the roles played by the 3-hydroxyl group and the 4,5-*trans* carbon-carbon double bond of the sphingosine backbone (Figure 2) were found to be essential in the fusion process. In additon, ceramide was the simplest SL to accomplish this significant contribution in mediating the fusion, independently of the length of the acyl chain. More recently, a Ca^{2+}-dependent pathway of infection by the Rubella virus (RuV, Rubivirus family, *Togaviridae*) was demonstrated to proceed through direct binding of the fusion loop in the viral E1 protein to SM/cholesterol-enriched membranes [49]. However, the treatment of host cells with sphingomyelinase proved that SM is exclusively required for viral entry but is not required for the further steps of viral replication. SM in the host cell membrane and acid sphingomyelinase (ASMase) activity have also been shown to be required by the Ebola virus (EBOV), a negative single-stranded RNA virus belonging to the *Filoviridae* family, to get into the host cell. The ASMase activity renders ceramide that provoques raft enlargement and membrane invagination [50]. This study also showed that the virus was able to recruit both SM and ASMase to the raft where the viral attachment was happening. Conversely, Bovine herpesvirus 1 (BoHV-1, *Herpesviridae* family) seems to require SM in the virus envelope but does not in the host cell [51]. The role played by ceramides is contradictory as they may enhance or inhibit virus replication, but this SL action seems to be related to the viral replication phase rather than to the internalization phase [52–54]. In virus using the endocytic pathway, similar to the influenza virus or the Ebola virus, it has been shown that activity of glucosylceramidase (GBA) is required for viral entry and membrane fusion through the regulation of endocytosis, but in a virus-dependent manner. It was also shown that trafficking of the epidermal growth factor (EGF) to late endosomes was impaired in GBA-knockout cells, which negatively affects the virus entry through spoiling the endocytic pathway [55]. Indeed, co-clustering of the HA attachment factor and EGF in submicrometer domains that overlap partially has been reported recently [56]. Accordingly, there is evidence that SLs have a function in enveloped ssRNA viruses at the early stage of infection that accounts for the viral entry modulation, but further research is still necessary to unveil the exact mechanisms of SL reactions.

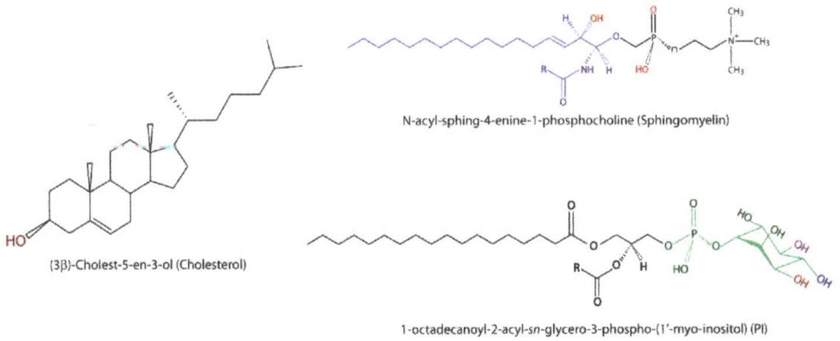

Figure 2. Structure of the most relevant lipids in virus infection is illustrated. Hydroxyl (HO) and oxygen (O) atoms potentially involved in the interaction with the fusion peptide or fusion loop are marked in red in cholesterol and sphingomyelin. The basic ceramide structure is marked in blue in the sphigomyelin structure. In phosphatidylinositol (PI), the hydroxyl groups that can be esterified with phosphate at the positions 3, 4, and 5 of the myo-inositol group to render PIP (PI3P or PI4P), PIP2 (PI(3,4)P or PI(4,5)P), and PIP3 (PI(3,4,5)P), which are marked in red, blue, and violet, respectively, are shown.

Some CoVs (HCoV-OC43 and HCoVHKU1), as well as influenza A virus (whose fusion loop is hemagglutinin, HA) and other non-related viruses (i.e., non-enveloped simian virus 40 SV-40, of polyomavirus family), use the sialoglycan moiety (9-*O*-acetyl-sialic acid) of gangliosides or glycoproteins located in membrane lipid rafts as receptors for the spike protein. The amino acid Trp90 in the domain A of the HCoV-OC43 S protein was shown to be essential for receptor binding. However, despite the fact that binding to 9-*O*-acetyl-sialic acid is required for membrane fusion, further interaction of the virus protein with other host membrane sialoglycans or proteins is also necessary to induce the conformational changes leading to membrane fusion [57,58]. Conversely, formation of the complex SV40 protein with the host cell ganglioside GM1 was found to be enough to induce the membrane curvature and invaginations required for membrane fusion [59].

As already discussed above, some studies have depicted the possibility that interaction of the fusion peptide or fusion loop with negatively charged phospholipids on the host membrane might be required for an efficient membrane fusion [25]. In this regard, phosphatidylserine (PS) contained in the virus envelope has been demonstrated to serve after externalization as a virus co-receptor through the T cell immunoglobulin mucin domain 1 (TIM-1) receptor in EBOV and other viruses, even in an indispensable fashion [60–63]. In the study of Nanbo et al. [63], flipping of PS from the inner leaflet to the outer leaflet of the cell membrane for virion adquisition and incorporation to its envelope is proposed as a previous step to TIM1 binding. In herpes simplex virus (HSV), phospholipid scramblase-1 (PLSCR1), after activation by HSV exposure, flips both PS and Akt to the outside of the membrane in a Ca^{2+}-dependent mechanism. PS is restored to the inner leaflet 2 to 4 h after infection to avoid apoptotic triggering [62], suggesting a different role for PS in relation to the TIM-1 PS receptor. However, the function of TIM-1 as an essential receptor for HAV has been disputed [64] due to the finding that quasi-enveloped HA virions (eHAV) were able to infect TIM1-knockout Vero cells to a similar extent to naked HAV. Hence, the authors proposed TIM1 to be an accessory attachment factor by binding PS on the HAV envelope rather than an essential virus protein receptor. In spite of these contradictory data, PS seems to act in any way in virus attachment and entry in certain virus families, at least contributing to the process efficiency, but the exact role may depend on every virus or it may be complementary to other factors.

A phospholipid currently associated to the inner leaflet of the lipid rafts is phosphatidylinositol (PI), which is a negatively charged phospholipid with important and versatil signaling functions (Figure 2) [65,66]. Abundant data suggest that a derivative of PI, the phosphatidylinositol 4,5-biphosphate (PIP2), accumulates preferently in liquid-disordered phases (L_d) [7], where the cholesterol content is presumed to be low, and interplays with PS, which is rather localized in liquid-ordered phases (L_o). PIs play an essential role also in endosome maturation, which is a requisite for efficient virus infection of those using the endosomal pathway [56,66]. During HIV infection, PIP2 has been proposed to coordinate the actin cytoskeleton changes required for efficient virus entry in CD4+ T cells [67]; after virus attachment to the host cell receptor, PIP2 is recruited to the binding membrane microdomain, and in this way, PIP2 controls the protein reactions, leading to actin polymerization. As well in HIV-1, the requisite of PIP2 accumulation for the virus Gag protein to be properly anchored and stabilized in the inner leaflet of the cell plasma membrane has been pointed out [68,69]. Two isoforms, α and γ, of the phosphatidylinositol-4-phosphate 5-kinase family type 1 (PIP5K1) have recently been shown to participate in Gag stabilization by PIP2 through targeting the Gag precursor Pr55^{Gag} to the cell plasma membrane [70]. As commented above, interaction with the headgroup of negatively charged phospholipids such as PS or PI may also contribute to the dehydration process in the formation of the hemifusion stalk, with this contribution happening by promotion of the inverted hexagonal phase in the lipid bilayer and binding of Ca^{2+} [25]. In in vitro experiments with COS-7 cells and multilamellar vesicles (MLVs), unspecific binding of the Marburg virus (MARV) mVP40 protein to PIP, PIP2, and even PIP3 species present in the MLVs, both in the presence or absence of PS, has been reported. In this study, it was also found that with increasing PS concentration, the association of mVP40 to MLVs rose up to a threshold. Furthermore, the addition of sphingosine with the aim to reduce the negative charge load in the inner leaftet of the COS-7 cells led to a decrease in the binding level. These facts suggest that the electronic density, rather than the specific lipid species, is a determinant factor for binding [70]. Activation of the PI3K pathway for signaling is one of the most relevant features taking place for both entry and budding during infection by a number of viruses [58,71–73]. PI3K converts PIP2 into phosphatidylinositol 3,4,5-triphosphate (PIP3). In addition to stabilizing proteins or serving as a binding factor, PIP2 has been shown to collaborate with Akt through the signaling pathway PI3K/Akt on avoiding apoptotic events, and in this way, keeping the host cell metabolically active for virus replication and budding [71–73].

All these results clearly bring evidence that the lipid environment surrounding proteins involved in virus infection has a relevant function in the virus entry mechanism. Different lipids are essential for virus docking to the cell receptor either serving directly as (co)-receptors or providing the appropriate environment (lipid rafts) for the necessary reactions (e.g., membrane curvature). In addition, the virus, through specific protein conformational changes, takes advantage of several cell signaling pathways controlled by diverse membrane lipids. This process allows the virus to govern the cell metabolism following endocytosis of the viral genetic molecules.

3. Lipid Regulation in Virus Replication: Viral Factories

After the virus or its genome gets inside the infected cell, ssRNA+ viruses and other enveloped ones that replicate in the cytoplasm manage the cell metabolism to develop the replication scaffold, this membrane structure bolstering the so-called 'virus factory' [5,58,74–80]. There is consensus on that the functions of these structures are (i) to compartmentalize the diverse processes involved in viral genome replication, its envelopment, and structural protein assembly; (ii) to increase virion concentration during budding before infecting naïve cells; and (iii) to create a protected environment to escape the innate immune recognition of the viral components. Virus replication imposes an extra-energetic expenditure to the cell metabolism. Hence, cell central metabolism is orchestated by viral proteins to redirect toward the generation of enough energy and metabolites that are required for virus replication. In particular, building the scaffold demands a high rate of new lipid synthesis. Therefore, the lipid metabolism is hijacked by the virus proteins for the *de novo* synthesis of fatty acids in order to generate the scaffold membranes, the replication complexes (RCs), as well as for energy production in the β-oxidation pathway in the mitochondria. Concurrently, the cell metabolism needs to be kept above a threshold level to avoid exhaustion of the host cell. Full understanding of the mechanisms and related factors involved in virus–host interaction is a requisite for developing efficient antiviral infection therapies.

3.1. Viral Replication Complexes

The scaffold structure raised for building the viral factory varies between different virus in their morphology and possibly lipid composition. Flaviviruses develop a so-called 'membranous network' (MN) in a spherule/invagination type, while coronavirus does through a quarter-like type delimited by 'double membrane vesicles' (DMVs). Nonetheless, HCV (*Flaviviridae*) uses DMVs instead [77]; hence, this morphological separation may have exceptions or be somewhat diffuse. An extended review of the different virus family-related morphologies of the MNs as well as diverse factors influencing their formation can be found in [22,75,76]. It should be remarked that the exact lipid composition of the RCs' membranes is not known in detail yet, although there is evidence that their lipid profile differs from that of the organelles from which they are generated. The enrichment of typical lipids such as cholesterol, sphingomyelins, and glycosphingolipids in the lipid rafts seems to be a common feature of these MNs. The RCs' membranes may be originated from the endoplasmic reticulum (ER) in the perinuclear area, as for example in SARS-CoV and *Faviviridae* [75,78,80], from the Golgi, giving rise to cytopathic vesicles (CPVs) as in *Togaviridae* and *Picornaviridae* [75], from mitochondria (*Nodaviridae*) [79], or from the cell plasma membrane (CPVs in Alphaviruses) [75]. However, vesicle trafficking between the ER and the Golgi organelles may contribute to an undefinition in this regard. MNs, and in particular DMVs, are connected to the cytosol through a pore, which is believed to serve as the gate to the replication scaffold for the requiered metabolites, in particular nucleotides. This pore-mediated gate has not been detected up to date in SARS-CoV's DMVs, which raises the concern of how the required metabolites get inside the RCs. There is evidence from a number of studies that DMVs are the site of replication, but it has also been shown that DMVs can be developed irrespective of whether RNA replication takes place by the sole action of the viral proteins, at least for HCV [81,82]. Viral nonstructural proteins nsp3, nsp4, and nsp6 are involved in DMV development in SARS-CoV-1 in a time-dependent manner and correlating with RNA replication. Timecourse events have been shown to run with the initial formation of single membrane vesicles (SMVs) during the first 2–4 h after cell infection. These futher evolve to DMVs 16 h after infection, and they ultimately turn into multimembraneous vesicles (MMVs) close to the *cis*-Golgi at the budding stage 36–48 h after infection, this latter transformation being coincident with the formation of vesicle packets [75,78,79,83]. In HCV, NS5A seems to be enough for DMV formation, but the collaboration of NS3-5B is required for completing efficient DMVs, whereas NS4B is likely responsible for inducing the formation of SMVs [77,80,82]. Even though particular hints can be likely associated to every particular virus, there are common features shared by all ssRNA+ viruses regarding RCs' structure and buildup.

3.2. Lipid-Related Host Factors Associated to the RCs' Buildup

Enveloped viruses such as ssRNA+ viruses have a membrane lipid whose profile is different to that of the original organelle membrane when the envelope is created. Since the viral membrane is known to be enriched in cholesterol, sphingolipids, and phospholipids with saturated acyl chains, the DMV is believed to be also primarily composed of such classes of lipids. An unusual sphingolipid, dehydrosphingomyelin, along with PS and plasmalogens of PE were reported in the HIV envelope [84]. A role for sphingomyelin-to-ceramide conversion has been proposed in WNV budding, as its envelope was found to be highly enriched in sphingomyelin [85]. More recently, using multi-color super-resolution microscopy and mass spectrometry analysis, a substantial increase in PIP2 (from 11% to 51%) and PIP3 (from 0.01% to 0.13%) was reported in the HIV membrane as compared with the plasma membrane of the host cell [69]; this fact is related to the recruitment of Gag protein for efficient membrane fusion as aforementioned (Figure 1).

However, the most striking and known lipid-related factor associated to the MNs' development is the PI4KIII signaling pathway. The PI4Pα isoform, which is mainly expressed in the ER, has been shown to be a key factor for HCV replication, whereas the PI4KIIIβ is found in the Golgi and is required by Picornaviruses and some HCV strains [75]. This enzyme interacts with the viral protein NS5A, and disrupting this interaction prevents virus replication. The product of the PI4K enzyme is PIP4; enrichment in this PI has been shown to act in different processes regarding virus replication: membrane curvature, directly or indirectly through recluting cholesterol [86], glycosphingolipid transport to the RCs by the action of the FAPP2 protein [87], and protein concentration. However, conversely to these studies, it has been shown that currently used inhibitors of PI4KIIIα, enviroxime and BF738735, actually exert their inhibition against PI3K [88]. Thus, this result points out a genomic dependence on the PI kinases in HCV; otherwise, the action on PI3K is required only at the entry stage (see above). Enviroxime-like inhibitors have been shown to halt enterovirus replication through the action against PI4Kβ [89]. The *de novo* lipid synthesis has also been evidenced for WNV, from the *Flaviviridae* family as HCV, to proceed in a PI4P-independent fashion and, concurrently, it is not related to PI4KIII signaling [90]. There is no clear evidence on the fact that the PI4K signaling pathway has a relevant function in MNs' development. Hence, while PI4KIIIβ was shown to be important for SARS-CoV's DMV formation [91], another study did not find its metabolite, PI4P, within the host factors involved in SARS-CoV replication, and the authors attribute to PI4P a function rather in virus entry. However, the authors of this latter study acknowledge that siRNA methodology may provide false negatives [92,93]. Since DMVs are not common in healthy cells but they can be observed during authophagy, it has been suggested that SARS-CoV and other coronaviruses use the autophagy pathway for development of the DMVs; indeed, it has been shown that nsp6 in MHV or the equivalent nsp5-7 in arteriviruses, which hits the ER, can activate such a pathway [79,94]. Nonetheless, DMVs are smaller than autophagosomes, and hence, they might be rather endoplasmic reticulum derived vesicles (EDEsomes) enriched in PI3P and not follow exactly the same synthetic route [94]. Further work on coronaviruses and autopahgy found that only the LC3-I protein, the microtubule-associated proteins 1A/1B light chain 3B, is localized on the replication membranes, but the active protein lipidated with phosphatidylethanolamine LC3-II inserted into the autophagosome membrane is absent. Accordingly, present knowledge on coronaviruses in regard to autophagy suggests that they take benefit of the autophagocytic components but do not develop autophagosomes per se [95].

The autophatocytic pathway has also been associated to the start of HCV infection, but it seems not to be necessary for the infection to go on [82]. Later on, it was shown that autophagy was key in RNA replication at the onset of HCV infection [96], but the virus life cycle can go ahead afterward without the autophagy system intervention. Further work has shown that HCV, and possibly DENV, uses the autophagy system to evade the innate immune system [97]. Using immortalized human hepatocytes defective of the autophagy-related proteins either beclin (BCN1) or ATG7, it was shown in the latter study that disruption of the autophagy machinery elicites activation of the interferon signaling pathway and leads to apoptosis of the infected cells. Triggering of the autophagy pathways takes place after binding of the virus to the cell surface via the downregulation of mTOR and inactivation of Akt signaling [95]. Conflicting results have been reported for the induction of autophagy by HCV in regard to the unfolded protein response (UPR) [95]. Recent work [98] has bound the induction of autophagy by HCV to Golgi membrane fragmentation to render vesicles that colocalize with the HCV replicons. The immunity-related GTPase M protein (IRGM) mediates the phosphorylation of the early autophagy initiator ULK1 as well as the Golgi membrane fragmentation in response to HCV infection. The protein LC3 has also been detected in the replication membranes of the HIV-1, and the association of LC3-II with Gag-derived proteins seems to be a requisite for the efficient maturation of the Gag subunit p24 [14,99]. Members of the *Picornaviridae* family, non-enveloped viruses, have been reported to subvert the autophagosome pathway as a means to exit the infected cell without membrane lysis; support for this spreading mechanism comes from the finding of numerous extracelular vesicles that are enriched in phosphatidylserine phospholipids [14]. The best studied virus regarding autophagy is the dengue virus (DENV). Even though it was initially suggested that the DENV replication complexes are developed from autophagosomes, further work pointed out that the replication of DENV took place on invaginations arising from the endoplasmic reticulum (ER), while autophagy was rather used by DENV to modify the lipid metabolism in a way that is known as lipophagy [100,101]. Lipophagy was first shown to be an active way to get energy under starvation [102] through the association of autophagic components with lipid droplets (LDs). Recently, lipophagy has been demonstrated to regulate the fatty acid availability for the β-oxidation through contact sites between the mitochondria and the ER [103]. Regarding virus-associated hijacking of the cell lipid metabolism, Heaton and Randall [100] early showed that increased β-oxidation and the depletion of triglycerides was concurrent with and necessary for DENV replication. Then, these features were linked to the action of autophagy through the association with lipid droplets. A recent study by Zhang et al. [104] has found that AUP1, a type III protein with signals for LDs and ER, plays a relevant role in lipophagy induced by DENV and other flaviviruses such as WNV. Unmodified AUP1 is required for lipophagy triggering. A 10-fold increase in the content of diacylglycerophosphocholines (PCs) was measured in this study in infected cells containing unmodified AUP1, this increase being concomitant with a depletion of triacylglycerols and cholesterol esters, whereas the contents of free fatty acids and unesterified cholesterol rose. Conversely, smaller LDs, but not a reduction of their abundance, were observed in AUP1-knocked-out cells. Thus, these data point to an augmented consumption of LDs in the infected cells. This study unveils the mechanism that leads to the commented results; after the DENV protein NS4A associates with AUP1, the complex is relocalized from LDs to autophagosomes, where the acyltransferase domain of AUP1 is activated for the generation of phospholipids. This process was found to be dependent on the AUP1 ubiquitylation status, with NS4A inhibiting the ubiquitylation of AUP1.

Similar to viral entry, cholesterol has been found to be also relevant in the RCs' membranes [79,82]. Up to a c.a. 9-fold enrichment of cholesterol was found in HCV-developed DMVs as compared to its content in the ER membranes from which DMVs were originated [77]. A key protein in cholesterol metabolism associated to non-vesicular transport is the oxysterol-binding protein (OSBP). This protein has been described to transport cholesterol to PI4P-enriched membranes, which would agree with its collaboration in delivering cholesterol to DMVs with an abundant content of this PI [77]. The ceramide transfer protein (CERT) and the four-phosphate adaptor protein 2 (FAPP2) are known to undergo a similar fate in HCV infection [82]. An important protein involved in cellular lipid homeostasis is the sterol regulatory element binding protein (SREBP), a bHLH-zip transcription factor with three isoforms; SREBP1c regulates the expression of fatty acid (FA) biosynthesis genes [105,106], whereas SREBP2 transactivates genes implied in cholesterol biosynthesis, intracellular lipid transport, and lipoprotein import [107]. A recent study shows that the inhibition of SREBP with the retinoid derivative and RAR-α agonist AM580 prevents MERS-CoV infection by avoiding the formation of functional DMVs [105]. In this study, the lipid metabolism was the most affected pathway, with sterol biosynthesis being strengthened at expense of the glycerophospholipid metabolic pathways. Fast activation of the lipid biosynthesis enzymes Acetyl-CoA carboxylase (ACC), fatty acid synthase (FAS), and HMG-CoA synthase (HMGCS) was observed in such study, whose activity was partially blocked by AM580 inhibition of SREBP enzymes. Promotion of lipid biosyntheis after infection had already been pointed out for HCV in an elegant proteomics and lipidomics study [108]. HCV infection elicites changes in the proteome of host cells that resembled the Warburg effect described in cancer cells toward lactate production and the support of continuous glycolysis; concurrently, the up-regulation of citrate synthase (CS) and other lipogenic enzymes 24 h after infection was interpreted by the authors of the latter study as indicative of re-routing of the tricarboxylic acid (TCA) cycle for cytosolic accumulation of citrate, which would be used in FA synthesis. The up-regulation of peroxisomal and mitochondrial FA oxidation pathways is concurrent with the other metabolic changes. An increase in pro-apoptotic ceramides was observed in the latter study as well; two possible interpretations were attributed to this finding, either a cytopathic effect after cell cycle arrest over time enough to complete virus offspring or a defense response of the host cell to avoid infection spread.

Blocking cholesterol suitability for the membraneous network or endosomes used for the virus replication and internalization has been demonstrated to inhibit the virus life cycle in a number of unrelated viruses. Disruption of the SREBP pathway restrains the Andes virus (ANDV), an ssRNA-virus, internalization, although it does not bind to the cell surface receptor [109]. In addition to SREBP2, other components of this pathway were found to be necessary. The dependence of viral entry on the sterol regulatory element binding protein cleavage activating protein (SCAP) and the site 1 protease (S1P) was evidenced in cells null for these proteins. Thus, in the study of Petersen et al. [109], the virus was not internalized in cells lacking S1P, this result pointing out that a complete cholesterol biosynthesis pathway is required. Infectivity was also reduced 10-fold when the cells were treated with methyl-β-cyclodextrin (MβCD), a cholesterol sequestering agent, and comparable results were obtained after cell treatment with mevastatin or the S1P inhibitor PF-429242. However, the S1P dependence of virus infectivity does not seem to affect other viruses, thus this route being likely selective for hantaviruses [110]. In this study, the genetic or pharmacological disruption of the SREBP pathway at the site of the regulatory element membrane-bound transcription factor peptidase/site 1 protesase (MBTPS1/S1P) dramatically reduced viral infection, which is a feature that confirms the essential dependence of hantavirus on the high membrane cholesterol content for membrane fusion and effective infection. The down-regulation of sterol synthesis at the gene level after infection was found to be controlled by an interferon regulatory loop, in which a type I interferon-dependent mechanism down-regulates the expression of SREBP2 [111], this result showing a link between the innate immune response and cholesterol biosynthesis after viral infection. This type I interferon response toward cholesterol synthesis down-regulation was dependent on the mevalonate-isoprenoid branch as a supply of mevalonate completely blocked the cholesterol synthesis, whereas a supply of cholesterol did not. Additionally, in the presence of geranylgeraniol, the type I interferon inhibition of sterol biosynthesis was severely diminished. Further research has shown that interferon may regulate the sterol synthesis pathway in multiple forms through microRNAs [112]. In particular, miR-342-5p was found to hit multiple SREBP-independent targets of the mevalonate–sterol synthesis pathway after viral infection. The type I interferon response was also observed in regard to the impairment of the formation of double membrane structures induced by arteriviruses as replication sites [113]. Host cell fight against viral infection by a reduction of cholesterol availability has been also pointed out to come from the antiviral effector protein interferon-inducible transmembrane protein 3 (IFITM3). This protein interacts with vesicle-membrane-protein-associated protein A (VAPA), impeding its association with the oxysterol binding protein (OSBP), and consequently, altering the normal function of OSBP. As a result of the IFITM3 action, virus release into the cytosol is blocked by the accumulation of cholesterol in multivesicular bodies and endosomes. This effect restrains the membrane fusion of the intraluminal vesicles and that of the multivesicular bodies, which is a requisite for virus budding and release to the cytosol [114]. The viral accesory protein of HIV Nef competes with the cholesterol transporter ABCA1 to prime the transport of cholesterol to lipid rafts as a viral strategy to raise the replication membranes, thus overcoming the antiviral properties of ABCA1 [115].

The replication of Rabies virus (RABV), an ssRNA virus, is halted by the action of viperin (virus inhibitory protein, endoplasmic reticulum-associated, IFN-inducible) in RAW264.7 cells. This protein is induced by the RABV, IFV, HIV, or HCV infection through promotion of the innate immune response bound to the TLR4 signaling pathway. The inhibitory activity of viperin on virus budding is related to its capability to substantially drop the contents of cholesterol and sphingomyelin in the replication membranes [116], thus pointing out the relevance of the membrane lipid composition for efficient virus replication. The induction of viperin has also been proven for HCV and IFAV [111]. However, viperin does not intervene in the inhibition of arterivirus-induced double membrane formation [113].

4. Additional Pathways of Lipid Metabolism Affected in Virus Infection

Remodeling of the lipid metabolism by virus infection may leave signals at the organism level even some years after healing. The metabolome profile of patients undergoing SARS-CoV-1 infection during the outbreak of 2002–2003 was assessed 12 years after overcoming the pathology [117]. An outstanding result of this study regarding disturbed lipid metabolism was the elevation of phosphatidylinositol (PI) and lysophosphatidylinositol (LPI) species concentrations in serum, which in turn correlated positively with the levels of very low-density lipoproteins (VLDL); higher concentrations of products of the phospholipase A_2 (PLA$_2$) such as lysophospholipids (LPPLs) and free arachidonic acid (AA) were also found in patients as compared to healthy voluntiers, with a correlation between the level of AA and the ratio of LPI(18:0) to total 18:0-PIs being observed as well. These results show a potential high sensitivity of SARS-CoV patients to PLA$_2$ activity. In the general context, the metabolome of these patients pointed to hyperlipidemia, cardiovascular abnormalities, and glucose metabolism alteration as a delayed efffect of the viral infection. Nonetheless, the authors acknowledge that some of the related metabolic disturbations are likely owed to the pharmacological treatment. High levels of PLA$_2$ group IID (PLA2G2D) in lungs of middle-aged mice as compared to young mice had previously been associated to a fatal or worse outcome [118]. The authors of this study conclude that the negative influence of this enzyme in SARS-CoV infection was to increase the concentration of anti-inflammatory lipid mediators, mainly protaglandin D_2 (PGD$_2$), which impaired the efficient function of the immune system [119]. In the recent SARS-CoV-2 outbreak (COVID-19), mortality has mostly affected aged people above 60 years old, thus showing an age-related fatality as for SARS-CoV-1 and MERS-CoV [120]. Using a lipidomics approach, the effect of HCoV-229E and MERS-CoV infection on the host cell lipid profile was recently investigated in cell culture [121]. The main conclusions of this study agree with the raised content of AA and LPPLs through PLase activity, which indicates that the possible virus-induced activation of cPLA$_2$ favors virus replication as a factor required for DMVs' formation. In this study, linoleic acid (LA) or AA supplementation to the culture cells suppressed replication, which is a result that may be interpreted as a demonstration of the perturbation of the LA/AA axis of the lipid metabolism.

In the COVID-19 outbreak, it has been suggested that increasing the levels of vitamin D could help fighting against the SARS-CoV infection [122]. This suggestion is based on the fact that 25-hydroxyvitamin D3 was found to protect Huh7 cells against MERS-CoV [105]. Vitamin D is a lipid-related compound belonging to the group of fat-soluble secosteroids, with the most important form in humans being vitamin D3 (cholecalciferol) [123]. In a recent study, high doses of vitamin D have shown protective effects against DENV infection through regulation of the Toll-like receptor expression as well as the modulation of pro-inflammatory cytokines release, which suggests that its action is focused toward the immune system modulation rather than to lipid metabolism [124]. However, evidence on the beneficial effects of vitamin D uptake is still poor, and more studies are devoted to this issue.

Lipids, as components of membranes, are related to viroporins, which are specific viral proteins that are known to create ion channels for ion trafficking [125–127]. The effect on cell metabolism of diverse viroporins differs among them, but there is evidence that they are closely related to viral pathogenity [125]. Viroporins may play a relevant role during virus infection, as they are involved in membrane permeability and calcium homeostasis. Their participation in the development of vacuoles from the ER during the DMVs' formation has been suggested, but data on this issue are still scarce. The regulation of Ca^{2+} flux by viroporins might favor the membrane fusion through the interaction of this cation with the phospholipid headgroups and concurrently facilitate the required dehydratation reaction. Viroporins are not required for virus replication with the exception of rotaviruses and picornaviruses; thus, whether this function is exerted through the ion channels or another property of viroporins remains an issue still unknown [125]. The lipid composition of the membrane may influence the viroporin activity, leading to different versions of ion channels, which depends on the electric charge that the phospholipids confer to the membrane and curvature [127]. A viroporin from rotavirus, NSP4, was shown to co-localize with the autophagy marker protein LC3 in membranes accomodating virus replication; this viroporin is implicated in the sequestering of autophagy for the transport of proteins from the ER to the replication sites [128]. Further research is necessary to understand the role played by viroporins in virus infection in order to consider them as potential therapeutic targets.

5. Conclusions

Remodeling of the virus-induced host cell lipid metabolism is a remarkable feature of the viral infection that affects viral entry, replication of the genomic material, and the releasing of progeny. A comperhensive view of the process is illustrated in Figure 3. The main actors are well known to be cholesterol, sphingolipids, and PIs, but other lipid species and their related pathways such as the LA/AA axis are also relevant. How to target the lipid metabolism in a safe manner to avoid virus infection or reduce its pathogenity is a promising therapeutic tool, but it demands improving the knowledge on the actual pathways that are affected over the virus life cycle. The exact mechanism through which the enzyme inhibitors act on the key enzymes of the lipid metabolism is additionally required to develop more efficient and safe therapeutic drugs. Since the lipid metabolism is essential for proper cell function, selective drugs targeting the virus or exclusively the infected cells have to be used to avoid harmful side effects.

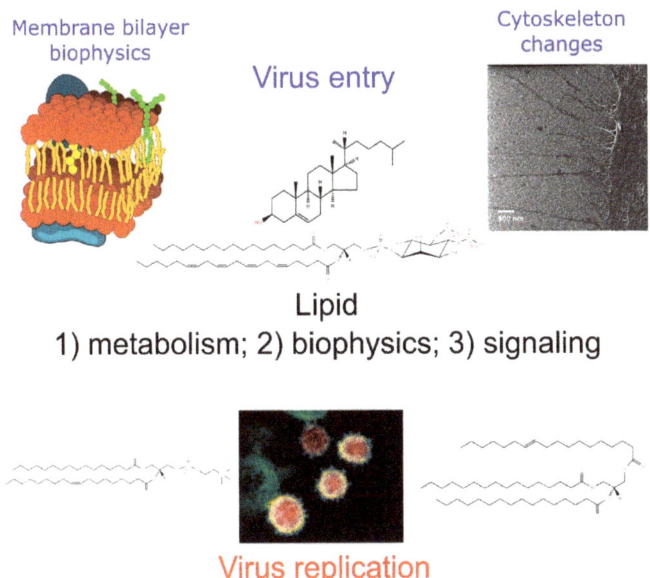

Figure 3. Comprehensive view of the virus replication process and the main lipids involved in every step.

Author Contributions: O.M. wrote the main draft. D.B. drew the figures and contributed with L.G.-d-G. to writing, reviewing and editing the manuscript. All authors have read and agreed to the published version of the manuscript.

Funding: Thanks are given to Spanish Government (project SAF2017-83079-R) and Fundación Dominguez Martínez for financial support to OM.

Acknowledgments: The authors thank M. S. Crespo for helpful reading of the manuscript.

Conflicts of Interest: The authors declare no conflict of interest.

References

1. Wu, F.; Zhao, S.; Yu, B.; Chen, Y.-M.; Wang, W.; Song, Z.-G.; Hu, Y.; Tao, Z.-W.; Tian, J.-H.; Pei, Y.-Y.; et al. A new coronavirus associated with human respiratory disease in China. *Nature* **2020**, *579*, 265–269. [CrossRef]
2. World Health Organization. WHO Director-General's Opening Remarks at the Media Briefing on COVID-19. Available online: https://www.who.int/emergencies/diseases/novel-coronavirus-2019 (accessed on 9 June 2020).
3. Fehr, A.R.; Perlman, S. Coronaviruses: An Overview of Their Replication and Pathogenesis. In *Methods in Molecular Biology*; Maier, H.J., Bickerton, E., Britton, P., Eds.; Springer: New York, NY, USA, 2015; Volume 1282, pp. 1–23. ISBN 978-1-4939-2437-0.
4. Reid, C.R.; Airo, A.M.; Hobman, T.C. The Virus-Host Interplay: Biogenesis of +RNA Replication Complexes. *Viruses* **2015**, *7*, 4385–4413. [CrossRef] [PubMed]
5. Novoa, R.R.; Calderita, G.; Arranz, R.; Fontana, J.; Granzow, H.; Risco, C. Virus factories: Associations of cell organelles for viral replication and morphogenesis. *Biol. Cell* **2005**, *97*, 147–172. [CrossRef]
6. Marsh, M.; Helenius, A. Virus Entry: Open Sesame. *Cell* **2006**, *124*, 729–740. [CrossRef] [PubMed]
7. Yang, S.-T.; Kreutzberger, A.J.B.; Lee, J.; Kiessling, V.; Tamm, L.K. The role of cholesterol in membrane fusion. *Chem. Phys. Lipids* **2016**, *199*, 136–143. [CrossRef] [PubMed]
8. Ou, X.; Zheng, W.; Shan, Y.; Mu, Z.; Dominguez, S.R.; Holmes, K.V.; Qian, Z. Identification of the Fusion Peptide-Containing Region in Betacoronavirus Spike Glycoproteins. *J. Virol.* **2016**, *90*, 5586–5600. [CrossRef]

9. Zhang, W.; Zheng, Q.; Yan, M.; Chen, X.-B.; Yang, H.; Zhou, W.; Rao, Z. Structural characterization of the HCoV-229E fusion core. *Biochem. Biophys. Res. Commun.* **2018**, *497*, 705–712. [CrossRef]

10. Zeisel, M.B.; Felmlee, D.J.; Baumert, T.F. Hepatitis C Virus Entry. In *Hepatitis C Virus: From Molecular Virology to Antiviral Therapy*; Bartenschlager, R., Ed.; Current Topics in Microbiology and Immunology; Springer: Berlin/Heidelberg, Germany, 2013; Volume 369, pp. 87–112. ISBN 978-3-642-27339-1.

11. Zhu, Y.-Z.; Qian, X.-J.; Zhao, P.; Qi, Z.-T. How hepatitis C virus invades hepatocytes: The mystery of viral entry. *World J. Gastroenterol.* **2014**, *20*, 3457. [CrossRef]

12. Lakadamyali, M.; Rust, M.J.; Zhuang, X. Endocytosis of influenza viruses. *Microbes Infect.* **2004**, *6*, 929–936. [CrossRef]

13. Mercer, J.; Schelhaas, M.; Helenius, A. Virus Entry by Endocytosis. *Annu. Rev. Biochem.* **2010**, *79*, 803–833. [CrossRef]

14. Robinson, M.; Schor, S.; Barouch-Bentov, R.; Einav, S. Viral journeys on the intracellular highways. *Cell. Mol. Life Sci.* **2018**, *75*, 3693–3714. [CrossRef] [PubMed]

15. Wang, H.; Yuan, X.; Sun, Y.; Mao, X.; Meng, C.; Tan, L.; Song, C.; Qiu, X.; Ding, C.; Liao, Y. Infectious bronchitis virus entry mainly depends on clathrin mediated endocytosis and requires classical endosomal/lysosomal system. *Virology* **2019**, *528*, 118–136. [CrossRef] [PubMed]

16. Jin, C.; Che, B.; Guo, Z.; Li, C.; Liu, Y.; Wu, W.; Wang, S.; Li, D.; Cui, Z.; Liang, M. Single virus tracking of Ebola virus entry through lipid rafts in living host cells. *Biosaf. Health* **2020**, *2*, 25–31. [CrossRef] [PubMed]

17. Wang, H.; Yang, P.; Liu, K.; Guo, F.; Zhang, Y.; Zhang, G.; Jiang, C. SARS coronavirus entry into host cells through a novel clathrin- and caveolae-independent endocytic pathway. *Cell Res.* **2008**, *18*, 290–301. [CrossRef]

18. Hamming, I.; Timens, W.; Bulthuis, M.; Lely, A.T.; Navis, G.; Van Goor, H. Tissue distribution of ACE2 protein, the functional receptor for SARS coronavirus. A first step in understanding SARS pathogenesis. *J. Pathol.* **2004**, *203*, 631–637. [CrossRef]

19. Hoffmann, M.; Kleine-Weber, H.; Schroeder, S.; Krüger, N.; Herrler, T.; Erichsen, S.; Schiergens, T.S.; Herrler, G.; Wu, N.-H.; Nitsche, A.; et al. SARS-CoV-2 Cell Entry Depends on ACE2 and TMPRSS2 and Is Blocked by a Clinically Proven Protease Inhibitor. *Cell* **2020**, *181*, 271–280.e8. [CrossRef]

20. Chigbu, D.G.I.; Loonawat, R.; Sehgal, M.; Patel, D.; Jain, P. Hepatitis C Virus Infection: Host–Virus Interaction and Mechanisms of Viral Persistence. *Cells* **2019**, *8*, 376. [CrossRef]

21. Plemper, R.K. Cell entry of enveloped viruses. *Curr. Opin. Virol.* **2011**, *1*, 92–100. [CrossRef]

22. Harrison, S.C. Viral membrane fusion. *Virology* **2015**, *479–480*, 498–507. [CrossRef]

23. Benhaim, M.A.; Lee, K.K. New Biophysical Approaches Reveal the Dynamics and Mechanics of Type I Viral Fusion Machinery and Their Interplay with Membranes. *Viruses* **2020**, *12*, 413. [CrossRef]

24. Ivanovic, T.; Choi, J.L.; Whelan, S.P.; Van Oijen, A.M.; Harrison, S.C. Influenza-virus membrane fusion by cooperative fold-back of stochastically induced hemagglutinin intermediates. *eLife* **2013**, *2*, e00333. [CrossRef] [PubMed]

25. Chao, L.H.; Klein, D.E.; Schmidt, A.G.; Peña, J.M.; Harrison, S.C. Sequential conformational rearrangements in flavivirus membrane fusion. *eLife* **2014**, *3*, e04389. [CrossRef]

26. Aeffner, S.; Reusch, T.; Weinhausen, B.; Salditt, T. Energetics of stalk intermediates in membrane fusion are controlled by lipid composition. *Proc. Natl. Acad. Sci. USA* **2012**, *109*, E1609–E1618. [CrossRef]

27. Gui, L.; Ebner, J.L.; Mileant, A.; Williams, J.A.; Lee, K.K. Visualization and Sequencing of Membrane Remodeling Leading to Influenza Virus Fusion. *J. Virol.* **2016**, *90*, 6948–6962. [CrossRef] [PubMed]

28. Ge, M.; Freed, J.H. Two Conserved Residues Are Important for Inducing Highly Ordered Membrane Domains by the Transmembrane Domain of Influenza Hemagglutinin. *Biophys. J.* **2011**, *100*, 90–97. [CrossRef] [PubMed]

29. Basso, L.G.M.; Vicente, E.F.; Crusca, E.; Cilli, E.M.; Costa-Filho, A. SARS-CoV fusion peptides induce membrane surface ordering and curvature. *Sci. Rep.* **2016**, *6*, 37131. [CrossRef]

30. Kuhl, T.; Guo, Y.; Alderfer, J.L.; Berman, A.D.; Leckband, D.; Israelachvili, J.; Hui, S.W. Direct Measurement of Polyethylene Glycol Induced Depletion Attraction between Lipid Bilayers. *Langmuir* **1996**, *12*, 3003–3014. [CrossRef]

31. Mahajan, M.; Bhattacharjya, S. NMR structures and localization of the potential fusion peptides and the pre-transmembrane region of SARS-CoV: Implications in membrane fusion. *Biochim. Biophys. Acta (BBA) Biomembr.* **2015**, *1848*, 721–730. [CrossRef]

32. Carravilla, P.; Nieva, J.L.; Eggeling, C. Fluorescence Microscopy of the HIV-1 Envelope. *Viruses* **2020**, *12*, 348. [CrossRef]

33. Lu, Y.; Liu, D.X.; Tam, J.P. Lipid rafts are involved in SARS-CoV entry into Vero E6 cells. *Biochem. Biophys. Res. Commun.* **2008**, *369*, 344–349. [CrossRef]

34. Earnest, J.T.; Hantak, M.P.; Park, J.-E.; Gallagher, T. Coronavirus and Influenza Virus Proteolytic Priming Takes Place in Tetraspanin-Enriched Membrane Microdomains. *J. Virol.* **2015**, *89*, 6093–6104. [CrossRef] [PubMed]

35. Guo, H.; Huang, M.; Yuan, Q.; Wei, Y.; Gao, Y.; Mao, L.; Gu, L.; Tan, Y.W.; Zhong, Y.; Liu, D.; et al. The Important Role of Lipid Raft-Mediated Attachment in the Infection of Cultured Cells by Coronavirus Infectious Bronchitis Virus Beaudette Strain. *PLoS ONE* **2017**, *12*, e0170123. [CrossRef] [PubMed]

36. Zhang, J.; Pekosz, A.; Lamb, R.A. Influenza Virus Assembly and Lipid Raft Microdomains: A Role for the Cytoplasmic Tails of the Spike Glycoproteins. *J. Virol.* **2000**, *74*, 4634–4644. [CrossRef] [PubMed]

37. Takeda, M.; Leser, G.P.; Russell, C.J.; Lamb, R.A. Influenza virus hemagglutinin concentrates in lipid raft microdomains for efficient viral fusion. *Proc. Natl. Acad. Sci. USA* **2003**, *100*, 14610–14617. [CrossRef] [PubMed]

38. Van Meer, G.; Voelker, D.R.; Feigenson, G.W. Membrane lipids: Where they are and how they behave. *Nat. Rev. Mol. Cell Biol.* **2008**, *9*, 112–124. [CrossRef]

39. Sezgin, E.; Levental, I.; Mayor, S.; Eggeling, C. The mystery of membrane organization: Composition, regulation and roles of lipid rafts. *Nat. Rev. Mol. Cell Biol.* **2017**, *18*, 361–374. [CrossRef]

40. Levental, I.; Levental, K.R.; Heberle, F.A. Lipid Rafts: Controversies Resolved, Mysteries Remain. *Trends Cell Biol.* **2020**, *30*, 341–353. [CrossRef]

41. Biswas, S.; Yin, S.-R.; Blank, P.S.; Zimmerberg, J. Cholesterol Promotes Hemifusion and Pore Widening in Membrane Fusion Induced by Influenza Hemagglutinin. *J. Gen. Physiol.* **2008**, *131*, 503–513. [CrossRef]

42. Chlanda, P.; Mekhedov, E.; Waters, H.; Schwartz, C.L.; Fischer, E.R.; Ryham, R.J.; Cohen, F.S.; Blank, P.S.; Zimmerberg, J. The hemifusion structure induced by influenza virus haemagglutinin is determined by physical properties of the target membranes. *Nat. Microbiol.* **2016**, *1*, 16050. [CrossRef]

43. Zaitseva, E.; Yang, S.-T.; Melikov, K.; Pourmal, S.; Chernomordik, L.V. Dengue Virus Ensures Its Fusion in Late Endosomes Using Compartment-Specific Lipids. *PLoS Pathog.* **2010**, *6*, e1001131. [CrossRef]

44. Meher, G.; Bhattacharjya, S.; Chakraborty, H. Membrane Cholesterol Modulates Oligomeric Status and Peptide-Membrane Interaction of Severe Acute Respiratory Syndrome Coronavirus Fusion Peptide. *J. Phys. Chem. B* **2019**, *123*, 10654–10662. [CrossRef] [PubMed]

45. Liu, K.N.; Boxer, S.G. Target Membrane Cholesterol Modulates Single Influenza Virus Membrane Fusion Efficiency but Not Rate. *Biophys. J.* **2020**, *118*, 2426–2433. [CrossRef] [PubMed]

46. Masters, P.S. The Molecular Biology of Coronaviruses. In *Advances in Applied Microbiology*; Elsevier: Amsterdam, The Netherlands, 2006; Volume 66, pp. 193–292. ISBN 978-0-12-039869-0.

47. Munro, S.P. Lipid Rafts. *Cell* **2003**, *115*, 377–388. [CrossRef]

48. Garcia-Arribas, A.B.; Alonso, A.; Goñi, F.M. Cholesterol interactions with ceramide and sphingomyelin. *Chem. Phys. Lipids* **2016**, *199*, 26–34. [CrossRef] [PubMed]

49. Corver, J.; Moesby, L.; Erukulla, R.K.; Reddy, K.C.; Bittman, R.; Wilschut, J. Sphingolipid-dependent fusion of Semliki Forest virus with cholesterol-containing liposomes requires both the 3-hydroxyl group and the double bond of the sphingolipid backbone. *J. Virol.* **1995**, *69*, 3220–3223. [CrossRef] [PubMed]

50. Otsuki, N.; Sakata, M.; Saito, K.; Okamoto, K.; Mori, Y.; Hanada, K.; Takeda, M. Both Sphingomyelin and Cholesterol in the Host Cell Membrane Are Essential for Rubella Virus Entry. *J. Virol.* **2017**, *92*, e01130-17. [CrossRef] [PubMed]

51. Miller, M.E.; Adhikary, S.; Kolokoltsov, A.A.; Davey, R.A. Ebolavirus Requires Acid Sphingomyelinase Activity and Plasma Membrane Sphingomyelin for Infection. *J. Virol.* **2012**, *86*, 7473–7483. [CrossRef]

52. Pastenkos, G.; Miller, J.L.; Pritchard, S.M.; Nicola, A.V. Role of Sphingomyelin in Alphaherpesvirus Entry. *J. Virol.* **2018**, *93*, e01547-18. [CrossRef]

53. Soudani, N.; Hage-Sleiman, R.; Karam, W.; Dbaibo, G.; Zaraket, H. Ceramide Suppresses Influenza A Virus Replication In Vitro. *J. Virol.* **2019**, *93*, e00053-19. [CrossRef]

54. Grassmé, H.; Riehle, A.; Wilker, B.; Gulbins, E. Rhinoviruses Infect Human Epithelial Cells via Ceramide-enriched Membrane Platforms. *J. Biol. Chem.* **2005**, *280*, 26256–26262. [CrossRef]

55. Aktepe, T.E.; Pham, H.; MacKenzie, J.M. Differential utilisation of ceramide during replication of the flaviviruses West Nile and dengue virus. *Virology* **2015**, *484*, 241–250. [CrossRef] [PubMed]

56. Drews, K.; Calgi, M.P.; Harrison, W.C.; Drews, C.M.; Costa-Pinheiro, P.; Shaw, J.J.P.; Jobe, K.A.; Nelson, E.A.; Han, J.D.; Fox, T.; et al. Glucosylceramidase Maintains Influenza Virus Infection by Regulating Endocytosis. *J. Virol.* **2019**, *93*, e00017-19. [CrossRef]

57. Sieben, C.; Sezgin, E.; Eggeling, C.; Manley, S. Influenza A viruses use multivalent sialic acid clusters for cell binding and receptor activation. *PLoS Pathog.* **2020**, *16*, e1008656. [CrossRef] [PubMed]

58. Tortorici, M.A.; Walls, A.C.; Lang, Y.; Wang, C.; Li, Z.; Koerhuis, D.; Boons, G.-J.; Bosch, B.-J.; Rey, F.A.; De Groot, R.J.; et al. Structural basis for human coronavirus attachment to sialic acid receptors. *Nat. Struct. Mol. Biol.* **2019**, *26*, 481–489. [CrossRef]

59. Mazzon, M.; Mercer, J. Lipid interactions during virus entry and infection. *Cell Microbiol.* **2014**, *16*, 1493–1502. [CrossRef]

60. Ewers, H.; Römer, W.; Smith, A.E.; Bacia, K.; Dmitrieff, S.; Chai, W.; Mancini, R.; Kartenbeck, J.; Chambon, V.; Berland, L.; et al. GM1 structure determines SV40-induced membrane invagination and infection. *Nat. Cell Biol.* **2010**, *12*, 11–18. [CrossRef] [PubMed]

61. Moller-Tank, S.; Kondratowicz, A.S.; Davey, R.A.; Rennert, P.D.; Maury, W.J. Role of the Phosphatidylserine Receptor TIM-1 in Enveloped-Virus Entry. *J. Virol.* **2013**, *87*, 8327–8341. [CrossRef] [PubMed]

62. Jemielity, S.; Wang, J.J.; Chan, Y.K.; Ahmed, A.A.; Li, W.; Monahan, S.; Bu, X.; Farzan, M.; Freeman, G.J.; Umetsu, D.T.; et al. TIM-family Proteins Promote Infection of Multiple Enveloped Viruses through Virion-associated Phosphatidylserine. *PLoS Pathog.* **2013**, *9*, e1003232. [CrossRef]

63. Nanbo, A.; Maruyama, J.; Imai, M.; Ujie, M.; Fujioka, Y.; Nishide, S.; Takada, A.; Ohba, Y.; Kawaoka, Y. Ebola virus requires a host scramblase for externalization of phosphatidylserine on the surface of viral particles. *PLoS Pathog.* **2018**, *14*, e1006848. [CrossRef]

64. Cheshenko, N.; Pierce, C.; Herold, B.C. Herpes simplex viruses activate phospholipid scramblase to redistribute phosphatidylserines and Akt to the outer leaflet of the plasma membrane and promote viral entry. *PLoS Pathog.* **2018**, *14*, e1006766. [CrossRef]

65. Das, A.; Maury, W.; Lemon, S.M. TIM1 (HAVCR1): An Essential "Receptor" or an "Accessory Attachment Factor" for Hepatitis A Virus? *J. Virol.* **2019**, *93*, e01793-18. [CrossRef] [PubMed]

66. Van den Bout, I.; Divecha, N. PIP5K-driven PtdIns(4,5)P2 synthesis: Regulation and cellular functions. *J. Cell Sci.* **2009**, *122*, 3837–3850. [CrossRef]

67. Vicinanza, M.; D'Angelo, G.; Di Campli, A.; De Matteis, M.A. Function and dysfunction of the PI system in membrane trafficking. *EMBO J.* **2008**, *27*, 2457–2470. [CrossRef]

68. Perugini, V.R.; Gordón-Alonso, M.; Sánchez-Madrid, F. PIP2: Choreographer of actin-adaptor proteins in the HIV-1 dance. *Trends Microbiol.* **2014**, *22*, 379–388. [CrossRef]

69. Muecksch, F.; Laketa, V.; Muller, B.; Schultz, C.; Kräusslich, H.-G. Synchronized HIV assembly by tunable PIP2 changes reveals PIP2 requirement for stable Gag anchoring. *eLife* **2017**, *6*, e25287. [CrossRef] [PubMed]

70. Mücksch, F.; Citir, M.; Lüchtenborg, C.; Glass, B.; Traynor-Kaplan, A.; Schultz, C.; Brügger, B.; Kräusslich, H.-G. Quantification of phosphoinositides reveals strong enrichment of PIP2 in HIV-1 compared to producer cell membranes. *Sci. Rep.* **2019**, *9*, 1–13. [CrossRef] [PubMed]

71. Wijesinghe, K.J.; Stahelin, R.V. Investigation of the Lipid Binding Properties of the Marburg Virus Matrix Protein VP40. *J. Virol.* **2016**, *90*, 3074–3085. [CrossRef]

72. Gonzales, B.; De Rocquigny, H.; Beziau, A.; Durand, S.; Burlaud-Gaillard, J.; Lefebvre, A.; Krull, S.; Emond, P.; Brand, D.; Piver, E. Type I phosphatidylinositol-4-phosphate 5-kinase α and γ play a key role in targeting HIV-1 Pr55Gag to the plasma membrane. *J. Virol.* **2020**, *94*, e00189-20. [CrossRef]

73. Diehl, N.; Schaal, H. Make Yourself at Home: Viral Hijacking of the PI3K/Akt Signaling Pathway. *Viruses* **2013**, *5*, 3192–3212. [CrossRef]

74. Wu, S.-X.; Chen, W.-N.; Jing, Z.-T.; Liu, W.; Lin, X.-J.; Lin, X. Hepatitis B Spliced Protein (HBSP) Suppresses Fas-Mediated Hepatocyte Apoptosis via Activation of PI3K/Akt Signaling. *J. Virol.* **2018**, *92*, e01273-18. [CrossRef]

75. Heaton, N.S.; Randall, G. Multifaceted roles for lipids in viral infection. *Trends Microbiol.* **2011**, *19*, 368–375. [CrossRef] [PubMed]

76. Harak, C.; Lohmann, V. Ultrastructure of the replication sites of positive-strand RNA viruses. *Virology* **2015**, *479–480*, 418–433. [CrossRef] [PubMed]

77. Neuman, B.W.; Angelini, M.M.; Buchmeier, M.J. Does form meet function in the coronavirus replicative organelle? *Trends Microbiol.* **2014**, *22*, 642–647. [CrossRef] [PubMed]

78. Paul, D.; Hoppe, S.; Saher, G.; Krijnse-Locker, J.; Bartenschlager, R. Morphological and Biochemical Characterization of the Membranous Hepatitis C Virus Replication Compartment. *J. Virol.* **2013**, *87*, 10612–10627. [CrossRef] [PubMed]

79. Knoops, K.; Kikkert, M.; Van Den Worm, S.H.E.; Zevenhoven-Dobbe, J.C.; Van Der Meer, Y.; Koster, A.J.; Mommaas, A.M.; Snijder, E.J. SARS-Coronavirus Replication Is Supported by a Reticulovesicular Network of Modified Endoplasmic Reticulum. *PLoS Biol.* **2008**, *6*, e226. [CrossRef]

80. Netherton, C.L.; Wileman, T. Virus factories, double membrane vesicles and viroplasm generated in animal cells. *Curr. Opin. Virol.* **2011**, *1*, 381–387. [CrossRef]

81. Romero-Brey, I.; Merz, A.; Chiramel, A.; Lee, J.-Y.; Chlanda, P.; Haselman, U.; Santarella-Mellwig, R.; Habermann, A.; Hoppe, S.; Kallis, S.; et al. Three-Dimensional Architecture and Biogenesis of Membrane Structures Associated with Hepatitis C Virus Replication. *PLoS Pathog.* **2012**, *8*, e1003056. [CrossRef] [PubMed]

82. Egger, D.; Wölk, B.; Gosert, R.; Bianchi, L.; Blum, H.E.; Moradpour, D.; Bienz, K. Expression of Hepatitis C Virus Proteins Induces Distinct Membrane Alterations Including a Candidate Viral Replication Complex. *J. Virol.* **2002**, *76*, 5974–5984. [CrossRef]

83. Wang, H.; Tai, A.W. Mechanisms of Cellular Membrane Reorganization to Support Hepatitis C Virus Replication. *Viruses* **2016**, *8*, 142. [CrossRef]

84. Angelini, M.M.; Akhlaghpour, M.; Neuman, B.W.; Buchmeier, M.J. Severe Acute Respiratory Syndrome Coronavirus Nonstructural Proteins 3, 4, and 6 Induce Double-Membrane Vesicles. *mBio* **2013**, *4*, e00524-13. [CrossRef]

85. Brügger, B.; Glass, B.; Haberkant, P.; Leibrecht, I.; Wieland, F.T.; Kräusslich, H.-G. The HIV lipidome: A raft with an unusual composition. *Proc. Natl. Acad. Sci. USA* **2006**, *103*, 2641–2646. [CrossRef] [PubMed]

86. Martín-Acebes, M.A.; Merino-Ramos, T.; Blazquez, A.; Casas, J.; Escribano-Romero, E.; Sobrino, F.; Sáiz, J.-C. The Composition of West Nile Virus Lipid Envelope Unveils a Role of Sphingolipid Metabolism in Flavivirus Biogenesis. *J. Virol.* **2014**, *88*, 12041–12054. [CrossRef] [PubMed]

87. Wang, H.; Perry, J.W.; Lauring, A.S.; Neddermann, P.; De Francesco, R.; Tai, A.W. Oxysterol-binding protein is a phosphatidylinositol 4-kinase effector required for HCV replication membrane integrity and cholesterol trafficking. *Gastroenterology* **2014**, *146*, 1373–1385. [CrossRef]

88. Khan, I.; Katikaneni, D.S.; Han, Q.; Sanchez-Felipe, L.; Hanada, K.; Ambrose, R.L.; MacKenzie, J.M.; Konan, K. Modulation of Hepatitis C Virus Genome Replication by Glycosphingolipids and Four-Phosphate Adaptor Protein 2. *J. Virol.* **2014**, *88*, 12276–12295. [CrossRef]

89. Delang, L.; Harak, C.; Benkheil, M.; Khan, H.; Leyssen, P.; Andrews, M.; Lohmann, V.; Neyts, J. PI4KIII inhibitor enviroxime impedes the replication of the hepatitis C virus by inhibiting PI3 kinases. *J. Antimicrob. Chemother.* **2018**, *73*, 3375–3384. [CrossRef]

90. Arita, M.; Kojima, H.; Nagano, T.; Okabe, T.; Wakita, T.; Shimizu, H. Phosphatidylinositol 4-Kinase III Beta Is a Target of Enviroxime-Like Compounds for Antipoliovirus Activity. *J. Virol.* **2011**, *85*, 2364–2372. [CrossRef]

91. Martín-Acebes, M.A.; Blazquez, A.; De Oya, N.J.; Escribano-Romero, E.; Sáiz, J.-C. West Nile Virus Replication Requires Fatty Acid Synthesis but Is Independent on Phosphatidylinositol-4-Phosphate Lipids. *PLoS ONE* **2011**, *6*, e24970. [CrossRef]

92. Yang, N.; Ma, P.; Lang, J.; Zhang, Y.; Deng, J.; Ju, X.; Zhang, G.; Jiang, C. Phosphatidylinositol 4-Kinase IIIβ Is Required for Severe Acute Respiratory Syndrome Coronavirus Spike-mediated Cell Entry. *J. Biol. Chem.* **2012**, *287*, 8457–8467. [CrossRef]

93. De Wilde, A.H.; Wannee, K.F.; Scholte, F.E.M.; Goeman, J.J.; Dijke, P.T.; Snijder, E.J.; Kikkert, M.; Van Hemert, M.J. A Kinome-Wide Small Interfering RNA Screen Identifies Proviral and Antiviral Host Factors in Severe Acute Respiratory Syndrome Coronavirus Replication, Including Double-Stranded RNA-Activated Protein Kinase and Early Secretory Pathway Proteins. *J. Virol.* **2015**, *89*, 8318–8333. [CrossRef]

94. De Wilde, A.H.; Snijder, E.J.; Kikkert, M.; Van Hemert, M.J. Host Factors in Coronavirus Replication. In *Roles of Host Gene and Non-Coding RNA Expression in Virus Infection*; Tripp, R.A., Tompkins, S.M., Eds.; Current Topics in Microbiology and Immunology; Springer International Publishing: Cham, Switzerland, 2017; Volume 419, pp. 1–42. ISBN 978-3-030-05368-0.

95. Chiramel, A.I.; Brady, N.R.; Bartenschlager, R. Divergent Roles of Autophagy in Virus Infection. *Cells* **2013**, *2*, 83–104. [CrossRef]

96. Dreux, M.; Gastaminza, P.; Wieland, S.F.; Chisari, F.V. The autophagy machinery is required to initiate hepatitis C virus replication. *Proc. Natl. Acad. Sci. USA* **2009**, *106*, 14046–14051. [CrossRef]

97. Shrivastava, S.; Raychoudhuri, A.; Steele, R.; Ray, R.; Ray, R.B. Knockdown of autophagy enhances the innate immune response in hepatitis C virus-infected hepatocytes. *Hepatology* **2011**, *53*, 406–414. [CrossRef] [PubMed]

98. Hansen, M.D.; Johnsen, I.B.; Stiberg, K.A.; Sherstova, T.; Wakita, T.; Richard, G.M.; Kandasamy, K.; Meurs, E.F.; Anthonsen, M.W. Hepatitis C virus triggers Golgi fragmentation and autophagy through the immunity-related GTPase M. *Proc. Natl. Acad. Sci. USA* **2017**, *114*, E3462–E3471. [CrossRef] [PubMed]

99. Choi, Y.; Bowman, J.W.; Jung, J.U. Autophagy during viral infection—A double-edged sword. *Nat. Rev. Microbiol.* **2018**, *16*, 341–354. [CrossRef] [PubMed]

100. Heaton, N.S.; Randall, G. Dengue Virus-Induced Autophagy Regulates Lipid Metabolism. *Cell Host Microbe* **2010**, *8*, 422–432. [CrossRef]

101. Heaton, N.S.; Randall, G. Dengue Virus and Autophagy. *Viruses* **2011**, *3*, 1332–1341. [CrossRef]

102. Singh, R.; Kaushik, S.; Wang, Y.; Xiang, Y.; Novak, I.; Komatsu, M.; Tanaka, K.; Cuervo, A.M.; Czaja, M.J. Autophagy regulates lipid metabolism. *Nature* **2009**, *458*, 1131–1135. [CrossRef]

103. Bosc, C.; Broin, N.; Fanjul, M.; Saland, E.; Farge, T.; Courdy, C.; Batut, A.; Masoud, R.; Larrue, C.; Skuli, S.; et al. Autophagy regulates fatty acid availability for oxidative phosphorylation through mitochondria-endoplasmic reticulum contact sites. *Nat. Commun.* **2020**, *11*, 1–14. [CrossRef]

104. Zhang, J.; Lan, Y.; Li, M.Y.; Lamers, M.M.; Fusade-Boyer, M.; Klemm, E.; Thiele, C.; Ashour, J.; Sanyal, S. Flaviviruses Exploit the Lipid Droplet Protein AUP1 to Trigger Lipophagy and Drive Virus Production. *Cell Host Microbe* **2018**, *23*, 819–831. [CrossRef]

105. Horton, J.D.; Goldstein, J.L.; Brown, M.S. SREBPs: Activators of the complete program of cholesterol and fatty acid synthesis in the liver. *J. Clin. Investig.* **2002**, *109*, 1125–1131. [CrossRef]

106. Balgoma, D.; Pettersson, C.; Hedeland, M. Common Fatty Markers in Diseases with Dysregulated Lipogenesis. *Trends Endocrinol. Metab.* **2019**, *30*, 283–285. [CrossRef] [PubMed]

107. Cottam, E.M.; Maier, H.J.; Manifava, M.; Vaux, L.; Chandra-Schoenfelder, P.; Gerner, W.; Britton, P.; Ktistakis, N.T.; Wileman, T. Coronavirus nsp6 proteins generate autophagosomes from the endoplasmic reticulum via an omegasome intermediate. *Autophagy* **2011**, *7*, 1335–1347. [CrossRef] [PubMed]

108. Yuan, S.; Chu, H.; Chan, J.F.-W.; Ye, Z.-W.; Wen, L.; Yan, B.; Lai, P.-M.; Tee, K.-M.; Huang, J.; Chen, D.; et al. SREBP-dependent lipidomic reprogramming as a broad-spectrum antiviral target. *Nat. Commun.* **2019**, *10*, 120. [CrossRef]

109. Petersen, J.; Drake, M.J.; Bruce, E.A.; Riblett, A.M.; Didigu, C.A.; Wilen, C.B.; Malani, N.; Male, F.; Lee, F.-H.; Bushman, F.D.; et al. The Major Cellular Sterol Regulatory Pathway Is Required for Andes Virus Infection. *PLoS Pathog.* **2014**, *10*, e1003911. [CrossRef]

110. Kleinfelter, L.M.; Jangra, R.K.; Jae, L.T.; Herbert, A.S.; Mittler, E.; Stiles, K.M.; Wirchnianski, A.S.; Kielian, M.; Brummelkamp, T.R.; Dye, J.M.; et al. Haploid Genetic Screen Reveals a Profound and Direct Dependence on Cholesterol for Hantavirus Membrane Fusion. *mBio* **2015**, *6*, e00801-15. [CrossRef]

111. Blanc, M.; Hsieh, W.Y.; Robertson, K.A.; Watterson, S.; Shui, G.; Lacaze, P.; Khondoker, M.; Dickinson, P.; Sing, G.; Rodríguez-Martín, S.; et al. Host Defense against Viral Infection Involves Interferon Mediated Down-Regulation of Sterol Biosynthesis. *PLoS Biol.* **2011**, *9*, e1000598. [CrossRef]

112. Robertson, K.A.; Hsieh, W.Y.; Forster, T.; Blanc, M.; Lu, H.; Crick, P.J.; Yutuc, E.; Watterson, S.; Martin, K.; Griffiths, S.J.; et al. An Interferon Regulated MicroRNA Provides Broad Cell-Intrinsic Antiviral Immunity through Multihit Host-Directed Targeting of the Sterol Pathway. *PLoS Biol.* **2016**, *14*, e1002364. [CrossRef]

113. Oudshoorn, D.; Van Der Hoeven, B.; Limpens, R.W.A.L.; Beugeling, C.; Snijder, E.J.; Bárcena, M.; Kikkert, M. Antiviral Innate Immune Response Interferes with the Formation of Replication-Associated Membrane Structures Induced by a Positive-Strand RNA Virus. *mBio* **2016**, *7*, e01991-16. [CrossRef]

114. Amini-Bavil-Olyaee, S.; Choi, Y.J.; Lee, J.H.; Shi, M.; Huang, I.-C.; Farzan, M.; Jung, J.U. The antiviral effector IFITM3 disrupts intracellular cholesterol homeostasis to block viral entry. *Cell Host Microbe* **2013**, *13*, 452–464. [CrossRef]

115. Cui, H.L.; Grant, A.; Mukhamedova, N.; Pushkarsky, T.; Jennelle, L.; Dubrovsky, L.; Gaus, K.; Fitzgerald, M.L.; Sviridov, D.; Bukrinsky, M. HIV-1 Nef mobilizes lipid rafts in macrophages through a pathway that competes with ABCA1-dependent cholesterol efflux. *J. Lipid Res.* **2012**, *53*, 696–708. [CrossRef]

116. Tang, H.-B.; Lu, Z.-L.; Wei, X.-K.; Zhong, T.-Z.; Zhong, Y.-Z.; Ouyang, L.-X.; Luo, Y.; Xing, X.-W.; Liao, F.; Peng, K.-K.; et al. Viperin inhibits rabies virus replication via reduced cholesterol and sphingomyelin and is regulated upstream by TLR4. *Sci. Rep.* **2016**, *6*, 30529. [CrossRef] [PubMed]

117. Diamond, D.L.; Syder, A.J.; Jacobs, J.M.; Sorensen, C.M.; Walters, K.-A.; Proll, S.C.; McDermott, J.E.; Gritsenko, M.A.; Zhang, Q.; Zhao, R.; et al. Temporal Proteome and Lipidome Profiles Reveal Hepatitis C Virus-Associated Reprogramming of Hepatocellular Metabolism and Bioenergetics. *PLoS Pathog.* **2010**, *6*, e1000719. [CrossRef]

118. Wu, Q.; Zhou, L.; Sun, X.; Yan, Z.; Hu, C.; Wu, J.; Xu, L.; Li, X.; Liu, H.; Yin, P.; et al. Altered Lipid Metabolism in Recovered SARS Patients Twelve Years after Infection. *Sci. Rep.* **2017**, *7*, 1–12. [CrossRef]

119. Vijay, R.; Hua, X.; Meyerholz, D.K.; Miki, Y.; Yamamoto, K.; Gelb, M.; Murakami, M.; Perlman, S. Critical role of phospholipase A2 group IID in age-related susceptibility to severe acute respiratory syndrome–CoV infection. *J. Exp. Med.* **2015**, *212*, 1851–1868. [CrossRef]

120. Vijay, R.; Fehr, A.R.; Janowski, A.M.; Athmer, J.; Wheeler, D.L.; Grunewald, M.; Sompallae, R.; Kurup, S.P.; Meyerholz, D.K.; Sutterwala, F.S.; et al. Virus-induced inflammasome activation is suppressed by prostaglandin D2/DP1 signaling. *Proc. Natl. Acad. Sci. USA* **2017**, *114*, E5444–E5453. [CrossRef]

121. Tay, M.Z.; Poh, C.M.; Rénia, L.; Macary, P.A.; Ng, L.F. The trinity of COVID-19: Immunity, inflammation and intervention. *Nat. Rev. Immunol.* **2020**, *20*, 363–374. [CrossRef]

122. Yan, B.; Chu, H.; Yang, D.; Sze, K.H.; Lai, P.-M.; Yuan, S.; Shuai, H.; Wang, Y.; Kao, R.Y.; Chan, J.F.-W.; et al. Characterization of the Lipidomic Profile of Human Coronavirus-Infected Cells: Implications for Lipid Metabolism Remodeling upon Coronavirus Replication. *Viruses* **2019**, *11*, 73. [CrossRef]

123. Grant, W.B.; Lahore, H.; McDonnell, S.; Baggerly, C.A.; French, C.B.; Aliano, J.L.; Bhattoa, H.P. Evidence that Vitamin D Supplementation Could Reduce Risk of Influenza and COVID-19 Infections and Deaths. *Nutrients* **2020**, *12*, 988. [CrossRef] [PubMed]

124. Holick, M.F. Sunlight and vitamin D for bone health and prevention of autoimmune diseases, cancers, and cardiovascular disease. *Am. J. Clin. Nutr.* **2004**, *80*, 1678S–1688S. [CrossRef]

125. Martínez-Moreno, J.; Hernández, J.C.; Urcuqui-Inchima, S. Effect of high doses of vitamin D supplementation on dengue virus replication, Toll-like receptor expression, and cytokine profiles on dendritic cells. *Mol. Cell. Biochem.* **2020**, *464*, 169–180. [CrossRef] [PubMed]

126. Nieva, J.L.; Madan, V.; Carrasco, L. Viroporins: Structure and biological functions. *Nat. Rev. Microbiol.* **2012**, *10*, 563–574. [CrossRef] [PubMed]

127. Sze, C.W.; Tan, Y.-J. Viral Membrane Channels: Role and Function in the Virus Life Cycle. *Viruses* **2015**, *7*, 3261–3284. [CrossRef] [PubMed]

128. Crawford, S.E.; Hyser, J.M.; Utama, B.; Estes, M.K. Autophagy hijacked through viroporin-activated calcium/calmodulin-dependent kinase kinase- signaling is required for rotavirus replication. *Proc. Natl. Acad. Sci. USA* **2012**, *109*, E3405–E3413. [CrossRef] [PubMed]

 *metabolites* 

Review

Fatty Acids and Membrane Lipidomics in Oncology: A Cross-Road of Nutritional, Signaling and Metabolic Pathways

Carla Ferreri [1,*], Anna Sansone [1], Rosaria Ferreri [2], Javier Amézaga [3] and Itziar Tueros [3]

[1] Istituto per la Sintesi Organica e la Fotoreattività, Consiglio Nazionale delle Ricerche, Via Piero Gobetti 101, 40129 Bologna, Italy; anna.sansone@isof.cnr.it
[2] Department of Integrated Medicine, Tuscany Reference Centre for Integrated Medicine in the hospital pathway, Pitigliano Hospital, Via Nicola Ciacci, 340, 58017 Pitigliano, Italy; rosariaferreri1957@gmail.com
[3] AZTI, Food and Health, Parque Tecnológico de Bizkaia, Astondo Bidea, Edificio 609, 48160 Derio, Spain; jamezaga@azti.es (J.A.); itueros@azti.es (I.T.)
* Correspondence: carla.ferreri@isof.cnr.it

Received: 1 August 2020; Accepted: 23 August 2020; Published: 25 August 2020

Abstract: Fatty acids are closely involved in lipid synthesis and metabolism in cancer. Their amount and composition are dependent on dietary supply and tumor microenviroment. Research in this subject highlighted the crucial event of membrane formation, which is regulated by the fatty acids' molecular properties. The growing understanding of the pathways that create the fatty acid pool needed for cell replication is the result of lipidomics studies, also envisaging novel fatty acid biosynthesis and fatty acid-mediated signaling. Fatty acid-driven mechanisms and biological effects in cancer onset, growth and metastasis have been elucidated, recognizing the importance of polyunsaturated molecules and the balance between omega-6 and omega-3 families. Saturated and monounsaturated fatty acids are biomarkers in several types of cancer, and their characterization in cell membranes and exosomes is under development for diagnostic purposes. Desaturase enzymatic activity with unprecedented *de novo* polyunsaturated fatty acid (PUFA) synthesis is considered the recent breakthrough in this scenario. Together with the link between obesity and cancer, fatty acids open interesting perspectives for biomarker discovery and nutritional strategies to control cancer, also in combination with therapies. All these subjects are described using an integrated approach taking into account biochemical, biological and analytical aspects, delineating innovations in cancer prevention, diagnostics and treatments.

Keywords: cancer cell membranes; fatty acid biosynthesis; essential fatty acids; desaturase enzymes; fatty acid signaling; fatty acid biomarker; sapienic acid; sebaleic acid; molecular nutrition; inflammation

1. Introduction

The development of lipid research in the last two decades has brought a fundamental contribution to the understanding of the main processes for cellular life, in all types of organisms as well as in plants [1]. In particular, fatty acids are the building blocks of the large majority of lipid structures, differentiated from lipids that have steroid and isoprenoid scaffolds. Fatty acids are known for their multiple roles, ranging from energy providers and gene regulators to precursors of signaling molecules and other important metabolites, but it is worth noting that fatty acids in phospholipids have specific structural and functional roles in order to create the envelope of all types of cells, i.e., the cell membrane [2]. In eukaryotes, fatty acids display structural diversity and, as represented in Figure 1 with the most important molecules for the organization of membrane phospholipids, are characterized by specific chain length and number of unsaturations. First of all, the length of the hydrocarbon (hydrophobic) chains requires a certain number of carbon atoms (most often 16–22 carbon atoms) to create the membrane compartment and the thickness of the lipid bilayers. Biosynthesis is initiated with the formation of 16 carbon atoms containing palmitic acid, the first endogenous lipid which is a saturated fatty acid (SFA) (Figure 1) made by the enzymatic system of fatty acid synthase (FAS). Together with the chain length, another structural requirement present in unsaturated fatty acids is the geometry of cis double bonds. The enzymatic system of desaturases introduces the unsaturation in a precise position of the fatty chain (indicated with the carbon atom number; see Figure 1) and this creates a bend (angle of ca. 30 degrees), modifying profoundly the biophysical properties of the molecules [3]. The main endogenous formation of double bonds is due to delta-9 (Δ9) desaturase (also known as stearoyl CoA desaturase SCD-1) operating on palmitic and stearic acids, as shown in Figure 2A.

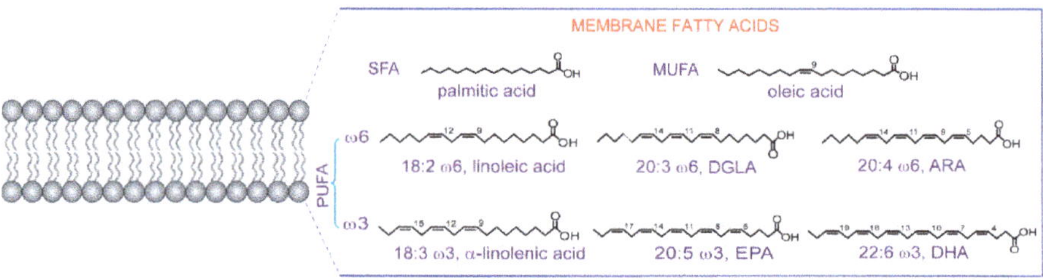

Figure 1. The fatty acid constituents of phospholipids: saturated fatty acids (SFA), monounsaturated fatty acids (MUFA) and polyunsaturated fatty acids (PUFA) are shown with their most present structures in eukaryotic membranes.

On the other hand, the polyunsaturated fatty acid (PUFA) structures are necessary to eukaryotic cells but are not biosynthesized *de novo*, and the precursors of the omega-6 and omega-3 families must be taken from the diet. The structures of the omega-6 and omega-3 precursors are shown in Figure 1 (linoleic acid and α-linolenic acid, respectively) and, after their uptake, other PUFAs are formed and enter into the membrane composition, as shown in Figure 1. In Figure 2B, the two pathways followed for long-chain PUFA biosynthesis are shown, with formation of omega-6 di-homo-gamma linolenic (DGLA) and arachidonic (ARA) acids and omega-3 eicosapentaenoic (EPA) and docosahexaenoic (DHA) acids. As a matter of fact, the transformation to mono- and polyunsaturated fatty acids (MUFA and PUFA) provides the precious building blocks of membrane phospholipids involved in the regulation of permeability and fluidity properties. MUFAs and PUFAs act in a manner opposite to SFA, which instead create the rigidity and the gel status of the lipid bilayer. The role of fats in cancer is generally recognized [4], and the SFA-MUFA pathway has been studied since it is one of the pieces of the puzzling scenario for tumoral cell development and invasion [5]. However, considering that the membrane is necessary for cell formation and reproduction, the ways in which the balance among SFAs, MUFAs and PUFAs influences these steps are still to be defined. MUFAs can be obtained totally by an endogenous process, whereas PUFAs cannot be biosynthesized in humans, as shown in Figure 2B. Due to this "dietary dependency", the effects of an impairment of both exogenous supply and endogenous metabolism needs a comprehensive approach in order to examine cellular metabolism, signaling and nutrition. This is why fatty acid-based membrane lipidomics drives important information in health and diseases and, in particular in cancer "-omics", it is needed for the comprehension of molecular mechanisms and for biomarker discovery.

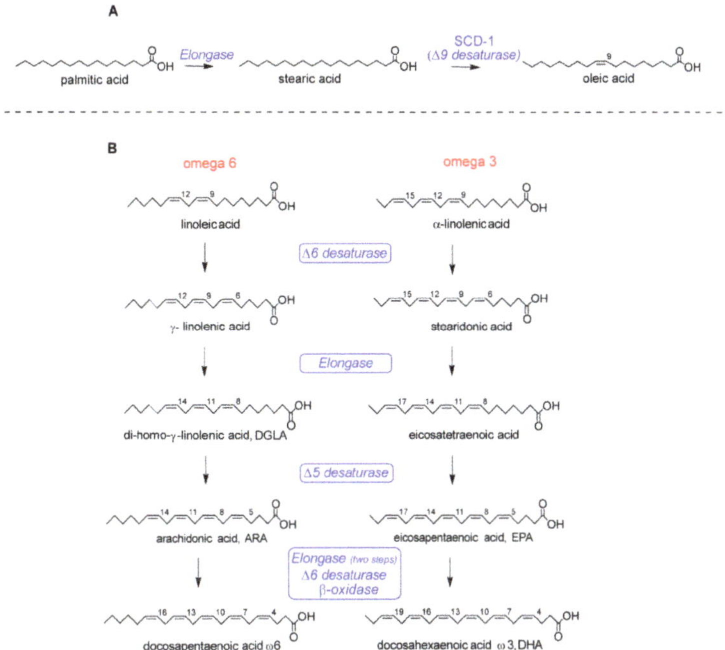

Figure 2. Some metabolic transformations of fatty acids: (**A**) the saturated fatty acid (SFA), palmitic acid, is transformed into stearic acid and the monounsaturated fatty acid (MUFA), oleic acid; (**B**) omega-6 and omega-3 precursors taken from the diet are transformed into the other polyunsaturated fatty acids (PUFA) members of the two families.

Here we wish to remark that a multidisciplinary approach is necessary, where chemical, biological and clinical skills are required all at once. Indeed, in membrane lipids research and medical applications, all these skills are also necessary to address critical issues in protocols: are we fully conscious of the difference in monitoring circulating lipids from those entering the cell membrane composition? Can we make crucial decisions about what is the best sampling procedure for cell membrane lipids? Finally, can we make an effort to unify protocols in one accredited procedure, so that big data can be collected and results can be compared in multicentric studies? In our opinion, analytical and chemical competences here come first, since they are required in order to build up accurate and reliable protocols: the recognition of fatty acid structures must be unambiguous [6], as will be shown in this review, taking into account that fatty acids are tissue-specific and each tissue has its own distribution of these molecular components [7]. Quality control must involve the exact separation of fatty acids from the sample to be analyzed, and if membranes are the target, the procedure must isolate them. This accuracy is fundamental because, after analysis, fatty acids are interpreted for their biological effects, as precursors to lipid mediators and contributors to membrane fluidity.

The contribution of fatty acids to membrane properties has been recognized for a long time, particularly in cancer development [8]. More recently, it has been discussed as evidence that the application of membrane modification and manipulation as part of cancer therapeutical strategies is still not developed [9].

An interplay between biosynthesis and diet regulates fatty acid availability. We gathered the literature on how fatty acids are implicated in tumor onset and progression and how the cancer lipidome reflects the activation of the *de novo* synthetic pathways. In this overview, we wish also to highlight our own work on the discovery of a family of MUFA positional isomers, the n-10 family, as new biomarkers of the metabolic shift that allows human cells to build up the first endogenous PUFA component, sebaleic acid [10]. The review also covers the link between obesity and cancer in order to understand why and when lipid supply causes health complications, highlighting specific fatty acids for their biological effects, signaling and contribution to the membrane properties that influence cell growth and death. From this scenario, several hints emerge for innovative strategies in cancer prevention (primary and secondary) using fatty acid-based membrane lipidomics and fatty acid balance.

2. Fatty Acids and Lipid Supply for Membrane Formation in Cancer

Cancer is a very complex disease due to the large number of factors involved. Cells develop a great capacity to grow, proliferate and survive under stress conditions. They modify several processes to achieve favorable environments, such as the metabolism of lipids, carbohydrates, proteins and nucleotides, being able to maintain the functionality of the structures and functions [11,12]. Adapted metabolic pathways allow cancer cells to obtain energy, form metabolic intermediates and synthesize fatty acids, even when the exogenous availability of these compounds is reduced. For example, the hyperactivation of the phosphatidylinositol-3 kinase and AKT (PI3K-AKT) transduces the signal from the hormone insulin to drive glucose uptake and is one of the most frequently mutated pathways in cancer [13]. In this case, glycolysis is favored, leading the cells to form pyruvate, which could be used for ATP synthesis or for *de novo* lipogenesis [14]. The hyperactivation of PI3K-AKT also activates the glutamate pyruvate transaminase 2 (GPT2), favoring glutamine anaplerosis to supply sufficient metabolites for FA synthesis and, finally, remodel the cellular lipidome [15]. In the latter case, it has been shown that such remodeling makes lipids an important hallmark of cancer [16]. The overexpression of FA transporters, such as fatty acid translocase CD36, plasma membrane fatty acid-binding proteins (FABP) and the fatty acid transport protein family (FATP), elevates the uptake of exogenous FAs with their subsequent storage in lipid droplets (LDs), as is known in ovarian cancer, and this is in connection also with adipose tissue, as will be explained in Section 5 [17]. To fully evaluate the lipid supply and understand their role in cancer, we must distinguish between *de novo* synthesized and dietary fatty acids, as explained below.

2.1. De Novo Synthesis of Saturated and Monounsaturated Fatty Acids

Combined with a greater capacity for the biosynthesis of lipids, cancer cells are not only able to maintain lipid homeostasis but also to provide ATP and NADPH in conditions of metabolic stress and sufficient precursors to deal with the formation of lipid rafts that are essential for protein dynamics in membranes and cell survival [18,19]. Since phospholipids are the basic units of membranes, in cancer disease different enzymes involved in their endogenous synthesis are highly expressed, such as ATP-citrate lyase (ACLY), acetyl CoA carboxylase (ACC) and fatty acid synthase (FAS) [20,21]. Each of them represents itself a target of study against cancer. Whereas in nutrient-unlimited and aerobic conditions, the glucose metabolism forms citrate through the tricarboxylic acid cycle (TCA), to later convert into acetyl CoA, being the key for *de novo* synthesis, cells also develop an alternative strategy to form FA when there is a lack of nutrients and hypoxia (Figure 3). Several groups have proposed that, in these mentioned cases, the TCA cycle can be modified to run in reverse and use glutamine from storage to act as source of acetyl CoA [22,23], whereas others describe that acetyl CoA can be obtained from histone deacetylation [24]. In the case of FAS, in addition, it responds to signals from the activation of the AKT and MAPK (mitogen-activated protein kinase) pathways, which, in turn, are also favored in cancer processes [25]. Besides the fact that the production of FAs is essential to sustain the structure and demands of membranes, their composition is also decisive in guaranteeing the functions of the dividing cells. The production of monounsaturated fatty acids (MUFA) from SFA provides fluidity, functionality and flexibility, which are essential for tumor cells. This step involves the action of delta-9 desaturase enzyme (known also as stearoyl CoA desaturase, SCD-1, and reported in Figure 3), which can act on both palmitic (16:0) and stearic (18:0) acids. As with the enzymatic complex from *de novo* synthesis, SCD-1 is overexpressed in cancer and regulated by different signaling cascades such as MAPK and AKT or systems such as p53 [26], attracting interest in their inhibition [4,5].

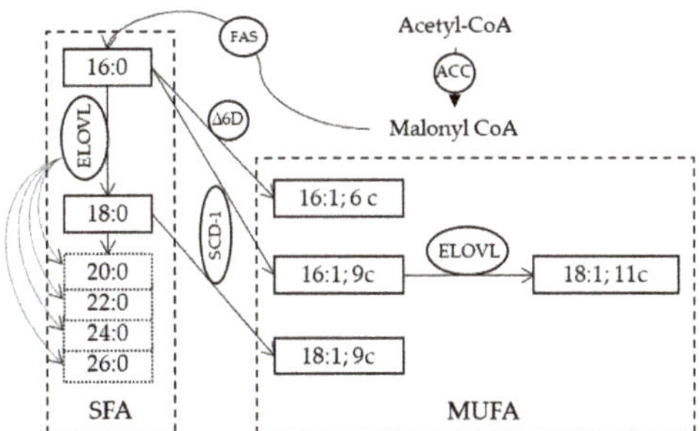

Figure 3. The *de novo* synthesis of saturated fatty acids (SFA) starting from acetyl CoA and the transformation to monounsaturated fatty acids (MUFA) by two desaturase enzymes. Structures of some of these fatty acids are shown in Figure 1. ACC: acetyl CoA: carboxylase; FAS: fatty acid synthase; ELOVL: elongase enzyme; Δ6D: delta-6 desaturase (Δ6); SCD-1: stearoyl CoA desaturase.

Cancer cells modify lipid metabolism in order to respond to environmental modifications. Hypoxia, for example, affects acetyl CoA formation from glucose and SCD-1 activity, as they are oxygen dependent. However, in this case, tumoral cells escape the need for fatty acid synthesis by increasing the uptake of lysophospholipids as a shortcut to prepare phospholipids. FABPs are transcriptional targets of hypoxia-inducible factors (HIFs) that facilitate extracellular scavenging of long-chain unsaturated lysophospholipids, which can be used as a nutrient source under conditions of metabolic stress [15]. Interestingly, this effect can occur even in aerobic conditions after oncogenic RAS activation, making it independent from SCD-1, to achieve sufficient MUFAs [27]. It is worth highlighting the existing debate about whether fatty acids used by cancer cells are of endogenous or exogenous (dietary) origin, since some studies did not find differences [28]. Lipidomic studies have a fundamental role in the elucidation of the decisive contribution of fatty acid biosynthesis, evidencing storage, lipolysis and membrane remodeling implied in tumor onset, progression and metastasis. Table 1 summarizes the most important fatty acid-driven mechanisms and related biological effects.

Table 1. The main fatty acid-driven mechanisms and biological effects in cancer onset, growth and metastasis.

Entry	Implicated Mechanism	Biological Effects	Lit
1	Desaturation from 16:0 to 6c-16:1 (sapienic acid)	Support of membrane biosynthesis during proliferation	[29,30]
2	mTORC2 regulation of lipid metabolism	Glycolysis and lipogenesis activation	[31,32]
3	Acetyl CoA synthetase 2 promotion of acetate utilization	Maintaining cancer cell growth under hypoxia and metabolic stress	[33]
4	Adipokines mediation of ovarian cancer metastasis	Induction of lipolysis and β-oxidation to provide energy	[34]
5	Enhanced uptake of exogenous lipoproteins	(a) Cholesteryl ester accumulation, induced by PTEN loss and PI3K/AKT activation, to sustain cancer aggressiveness (b) Increased amount of cholesterol and overexpression of low-density lipoprotein receptor to boost proliferation (c) Sustaining proliferation and aggressive potential of breast cancer tumors	[35] [36] [37]
6	Increase in lipid droplets in tumor cells	Increased COX-2 expression and storage in droplets, with effects on proliferation	[38]
7	Stearoyl CoA desaturase essentiality for cancer cell survival	Inhibition of FA desaturation, blocking the synthesis of lipids and impairing cell survival	[39]

Readers are directed to the original references cited in Table 1 to elaborate on each subject appropriately. Among these mechanisms, a recent one (Entry 1) was individuated by some of our group, investigating the analytical protocols for efficient separation of the MUFA positional isomers, which will be addressed in Section 4. Knowledge of different mechanisms is necessary for research of new therapeutic targets that can act in a synergic manner, to disturb organization of membrane lipids, destabilize lipid rafts and activate apoptosis signaling [18,19].

2.2. PUFA Intake and Omega-6/Omega-3 Balance for Membrane Fatty Acid-Mediated Signaling

On the basis of the importance of phospholipids for cell formation, the "membrane hypothesis" can be drawn, for which the initial steps of death or life of tumoral cells could be also driven by the quality of the membrane fatty acids. To create a fatty acid balance among SFA, MUFA and PUFA residues in the individual, it must be taken into account that the dietary intake of omega-6 and omega-3 regulates the presence of PUFA residues in lipid pools. Once the individual pool is formed, it exerts strong control upon the membrane composition and the types of fatty acids that will be detached from membrane phospholipids to determine the related cell fate. Indeed, the phospholipase A_2 (PLA$_2$)-induced release of fatty acids from membranes is a well-known process, involved in the membrane remodeling cycle, i.e., the Lands cycle [40]. It does not discriminate between omega-3 and omega-6 structures, thus highlighting the importance of the above-mentioned balance present in membranes for pro- and anti-inflammation signaling. Indeed, every time that the release in the cytoplasm of arachidonic acid from phospholipids occurs by PLA$_2$, causing the subsequent formation of its eicosanoid mediators, other omega-6 and omega-3 fatty acids are released as well, such as di-homo gamma-linolenic acid (DGLA), eicosapentaenoic and docosahexaenoic acids (EPA and DHA). They are, in their turn, precursors of other lipid mediators with mainly anti-inflammatory properties, thus integrating the final inflammation and resolution responses [41]. Obviously, the result depends on the presence and balance of these fatty acids in membranes. Since recent data suggest inflammation as an important aspect in activating cancer proliferation pathways and resistance, it is evident that the membrane predisposition through its fatty acid composition is a piece of information to gather in the puzzling scenario of the cancer disease. Cancer is generated not only by genetic alterations, as a result of intrinsic or exogenous mutagens, but also by long-term exposure to acute or chronic inflammation. It is now becoming clear that the proliferation of cells alone does not cause cancer. However, sustained cell proliferation in an environment rich in inflammatory cells, growth factors and DNA-damage-promoting agents is necessary in the neoplastic process, promoting survival and migration. In this way, the causal relationship that exists between inflammation, innate immunity and cancer is more widely accepted [42]. Many of the molecular and cellular mechanisms that mediate this relationship are still unresolved, but the role that FAs play in inflammation processes related to cancer is increasingly relevant. Indeed, the role of dietary PUFAs omega-6 and omega-3 is a matter for discussion of their effects on cancer incidence and evolution [43]. The negative impact of Western diets, rich in omega-6, has recently been described in societies in which the intake of omega-6 fatty acids was traditionally in balance with that of omega-3. The number of cases with diseases associated with inflammatory processes, as well as their worse prognosis, has increased [44]. The scientific debate on the importance of the PUFA intake for cancer risk has not yet reached a conclusion. Large population studies are needed to address this task. For example, in a recent population-based (100,881 participants) prospective cohort study, using self-reported dietary data from the Västerbotten Intervention Programme, statistically significant associations have been described between a more anti-inflammatory or healthier diet and reduced risk of cancer [45]. In the development of inflammation mediated by PUFAs, both omega-6 and omega-3 FAs play crucial roles, since these two families of FAs are in constant competition with each other and, broadly speaking, they develop opposite effects. In this sense, omega-6 FAs are more related to inflammation (through the arachidonic acid (AA) cascade) and omega-3 to anti-inflammatory effects. Figure 2B shows how both omega-3 and omega-6 are closely related, by sharing the same enzymes for each step of their transformations. This fact implies that, from the beginning, there is a need for balance between the families, since, if one is favored, it will hinder the synthesis of products from the other. It is worth adding that there are regulations also from the FA-derived mediators' formation and interactions: in the formation of prostaglandins (PG) and leukotrienes (LT) from omega-6 or omega-3 fatty acids, cyclooxygenases (COX-1, COX-2) respond more intensely for intermediates of omega-6 origin in the case of PG (PGD2, PGE2). Furthermore, not only enzymes but also receptors show different affinities, again being favored by PGs and some LTs of the omega-6 series [44]. Thus, in an environment in which omega-6 is biochemically favored, sufficient intake of omega-3 can have a key effect on the PUFA metabolism dynamics, as well as on the intensity of the action of the different eicosanoids. The involvement of PUFAs in cancer is demonstrated here by a few representative examples: regarding omega-6, AA modulates the activation of the nuclear factor kappa-light-chain-enhancer of activated B cells (NF-κB), involved in the immune response and altered in this disease. It also induces focal adhesion kinases, which promote progression and metastasis [46]. The signaling activity of AA is exerted through its transformation to PGE_2, and, indeed, the overexpression of COX-2 protein was highlighted in several types of cancer, whereas human breast cancers frequently have high PGE_2 levels, and breast tumors with high COX-2 protein levels are more likely to metastasize [47]. This led to consideration of COX inhibitors for cancer therapies, evaluating also their effects on angiogenesis. In this scenario, the PG interaction with EP receptors (E-series prostaglandin receptors), a family of G-protein coupled receptors designated as EP1–3 and EP4, was individuated, with the corresponding activation or deactivation of the c-AMP cascade or the extracellular signal-regulated kinase (ERK) 1 and ERK2 by way of PI3K [48]. As a matter of fact, not all the omega-6 FAs are equal in the tumor effects, and some studies suggest that, unlike the downstream omega-6 AA, the upstream omega-6s, such as linoleic acid (LA), γ-linolenic acid (GLA) and di-homo gamma-linolenic acid (DGLA), may possess anticancer effects. In fact, GLA and DGLA may exert anticancer properties via the production of PGE_1. Although more work is needed to clarify the molecular basis of the anticancer effects of GLA and DGLA, it has been demonstrated that they are able to regulate gene and protein expression, disrupting cell-cycle progression and inducing apoptosis, a mechanism which implies also a direct effect on the lipid composition in cell membranes [48]. Regarding omega-3 PUFAs, they have the opposite effect to that mentioned for AA. For example, the combination of EPA and DHA decreases the production of eicosanoids formed from AA, leading to the inactivation of NF-κB and hindering proliferation [46]. They are also able to inhibit the activity of AKT protein, which is involved in cell survival and the inhibition of apoptotic processes [49]. Furthermore, PUFAs are involved in different processes such as lipid peroxidation, cell oxidative stress [50,51] and regulation of gene expression for controlling growth factor mediated carcinogenesis [52]. Moreover, there are other mechanisms involving lipid-based events that affect human health. The interest in ethanolamides has increased, as they are biological compounds that may have important beneficial actions by controlling inflammatory responses without being classical steroidal and non-steroidal anti-inflammatory drugs, that act by inhibiting the cascade of arachidonic acid. Palmitoyl ethanolamide (PEA) is an endogenous lipid mediator that can be found in foods (tomato, soybean, peanut) formed by palmitic acid and ethanolamide, with anti-inflammatory, neuroprotective and analgesic activities [53]. The suggested mechanisms of action involve various metabolic pathways in which some receptors seem to be activated directly (peroxisome proliferator activated receptor alpha (PPAR-α) and orphan G-protein coupled receptor 55 (GPR55)) or indirectly (cannabinoid receptors, CB_1 and CB_2) and transient receptor potential vanilloid type-1 channel (capsaicin receptor or TRPV1)) [53,54] to modulate mast cell activation and degranulation [53]. In colitis-associated cancer, more specifically, PEA inhibits angiogenesis through PPAR-α, suggesting a protective effect in both inflammation and cancer, being able to reduce mucosal damage, disease progression and carcinogenesis [55]. To conclude with the PUFA scenario, a brief mention of PUFA peroxidation products and the oxidative-based pathways to induce apoptosis in cancer, including ferroptosis, is made here, directing readers to reported works in these fields [56,57].

3. The Membrane Fatty Acid-Based Profile in Cancer and the Relevance of Erythrocytes

Considering the importance of fatty acids in membranes, a "bottom-up" approach can reveal the main changes in fatty acid profile between healthy controls and cancer patients. A systematic study of membrane fatty acid-based profiles in large populations is lacking, as is the agreement regarding the biological compartment in which fatty acids are measured. Therefore, there is large heterogeneity of data that does not currently allow us to draw conclusions about the most significant changes occurring in fatty acids in the human body. However, it is important to remark that, from the emerging scenario of fatty acids' involvement in cancer metabolism, it is reasonable to focus the efforts towards the cell membrane compartment. In this respect, the significance of the erythrocyte cell membrane and its fatty acid composition is highlighted for several reasons: (a) the numerosity of erythrocytes and the predominance among other tissues of these cells, which constitute 70–80% of the total cells formed each day [58], rendering them the best representatives of the availability of fatty acids to construct membrane phospholipids; (b) the continuous exchange of the erythrocyte membrane phospholipids with lipoproteins and tissues in order to reshape the molecular content and satisfy homeostatic requirements [59,60]; (c) the biological mission to reach tissues and organs during the erythrocyte's average life span of four months in humans, which requires the best performance of membrane properties in order to efficiently exchange gases; (d) the presence of the most representative SFA, MUFA and PUFA molecules and the preferred storage of arachidonic acid, known to be present in membrane phospholipids by 13–17%, as well as of other precious PUFAs [6]. Based on these considerations, the fatty acid profile can give information on the balance of these molecular components in erythrocyte phospholipids and help to establish the changes occurring under healthy and unhealthy conditions. Indeed, the fatty acid-based membrane lipidome monitoring used in different human conditions revealed how the endogenously and exogenously-derived fatty acids of erythrocytes are affected [61–64]. In Table 2, the relevant data relating to erythrocyte fatty acid monitoring from studies on cancer patients are gathered, highlighting the cancer types, the country and the number of patients and detailing the most important conclusions of each study. It is interesting to note that the SFA-MUFA transformation emerges as an important biomarker of cancer status, as well as the ratio between omega-6 and omega-3 PUFAs.

Table 2. Fatty acids in cancer: collection of data from studies on erythrocyte membrane fatty acids in patients affected by different types of cancer and emerging biomarkers.

Cancer Type	Country	Human Cohort Size	Outcomes	Reference
Breast/Prostate/Liver/Pancreas/Colon/Lung	Puerto Rico	255 cancer patients 2800 non cancer patients 34 healthy volunteers	Lower levels of stearic acid and increased content of oleic acid. EPA and DHA/ALA ratio to estimate PUFA imbalances in cancer patients.	[65]
Colorectal	Japan	61 cases 42 controls	Less EPA and linoleic acid and high levels of arachidonic acid in cancer patients.	[66]
Breast	Italy	71 cases 141 controls	High oleic acid and low stearic acid in patients. Oleic acid and MUFA positively associated with breast cancer risk. Saturation index (stearic/oleic acids ratio) inversely correlated.	[67]
Colorectal	Italy	13 cancer patients 13 patients with no malignant diseases	Lower levels of n-3 PUFAs and higher n-6/n-3 PUFA ratio in cancer patients.	[68]
Breast/Colon/Lung	Spain	54 cases 34 controls	Less SFA (C16:0 and C18:0), high MUFA (9c-C18:1 and 11c-C18:1) compared to controls. In the PUFA families, increase in n-6 C18:2 and C20:3 (15.7% and 22.2%, respectively).	[69]
Colorectal	France	328 cases 619 controls	High levels of pentadecanoic and heptadecanoic acids; oleic acid and linoleic acid associated with the risk of advanced adenomas. EPA and DHA negatively associated with the risk of advanced adenomas.	[70]
Basal Cell Carcinoma	Iran	40 cases, 40 controls	Low palmitic and high oleic acid levels in cancer patients. Saturation index (stearic/oleic acids ratio) lower in cancer patients.	[71]
Basal Cell Carcinoma	Iran	40 cases, 40 controls	Higher AA, total omega-6 and LA in cancer patients, lower omega-3.	[72]
Colorectal	Japan	74 cases, 221 controls	Risk of colorectal cancer inversely associated with DHA, AA and PUFAs and positively associated with palmitic acid, SFAs and SFA/PUFA.	[73]
Breast	China	322 cases, 1030 controls	Significant direct association among palmitic, γ-linolenic, palmitoleic and vaccenic acids and risk of breast cancer. Total n-3 fatty acids, EPA and 16:0/16:1 saturation index associated with significantly lower risk of breast cancer.	[74]
Prostate	USA	127 cases, 183 controls	MUFA and α-linolenic/EPA ratio associated with reduced risk of prostate cancer.	[75]
Advanced squamous cell lung carcinoma (SCC), lung adenocarcinoma (ADC) and small cell lung cancer (SCLC)	Spain	63 patients, 50 controls	AA, EPA, palmitic, oleic acids biomarkers in diagnosis and in other aspects related to clinical disease management of cancer.	[76]

From these results, it is also clear that a large multicentric population study would definitely yield important results regarding the adoption of the fatty acid membrane profile for the follow-up of patients and therapies, assessing the importance of fatty acid biomarkers in primary and secondary prevention and discovering the molecular and clinical effects of personalized diets for cancer. We believe that the work in progress to map genetic alterations that control cell-cycle progression, apoptosis and cell growth in cancer [77] can be combined with molecular indicators such as membrane lipidomics in each tumor type to obtain more insights into the lipid pathways in cancer and clarify the epigenetic role of nutrition.

4. The Study of the Cancer Lipidome and the Discovery of *De Novo* Pathways: Fatty Acid Positional Isomers as New Biomarkers of Metabolic Shift

Lipidomics in cancer helps to clarify the connections between disease and lipidome, discovering novel lipid biomarkers for diagnosis as well as alternative and synergic strategies for therapy [78]. As mentioned before, the intake of essential fatty acids (EFA) with the omega-6/omega-3 PUFA balance is of crucial importance since membranes cannot be formed without this supply. The essentiality of PUFA derives from the fact that the insertion of a second double bond in the MUFA structure cannot occur in eukaryotic cells, which means that cells do not have desaturase enzymes able to convert oleic acid (as well as palmitoleic and vaccenic acids) into PUFA in the biosynthesis (see Figures 1 and 3). Since neither healthy nor cancerous cells can be formed without PUFA, it can be asked whether the dependence on dietary PUFA is a common feature and a limiting step of both types of cell metabolism. The answer to this question is not as straightforward as it seems, and in fact only recently have investigations been directed toward the study of the influence of metabolism and diet on the human lipidome. In lipidome analysis, it was also discovered that chemical skills are very important to create unambiguous protocols and distinguish fatty acid structures, especially those presenting unsaturations. A seminal example is provided by the report demonstrating for the first time of the presence of sapienic acid in various fractions of human plasma. This is a positional isomer of palmitoleic acid, which has the double bond in C6-C7 instead of C9-C10 [79]. The analytical approach for the unambiguous characterization and discrimination of positional and geometrical fatty acid isomers having the 16:1 structure is crucial for the determination of the sapienic acid presence. We described in detail the protocol of fatty acid analysis, which includes a crucial derivatization step to localize the double bond position, using the well-known dimethyl disulfide (DMDS) adducts and its diagnostic fragmentation in mass spectrometry [6,10,63,79]. It must be added that such derivatization procedure and mass spectra can be performed by regular equipment in chemical labs, and do not require specialized and expensive instrumentation. The quantitation of this fatty acid was performed in cholesteryl esters isolated from human plasma of healthy people (n = 5) (50.0 ± 4.0 ng/mL) and in commercially available human low density lipoprotein (LDL) samples (35.0 ± 2.0 ng/mL). How these levels are affected by health conditions in large cohorts remains to be thoroughly explored. These findings prompted us to understand in more detail the biosynthetic origin of sapienic acid. It is reported that, compared to all other types of cells that primarily form oleic acid (Figure 2A), sebocytes change their palmitic acid metabolism by the intervention of delta-6 (Δ6) desaturase enzyme (Figure 4) [80]. However, the systemic role of sapienic acid was not explored, and it was not highlighted the crucial step, that is the partition of palmitic acid between SCD-1 and delta-6 desaturase enzymes (see Figure 3). Whether this partition indicates a metabolic diversion with health significance is under current investigation. As a matter of fact, palmitic acid is an unusual substrate for delta-6 desaturase, which is an enzyme mostly involved with exogenous omega-6 and omega-3 EFA; therefore, the activation of sapienic acid biosynthesis could be attributed to several reasons, including (a) strong availability of the SFA substrate due to FAS activation, the latter well known in cancer [12,26,81]; (b) enzymatic activity competition or lack of normal intake/ presence of PUFA substrates [80,82]; (c) involvement of enzymatic polymorphism and competitive activity of desaturase for PUFA and SFA metabolisms [83]. Aiming at exploring sapienic acid and the other positional MUFA isomers in cell metabolism, we used the human colon carcinoma cell line Caco-2 to compare the results of supplementation of sapienic and palmitoleic acids (150 and 300 µM), discovering that both are rapidly incorporated into membrane phospholipids and also that the former is converted to 8cis-C18:1 and 5cis, 8cis-18:2, as depicted in Figure 4, bringing also these two fatty acids in the cell membrane phospholipid composition. The n-10 fatty acid family has a still unexplored meaning for cancer cells and we were the first to demonstrate in a cancer cell line that it involves a unique type of endogenous PUFA biosynthesis (i.e., sebaleic acid; Figure 4) leading, more importantly, to its incorporation into membranes.

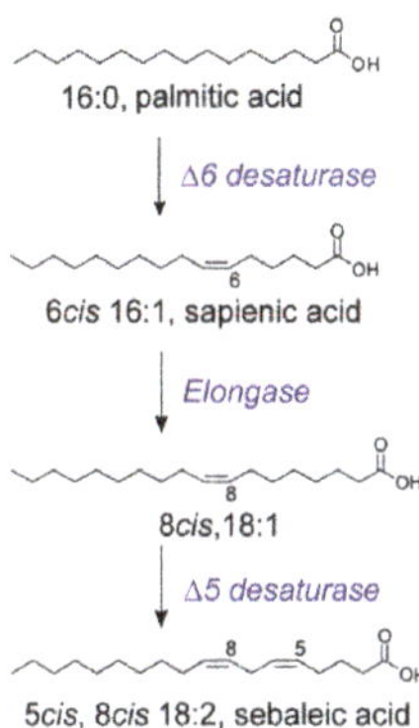

Figure 4. The metabolism of palmitic acid to sapienic acid (6cis-16:1) and its subsequent transformation to obtain the PUFA, sebaleic acid (5cis, 8cis-18:2).

The concomitant isolation of cholesteryl esters and triglycerides from the cell line demonstrated that the n-10 fatty acids "invade" all lipid classes, and, even at high concentrations (300 µM) and at long time exposures, they are not harmful (sapienic acid EC_{50} 232–265 µM for 96 h) [29]. In the same study, the biophysical properties of the cell membranes were monitored by two-photon fluorescent microscopy, using Laurdan as a dye, showing that the supplementation of sapienic acid, with respect to its positional isomer palmitoleic acid, increased fluidity in several regions, evidently correlated with the formation and distribution of n-10 MUFA and PUFA in lipid domains. Following the interest in extracellular vesicles EVs (exosomes) as relevant sites for cancer metabolism and diagnostics [84], we investigated the presence of the n-10 fatty acid family, comparing membrane phospholipids and EVs of prostate cancer cell lines with different degrees of aggressiveness: PC3 (prostate cancer) and LNCaP (prostate derived from metastatic site: left supraclavicular lymph node), the former being more aggressive [10]. We found that 12–13% of the membrane fatty acids of these cell lines were composed of n-10 fatty acids, with the sapienic acid content >7%. In EVs, n-10 fatty acids were 9% for PC3 EVs and 13% for LNCaP EVs, with statistically significant increases in 8cis-18:1 and 5cis, 8cis-18:2, which is relevant considering that the EV are involved also in the transport of biologically active lipids and lipid metabolites to feed cancer tissues. This discovery can have a strong impact also in cancer diagnostics and follow-up of intervention efficacy. We envisaged that the sapienic/pamitoleic ratio, found equal to 3.5 in prostate cancer cells, also provides a measure of the partition into two metabolic pathways, and, in these cell lines, the delta-6 desaturase transformation of palmitic acid was found to be unusually high. A parallel evaluation of gene expression for desaturase (FADS) and elongase (ELOVL) enzymes by qRT-PCR (quantitative real time polymerase chain reaction) evidenced significant increases in FADS expression in PC3 with respect to LNCaP cells, and the higher expression of ELOVL5 in PC3 compared to LNCaP cells with ELOVL6 significantly lower. We found interesting evidence of higher desaturase activity in the most aggressive PC3 cell line, and suggested deepening the study of FADS3 desaturase, which, so far, has an uncertain metabolic role [85]. Indeed, the role of desaturase enzymes represents an important aspect in cancer metabolism and is also considered as a target in anticancer therapy, as reported in reviews [86] showing these strategies applied in preclinical trials. Regarding elongases, most of them are tumor specific: for example, ELOVL1, ELOVL5, ELOVL6 and ELOVL2 are highly expressed in breast cancer [87,88] and ELOVL7 in prostate cancer [89]. In the new scenario of the n-10 fatty acid family, a recent work confirmed the presence of sapienic acid in different cancer cell lines, defining it as a contributor to cancer plasticity [30], and another paper reported an increase in the transformation of palmitic acid to sapienic acid induced by the increase in mammalian target of rapamycin (mTOR) and sterol regulatory element-binding protein 1 (SREBP-1) signaling in mouse embryonic fibroblasts (MEFs) and U87 glioblastoma cells [90]. In this report, the inhibition of the two signaling pathways led to a decrease in sapienic acid biosynthesis. On the other hand, it must be recalled that fatty acids' enzymatic activities can be influenced by dietary fats, as previously shown for the competition between palmitic acid and PUFA omega-6 and omega-3 precursors [91].

New pathways involving SFA and MUFA are going to be discovered, provided that analytical protocols are able to give satisfactory results, such as was recently shown by the transformation of oleic acid in MCF7 cell lines into an eicosanoic fatty acid (7cis, 11cis-20:2) obtained by the unusual activity of FADS1 desaturase introducing a double bond at the level of C7 and not C5 [92]. Considering all the published work on the subject, so far only our experiments with cancer cell lines demonstrated the new pathway that brings about endogenous PUFA synthesis (sebaleic acid) and determined the n-10 FA insertion at the level of membrane phospholipids. We believe that this outcome of the sapienate metabolism is the real contribution to cancer plasticity, strongly influencing fluidity changes that are deeply embedded in cancer signaling and metabolism. We envisage that the pathway of sapienic acid will have a strong development in metabolic, therapeutic and nutritional research; here, we have provided a careful literature summary of the various contributions available so far, that we hope will be useful to researchers interested in the field.

5. Link between Obesity and Cancer: When the Lipid Supply Becomes Dangerous

Despite the difficulty of definitively proving that obesity is one of the causes of cancer, it remains a recognized risk factor contributing to the development and progression of tumors [93]. Several observational studies evidenced that obese and overweight subjects have a higher risk of developing cancer than lean subjects; in 2016, the International Agency for Research on Cancer (IARC) declared that obesity was associated with an increased risk for 13 types of cancer, indicated in Table 3 with their corresponding epidemiological studies [94,95].

Table 3. Increased risk for 13 cancer types correlated to overweight/obesity (% increased risk OW/OB vs. lean) and their corresponding epidemiological studies.

Cancer Type	Increased Risk (OW/OB vs. Lean)	References
Endometrial	150–200%	[96,97]
Esophageal	200–400%	[98,99]
Gastric cardia	168–188%	[100,101]
Liver	17–89%	[102–104]
Kidney	200%	[105–107]
Multiple myeloma	10–20%	[108–110]
Meningioma	10–20%	[111,112]
Pancreatic	50–60%	[113,114]
Colorectal	30–60%	[115–117]
Gallbladder	20–60%	[118,119]
Breast	20–40%	[120–123]
Ovarian	10–30%	[97,124]
Thyroid	10–30%	[125–127]

The clarification of the mechanisms binding obesity to cancer is crucial for the diagnosis and implementation of effective therapies. Here, we have gathered the main molecular pathways connecting adipose tissue (AT) and adipocytes with cancer cells in the tumor microenvironment, as well as their impact on cancer growth, invasion and metastasis. We summarize relevant connections between adipose and cancer tissues in Figure 5.

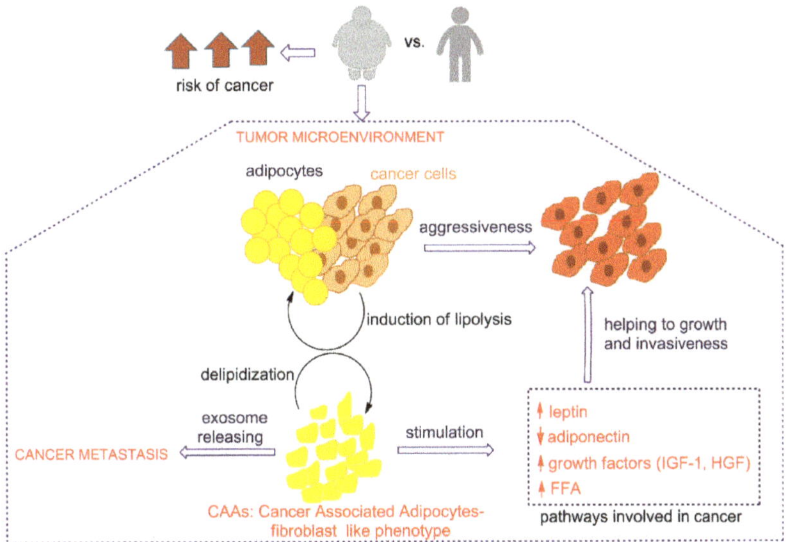

Figure 5. Relevant metabolic connections between adipose and cancer tissues; the arrow ↑ means increase, the arrow↓ means decrease.

The close localization between adipose tissue and cancer cells, immoderately increased in obese subjects due to the effect of excess of calories not consumed, induces a deep modification of the phenotype and functioning of adipocytes, which become cancer-associated adipocytes (CAA), promoted for the induction of lipolysis by cancer cells. Adipocytes are decreased in number and size, showing delipidization and de-differentiation to fibroblast-like phenotype [128–131]. It is well known that the exposure of adipocytes to cancer cells for long periods, with consequent fibroblast morphology, induces the formation of cancer cell fibroblast populations that are involved in tumor invasiveness [132]. In this context, the transformation of adipocytes provides a total alteration of their secretory function involved in endocrine, metabolic and immune systems. The ways in which the specific fatty acid status of adipocytes is involved in the support of cancer cells growth and metastasis are not yet well defined. The identification of several fatty acid unbalances in the erythrocyte membranes of obese patients certainly highlights derangements of lipid metabolism, including the above mentioned sapienic acid pathway [63]. Increased release of free fatty acids accompanies altered levels of adipokines and pro-inflammatory cytokines, growth factors and hormones [133]. The signaling in several types of cancer cells is sustained by adipokines secreted by adipocytes, mainly including leptin, adiponectin, oestrogens, insulin-like growth factor 1 (IGF-1) and hepatocyte growth factor (HGF). In particular, the leptin/adiponectin ratio can be an interesting value to examine, due to the opposite effects of these hormones. Leptin stimulates a cascade of signaling events inducing JAK2/STATs, MAPK/ERK 1/2, PI3K/AKT and PKC, JNK, p38 MAPK and AMPK pathways in diverse cellular types (see abbreviations) [134]. The mechanism implicates the interaction with transmembrane leptin receptor (LRb) that, if phosphorylated, mediates downstream LRb signaling controlling STAT3 (signal transducer and activator of transcription 3) and ERK activation [135,136]. Simultaneously, high levels of leptin induce the stimulation of monocytes into macrophages, leading to chronic, obesity-associated inflammation. Leptin increases the expression of anti-apoptotic proteins, inflammatory markers (tumor necrosis factor, TNF-α, interleukin IL-6) and angiogenic factors (vascular endothelial growth factor, VEGF), all processes involved in cancer cell survival, proliferation and migration [137,138]. On the other hand, adiponectin is inversely correlated to the body mass and cancer, inducing apoptosis and decreasing tumor vascularization. It modulates multiple signaling pathways, exerting its physiological and protective functions through the receptors AdipoR1 and AdipoR2 [139–141]. It is also able to block angiogenesis, inhibiting endothelial cell proliferation induced by FGF2 (fibroblast growth factor 2) as well as the migration of endothelial cells by VEGF. Furthermore, adiponectin inhibits cancer growth and proliferation, interfering with several pathways like AMPK, MAPK and PI3K/AKT, ERK1/2-MAPK pathway and GSK3/catenin, inducing G0/G1 cell-cycle arrest [142–145]. Adiponectin-induced cell death is also accompanied by an increase in intracellular reactive oxygen species (ROS). As a matter of fact, adiponectin pre-treatment suppresses leptin-induced ERK and AKT signaling [146]. Here, we only mention the roles of insulin and glucose levels and their interactions with specific receptors, such as insulin-like growth factor 1 (IGF-1) and hepatocyte growth factor (HGF), that correlate with increased risks of specific cancers, like ovarian and breast cancers, mainly through activation of PI3K/AKT and MAPK pathways [147], while the inhibition of IGF-IR kinase activity prevents the growth-promoting effect of adipocytes on breast cancer cells [148]. Additional factors are connected with the altered environment of adipose tissue and cancer for the increase in inflammatory conditions, with consequent liberation of pro-inflammatory mediators, among them TNF-α and IL-6, contributing to the growth and differentiation in tumors like lymphoma, pancreatic and liver cancers. TNF-α induces carcinogenesis, activating the nuclear transcription factor NF-κB that prevents apoptosis, allowing enhanced cell survival, growth and proliferation [134,149]. IL-6 is normally elevated in obesity, induces JAK-STAT3 signal transduction and, stimulates cell proliferation, differentiation and metastasis. It mediates cell proliferation through the MAPK pathway; in fact, in some studies, the inhibition of MAPK stopped proliferation in the presence of IL-6, evidencing the role of cytokines in cell proliferation connected with inflammation [150]. Inflammation signaling, as discussed in Section 3, is an important piece of information to acquire in order to estimate factors that trigger cancer and its progression. The need for an integrated metabolic scenario emerges, linking the balance of membrane fatty acid precursors of eicosanoids and other lipid mediators with the effects of fat accumulation and hormonal control.

The effect of transformation of adipocytes in CAAs is evidenced also by increased release of free fatty acids (FFA), with the immediate effect of generating energy to fuel tumor growth [151]. The mobilization of FFA from adipocytes is performed in three steps by lipase enzymes: ATGL (adipose triglycerides lipase), HSL (hormone sensitive lipase) and MAGL (mono acylglycerol lipase), enhancing their circulating levels [93]. Seminal experiments of co-culture of adipocytes with cancer cells showed that there is stimulation of lipolysis in adipocytes releasing FFA and glycerol, with a reduction in adipocyte size [152]. The amount of FFA promoted cancer progression, delivering building blocks for cancer cells but also stimulating lipid metabolism; in ovarian cancer cells, co-cultures with adipocytes induced upregulation of fatty acid β-oxidation (FAO), with a consequent large quantity of ATP [133,139,153,154], supporting the energy demand of the tumor mass. The transfer process of FFA between adipocytes and cancer cells is mediated by fatty acid-binding protein 4 (FABP4), which supplies energy to the cells and also active oncogenic pathways like IL-6/STAT3/ALDH1, leading to an enhanced stem cell-like phenotype and tumor progression [133]; FABP4 expression increased in cancer cells co-cultivated with adipocytes [155]. The interesting connections of disease development with the fatty acid structures and functions discussed in the previous sections should appear clear at this point, and this review has the scope of stimulating a constructive debate among scientists involved in cancer cell biology, metabolomics and lipidomics in order to use the substantial information available to develop lipid-based diagnostics and strategies for cancer.

The recent results obtained with EVs in cancer offer promising perspectives on mechanistic and diagnostic developments [156]. The transport of lipids by EV from AT can be an important player in the whole scenario. In a case study of melanoma, the biomolecular transfer process of EVs seems to increase in the presence of obesity. Incubation of melanoma cells with EVs deriving from AT caused the redistribution of lipid droplets close to mitochondria and the increase of fatty acid oxidation [157]. Research conducted on overweight subjects showed that exosomes derived from cancer cells were incorporated by adipocytes, modifying transcriptome and cytokine secretion; the exosomes obtained from adipocytes strongly helped tumor growth by angiogenesis and enhanced inflammation, recruiting macrophages, activating kinases and involving the NF-κB signaling pathway [158,159]. Together with the analysis of the fatty acid types contained in EVs, including sapienic and sebaleic acids, the EVs have enormous potential for unveiling new aspects of lipid supply to cancer.

Obesity also influences the effects of anticancer therapies, as shown in obese cancer patients compared with non-obese patients evaluated for the effects of the same drug treatment [160]. Besides the various aspects involved in these effects, it is important to highlight that the fatty acid constituents of adipose tissue assume a fundamental role in the modification of pharmacokinetics, conferring drug resistance [161,162].

In the scenario of lipid metabolism, the role of lipophagy (i.e., autophagic degradation of lipid droplets, the main lipid storage organelles of eukaryotic cells), discovered in 2009 to have important consequences on health [163], must be mentioned here, in connection with the ongoing debate concerning the role of fasting strategies in cancer treatment [164]. The ways in which calorie restriction/control impacts obesity and cancer treatment will be matter for further research and active debates from different perspectives [165].

6. Some Considerations of Fatty Acid-Based Membrane Lipidomics and Lipid Therapy

Tumoral cells develop accelerated *de novo* lipogenesis as well as strong lipid recruitment, also taking advantage of obesity, to sustain their needs. However, the quality of fatty acids contributes to their invasiveness, also due to their influence on the biophysical properties of membranes and signaling cascades. The proposal of the "membrane hypothesis" links the initial steps of death or life of tumoral cells with the moment of the phospholipid aggregation for membrane formation and the balance between the saturated and unsaturated fatty acid types present in the individual. This crucial balance is different from tissue to tissue, since each tissue has its own composition [7], and it is important to remark that the membrane formation is a completely spontaneous process of phospholipid aggregation, which in their turn are formed by the availability of fatty acids in the lipid pools. It could be said that, with respect to the adequate intake (AI) of lipids established by the main international agencies of health and food [166], the lipid pool should be able to reach a satisfactory balance, with scarce possibilities for impairment or excess. We are aware of the strong ongoing debate about the interplay between genetics and other causes of cancer [167–170]; however, we wish to highlight the importance of the environment (including nutrition), able to interfere with fatty acid levels and metabolic transformations, with strong impact on inflammatory responses and stress conditions, including on hormonal effects such as explained in obesity (Section 5), that can change the "normal" scenario and create unbalances. In our opinion, it is timely to introduce the monitoring of SFA, MUFA and PUFA membrane levels in clinical practice, in view of evaluating strategies that influence the formation of membranes in the individual. Fatty acid-based membrane lipidomics can give the necessary information to estimate the correctness of the molecular pool, which is the *conditio sine qua non* for the healthy behavior of this important compartment [61]. It is worth recalling that the erythrocyte membrane compositions of patients under parenteral nutrition reflected the lipid emulsions given to them. In particular, olive oil emulsion was able to induce statistically significantly higher levels of arachidonic acid and omega-6/omega-3 ratio compared to patients treated with a lipid emulsion containing a small percentage of fish oil [64]. This is an important message for those involved in patient care and nutrition and also for considering the exact dosage of fatty acid supplementations for therapeutic purposes. As a matter of fact, membrane homeostasis and related therapies are nowadays emerging, targeting cell membranes by dietary bioactive molecules able to obtain the remodeling of plasma membrane domains. The attenuation of oncogenic protein activity by modulating the membrane organization of essential proteins and lipids was proven and this is a promising way to use such an approach to manage cancer expansion. It is worth underlining that omega-3 has a very potent influence on membrane organization and this ability, combined with the anti-inflammatory activity, should be developed toward successful cancer treatments [171,172]. However, it is also evident that, without assessing the membrane status in the individual, the assignment of the lipid strategy cannot be precise, in types and doses, thus even bringing about contrasting clinical outcomes since the membrane unbalance results or remains altered. Therefore, it will be necessary to develop a multidisciplinary approach, involving also clinicians, for the understanding of membrane molecular profiles and for creating protocols of membrane lipidomics and lipid therapy, gathering evidence-based results. In this direction, lipid replacement therapy (LRT) is described as a natural medicine approach to replace damaged lipids in cellular membranes and organelles; however, no personalization is proposed [173]. In the context of membrane therapy, we must also mention natural fatty acids with a structure able to interfere with lipid enzymes, such as sterculic acid, a cyclopropane-containing derivative of oleic acid (9,10-methylene-9-octadecenoic acid) found in plants of the genus *Sterculia*. This is an inhibitor of SCD-1, and of the related cascades, as previously explained, which has attracted interest for application in various diseases, including cancer [174]. As previously described, lipid enzyme inhibitors (fatty acid synthesis and desaturation) are attracting interest for innovative cancer treatments, and readers are directed to reviews to deepen the state-of-the-art of such therapeutic strategies [15,21,174].

Focusing on mature erythrocytes, their membrane composition data can be gathered from cancer patients, accompanying the biological sample with an accurate food questionnaire. By this approach, we were able to highlight in a preliminary study of cancer patients that they have SFA-MUFA membrane levels which are significantly different from controls and independent of dietary intakes [69]. It is also straightforward that the analytical protocols used for membrane lipidomic analysis must be certified by international accreditation bodies, and it is advisable that such protocols are unified and automatized by high-throughput procedures, in order that clinical laboratories can gather reliable "big data" to depict cancer lipidomics in an incontrovertible manner.

7. Conclusions

The acquisition of a multidisciplinary vision of fatty acids' relevance to membrane formation and cancer development is necessary in order to go from the bench to the bedside and to the home of patients, associating nutrient choice with strategies to defeat cancer. The growing understanding of the response of cancer to diet will lead to new therapeutic opportunities but, at the same time, will have practical use in the everyday lives of patients, solving also contrasting effects reported in the literature for PUFA supplementation [175]. It is desirable to increase efforts for a larger understanding of molecular nutrition effects in combination with pharmacology and immunology to control this multifaceted disease [176]. Researchers of several disciplines are required in order to accomplish such goals. Specific effort is needed by clinical units to introduce fatty acid diagnostics tools and therapies to prove the validity of the concepts and translate them into medical practice. Indeed, previously reported clinical effects for some fatty acids, such as the omega-6 γ-linolenic acid (see Figure 2), of its antitumoral synergy with chemotherapy [177] must take into account its rare presence in foods and be evaluated in a personalized way, also determining the level of this fatty acid in the individual. Therefore, knowledge of molecular diagnostics, such as membrane lipidomics, is a fundamental step toward including endogenous and exogenous fatty acids in the cancer scenario.

Author Contributions: Contributed to data acquisition and writing of the initial version, C.F., A.S., I.T. and J.A.; revision performed, C.F. and R.F. All authors have read and agreed to the published version of the manuscript.

Funding: This work was financially supported by the Centre for the Development of Industrial Technology (CDTI) of the Spanish Ministry of Science and Innovation under the grant agreement TECNOMIFOOD project, CER-20191010. This is contribution number 990 from AZTI.

Conflicts of Interest: C.F. is co-founder and scientific director of the company Lipinutragen srl, born as spin-off recognized officially by the National Council of Research and involved in research and development of the membrane lipidomic analysis. The company had no role in the design of the study; in the collection and interpretation of data; in the writing of the manuscript, or in the decision to publish the results.

Abbreviations

AKT	Protein kinase B
AMPK	5′ adenosine monophosphate-activated protein kinase
AT	Adipose tissue
ATGL	Adipose triglycerides lipase
ATP	Adenosine triphosphate
CAAs	Cancer-associated adipocytes
DMDS	Dimethyl disulfide
EFA	Essential fatty acids
ERK	Extracellular signal-regulated kinases
EVs	Extracellular vesicles
FABP	Fatty acid binding protein
FADS	Fatty acid desaturase
FAO	Fatty acid oxidase
FFA	Free fatty acids
FGF2	Fibroblast growth factor 2
GSK3	Glycogen synthase kinase 3 beta
HGF	Hepatocyte growth factor
HSL	Hormone sensitive lipase
IGF-1	Insulin growth factor 1
IL-6	Interleukin 6
JAK2	Janus kinases 2
JNK	c-Jun N-terminal kinases
LDL	Low density lipoproteins
LNCAP	Prostate derived from metastatic site
LR	Leptin receptor
MAGL	Mono acylglycerol lipase
MAPK	Mitogen-activated protein kinases
MEFs	Mouse embryonic fibroblasts
PC3	Prostate cancer
PI3K	Phosphoinositide 3-kinases
PKC	Protein kinase C
PPAR	Peroxisome proliferator activated receptor
SREBP-1	Sterol regulatory element- binding protein 1
STAT	Signal transducer and activator of transcription protein
STAT3	Signal transducer and activator of transcription 3
TNF-α	Tumor necrosis factor alpha
VEGF	Vascular endothelial growth factor

References

1. Gross, R.W.; Han, X. Lipidomics at the interface of structure and function in systems biology. *Chem. Biol.* **2011**, *18*, 284–291. [CrossRef] [PubMed]
2. Abbott, S.K.; Else, P.L.; Atkins, T.A.; Hulbert, A.J. Fatty acid composition of membrane bilayers: Importance of diet polyunsaturated fat balance. *Biochim. Biophys. Acta* **2012**, *1818*, 1309–1317. [CrossRef] [PubMed]
3. Chatgilialoglu, C.; Ferreri, C.; Melchiorre, M.; Sansone, A.; Torreggiani, A. Lipid geometrical isomerism: From chemistry to biology and diagnostics. *Chem. Rev.* **2014**, *114*, 255–284. [CrossRef] [PubMed]
4. Rohrig, F.; Schulze, A. The multifaceted roles of fatty acid synthesis in cancer. *Nat. Rev. Cancer* **2016**, *16*, 732–749. [CrossRef]

5.   Baenke, F.; Peck, B.; Miess, H.; Schulze, A. Hooked on fat: The role of lipid synthesis in cancer metabolism and tumour development. *Dis. Model. Mech.* **2013**, *6*, 1353–1363. [CrossRef]

6.   Ferreri, C.; Masi, A.; Sansone, A.; Giacometti, G.; Larocca, A.V.; Menounou, G.; Scanferlato, R.; Tortorella, S.; Rota, D.; Conti, M.; et al. Fatty Acids in Membranes as Homeostatic, Metabolic and Nutritional Biomarkers: Recent Advancements in Analytics and Diagnostics. *Diagnostics* **2017**, *7*, 1. [CrossRef]

7.   Harayama, T.; Riezman, H. Understanding the diversity of membrane lipid composition [published correction appears in *Nat Rev Mol Cell Biol.* **2019**, *20*, 715]. *Nat. Rev. Mol. Cell Biol.* **2018**, *19*, 281–296. [CrossRef]

8.   Nakazawa, I.; Goto, Y. A role of the cellular phospholipid in the metastasis into the liver. *Cell. Mol. Biol.* **1981**, *27*, 23–26.

9.   Zalba, S.; Ten Hagen, T.L. Cell membrane modulation as adjuvant in cancer therapy. *Cancer Treat. Rev.* **2017**, *52*, 48–57. [CrossRef]

10.   Ferreri, C.; Sansone, A.; Buratta, S.; Urbanelli, L.; Costanzi, E.; Emiliani, C.; Chatgilialoglu, C. The n-10 Fatty Acids Family in the Lipidome of Human Prostatic Adenocarcinoma Cell Membranes and Extracellular Vesicles. *Cancers (Basel)* **2020**, *12*, 900. [CrossRef]

11.   Asgari, Y.; Zabihinpour, Z.; Salehzadeh-Yazdi, A.; Schreiber, F.; Masoudi-Nejad, A. Alterations in cancer cell metabolism: The Warburg effect and metabolic adaptation. *Genomics* **2015**, *105*, 275–281. [CrossRef] [PubMed]

12.   Long, J.; Zhang, C.J.; Zhu, N.; Du, K.; Yin, Y.F.; Tan, X.; Liao, D.F.; Qin, L. Lipid metabolism and carcinogenesis, cancer development. *Am. J. Cancer Res.* **2018**, *8*, 778–791.

13.   Rabinowitz, J.D.; Coller, H.A. Partners in the Warburg effect. *Elife* **2016**, *5*, e15938. [CrossRef]

14.   Yeung, S.J.; Pan, J.; Lee, M.H. Roles of p53, MYC and HIF-1 in regulating glycolysis—The seventh hallmark of cancer. *Cell. Mol. Life Sci.* **2008**, *65*, 3981–3999. [CrossRef] [PubMed]

15.   Koundouros, N.; Poulogiannis, G. Reprogramming of fatty acid metabolism in cancer. *Br. J. Cancer* **2020**, *122*, 4–22. [CrossRef]

16.   Hanahan, D.; Weinberg, R.A. Hallmarks of cancer: The next generation. *Cell* **2011**, *144*, 646–674. [CrossRef] [PubMed]

17.   Zhao, G.; Cardenas, H.; Matei, D. Ovarian Cancer—Why Lipids Matter. *Cancers* **2019**, *11*, 1870. [CrossRef] [PubMed]

18.   Mollinedo, F.; Gajate, C. Lipid rafts as signaling hubs in cancer cell survival/death and invasion: Implications in tumor progression and therapy. *J. Lipid Res.* **2020**, *61*, 611–635. [CrossRef]

19.   Sviridov, D.; Mukhamedova, N.; Miller, Y.I. Lipid rafts as a therapeutic target. *J. Lipid Res.* **2020**, *6*, 687–695. [CrossRef]

20.   Currie, E.; Schulze, A.; Zechner, R.; Walther, T.C.; Farese, R.V. Cellular Fatty Acid Metabolism and Cancer. *Cell Metabol.* **2013**, *18*, 153–161. [CrossRef]

21.   Zhang, J.S.; Lei, J.P.; Wei, G.Q.; Chen, H.; Ma, C.Y.; Jiang, H.Z. Natural fatty acid synthase inhibitors as potent therapeutic agents for cancers: A review. *Pharm. Biol.* **2016**, *54*, 1919–1925. [CrossRef] [PubMed]

22.   Leonardi, R.; Subramanian, C.; Jackowski, S.; Rock, C.O. Cancer-associated isocitrate dehydrogenase mutations inactivate NADPH-dependent reductive carboxylation. *J. Biol. Chem.* **2012**, *287*, 14615–14620. [CrossRef] [PubMed]

23.   Marin-Valencia, I.; Yang, C.; Mashimo, T.; Cho, S.; Baek, H.; Yang, X.L.; Rajagopalan, K.N.; Maddie, M.; Vemireddy, V.; Zhao, Z.; et al. Analysis of tumor metabolism reveals mitochondrial glucose oxidation in genetically diverse human glioblastomas in the mouse brain in vivo. *Cell Metab.* **2012**, *15*, 827–837. [CrossRef] [PubMed]

24.   Comerford, S.A.; Huang, Z.; Du, X.; Wang, Y.; Cai, L.; Witkiewicz, A.K.; Walters, H.; Tantawy, M.N.; Fu, A.; Manning, H.C.; et al. Acetate dependence of tumors. *Cell* **2014**, *159*, 1591–1602. [CrossRef] [PubMed]

25.   Flavin, R.; Peluso, S.; Nguyen, P.L.; Loda, M. Fatty acid synthase as a potential therapeutic target in cancer. *Future Oncol.* **2010**, *6*, 551–562. [CrossRef] [PubMed]

26.   Igal, R.A. Stearoyl CoA desaturase-1: New insights into a central regulator of cancer metabolism. *Biochim. Biophys. Acta* **2016**, *1861*, 1865–1880. [CrossRef]

27.   Kamphorst, J.J.; Cross, J.R.; Fan, J.; De Stanchina, E.; Mathew, R.; White, E.P.; Thompson, C.B.; Rabinowitz, J.D. Hypoxic and Ras-transformed cells support growth by scavenging unsaturated fatty acids from lysophospholipids. *Proc. Natl. Acad. Sci. USA* **2013**, *110*, 8882–8887. [CrossRef]

28. Hopperton, K.E.; Duncan, R.E.; Bazinet, R.P.; Archer, M.C. Fatty acid synthase plays a role in cancer metabolism beyond providing fatty acids for phospholipid synthesis or sustaining elevations in glycolytic activity. *Exp. Cell Res.* **2014**, *320*, 302–310. [CrossRef]

29. Scanferlato, R.; Bortolotti, M.; Sansone, A.; Chatgilialoglu, C.; Polito, L.; De Spirito, M.; Maulucci, G.; Bolognesi, A.; Ferreri, C. Hexadecenoic Fatty Acid Positional Isomers and De Novo PUFA Synthesis in Colon Cancer Cells. *Int. J. Mol. Sci.* **2019**, *20*, 832. [CrossRef]

30. Vriens, K.; Christen, S.; Parik, S.; Broekaert, D.; Yoshinaga, K.; Talebi, A.; Dehairs, J.; Escalona-Noguero, C.; Schmieder, R.; Cornfield, T.; et al. Evidence for an alternative fatty acid desaturation pathway increasing cancer plasticity. *Nature* **2019**, *566*, 403–406. [CrossRef]

31. Guri, Y.; Colombi, M.; Dazert, E.; Hindupur, S.K.; Roszik, J.; Moes, S.; Jenoe, P.; Heim, M.H.; Riezman, I.; Riezman, H.; et al. mTORC2 promotes tumorigenesis via lipid synthesis. *Cancer Cell* **2017**, *32*, 807–823. [CrossRef]

32. Hagiwara, A.; Cornu, M.; Cybulski, N.; Polak, P.; Betz, C.; Trapani, F.; Terracciano, L.; Heim, M.H.; Rüegg, M.A.; Hall, M.N. Hepatic mTORC2 activates glycolysis and lipogenesis through Akt, glucokinase, and SREBP1c. *Cell Metab.* **2012**, *15*, 725–738. [CrossRef]

33. Schug, Z.T.; Peck, B.; Jones, D.T.; Zhang, Q.; Grosskurth, S.; Alam, I.S.; Goodwin, L.M.; Smethurst, E.; Mason, S.; Blyth, K.; et al. Acetyl-CoA synthetase 2 promotes acetate utilization and maintains cancer cell growth under metabolic stress. *Cancer Cell* **2015**, *27*, 57–71. [CrossRef] [PubMed]

34. Nieman, K.M.; Kenny, H.A.; Penicka, C.V.; Ladanyi, A.; Buell-Gutbrod, R.; Zillhardt, M.R.; Romero, I.L.; Carey, M.S.; Mills, G.B.; Hotamisligil, G.S.; et al. Adipocytes promote ovarian cancer metastasis and provide energy for rapid tumor growth. *Nat. Med.* **2011**, *17*, 1498–1503. [CrossRef]

35. Yue, S.H.; Li, J.J.; Lee, S.Y.; Lee, H.J.; Shao, T.; Song, B.; Cheng, L.; Masterson, T.A.; Liu, X.Q.; Ratliff, T.L.; et al. Cholesteryl Ester Accumulation Induced by PTEN Loss and PI3K/AKT Activation Underlies Human Prostate Cancer Aggressiveness. *Cell Metab.* **2014**, *19*, 393–406. [CrossRef] [PubMed]

36. Guillaumond, F.; Bidaut, G.; Ouaissi, M.; Servais, S.; Gouirand, V.; Olivares, O.; Lac, S.; Borge, L.; Roques, J.; Gayet, O.; et al. Cholesterol uptake disruption, in association with chemotherapy, is a promising combined metabolic therapy for pancreatic adenocarcinoma. *Proc. Natl. Acad. Sci. USA* **2015**, *112*, 2473–2478. [CrossRef] [PubMed]

37. De Gonzalo-Calvo, D.; Lopez-Vilaro, L.; Nasarre, L.; Perez-Olabarria, M.; Vazquez, T.; Escuin, D.; Badimon, L.; Barnadas, A.; Lerma, E.; Llorente-Cortes, V. Intratumor cholesteryl ester accumulation is associated with human breast cancer proliferation and aggressive potential: A molecular and clinicopathological study. *BMC Cancer* **2015**, *15*, 460. [CrossRef]

38. Accioly, M.T.; Pacheco, P.; Maya-Monteiro, C.M.; Carrossini, N.; Robbs, B.K.; Oliveira, S.S.; Kaufmann, C.; Morgado-Diaz, J.A.; Bozza, P.T.; Viola, J.P.B. Lipid bodies are reservoirs of cyclooxygenase-2 and sites of prostaglandin-E-2 synthesis in colon cancer cells. *Cancer Res.* **2008**, *68*, 1732–1740. [CrossRef]

39. Peck, B.; Schug, Z.T.; Zhang, Q.F.; Dankworth, B.; Jones, D.T.; Smethurst, E.; Patel, R.; Mason, S.; Jiang, M.; Saunders, R.; et al. Inhibition of fatty acid desaturation is detrimental to cancer cell survival in metabolically compromised environments. *Cancer Metab.* **2016**, *4*, 6. [CrossRef]

40. Tontonoz, P.; Wang, B. Phospholipid remodeling in physiology and disease. *Ann. Rev. Physiol.* **2019**, *81*, 165–188.

41. Serhan, C.N. Novel pro-resolving lipid mediators in inflammation are leads for resolution physiology. *Nature* **2014**, *510*, 92–101. [CrossRef] [PubMed]

42. Coussens, L.M.; Werb, Z. Inflammation and cancer. *Nature* **2002**, *420*, 860–867. [CrossRef] [PubMed]

43. MacLean, C.H.; Newberry, S.J.; Mojica, W.A.; Khanna, P.; Issa, A.M.; Suttorp, M.J.; Lim, Y.W.; Traina, S.B.; Hilton, L.; Garland, R.; et al. Effects of omega-3 fatty acids on cancer risk: A systematic review. *JAMA* **2006**, *295*, 403–415. [CrossRef] [PubMed]

44. Lands, B. Omega-3 PUFAs lower the propensity for arachidonic acid cascade overreactions. *BioMed Res. Int.* **2015**, *2015*, 285135. [CrossRef]

45. Bodén, S.; Myte, R.; Wennberg, M.; Harlid, S.; Johansson, I.; Shivappa, N.; Hébert, J.R.; Van Guelpen, B.; Nilsson, L.M. The inflammatory potential of diet in determining cancer risk; A prospective investigation of two dietary pattern scores. *PLoS ONE* **2019**, *14*, e0214551. [CrossRef]

46. Calder, P.C. n-3 fatty acids, inflammation and immunity: New mechanisms to explain old actions. *Proc. Nutr. Soc.* **2013**, *72*, 326–336. [CrossRef]

47. Ristimaki, A.; Sivula, A.; Lundin, J.; Lundin, M.; Salminen, T.; Haglund, C.; Joensuu, H.; Isola, J. Prognostic significance of elevated cyclooxygenase-2 expression in breast cancer. *Cancer Res.* **2002**, *62*, 632–635.

48. Xu, Y.; Qian, S.Y. Anti-cancer activities of ω-6 polyunsaturated fatty acids. *Biomed. J.* **2014**, *37*, 112–119.

49. Zanoaga, O.; Jurj, A.; Raduly, L.; Cojocneanu-Petric, R.; Fuentes-Mattei, E.; Wu, O.; Braicu, C.; Gherman, C.D.; Berindan-Neagoe, I. Implications of dietary omega-3 and omega-6 polyunsaturated fatty acids in breast cancer (Review). *Exp. Ther. Med.* **2018**, *15*, 1167–1176. [CrossRef]

50. Larsson, S.C.; Kumlin, M.; Ingelman-Sundberg, M.; Wolk, A. Dietary long-chain n-3 fatty acids for the prevention of cancer: A review of potential mechanisms. *Am. J. Clin. Nutr.* **2004**, *79*, 935–945. [CrossRef]

51. Ferreri, C.; Chatgilialoglu, C. Lipidomics and Health: An Added Value to Olive Oil. In *Olives and Olive Oil as Functional Foods: Bioactivity, Chemistry and Processing*, 1st ed.; Shahidi, P., Kiritsakis, P., Eds.; Wiley: Chichester, UK, 2017; Chapter 27; pp. 505–520.

52. Schley, P.D.; Brindley, D.N.; Field, C.J. (n-3) PUFA alter raft lipid composition and decrease epidermal growth factor receptor levels in lipid rafts of human breast cancer cells. *J. Nutr.* **2007**, *137*, 548–553. [CrossRef] [PubMed]

53. Petrosino, S.; Di Marzo, V. The pharmacology of palmitoylethanolamide and first data on the therapeutic efficacy of some of its new formulations. *Br. J. Pharmacol.* **2017**, *174*, 1349–1365. [CrossRef] [PubMed]

54. De Petrocellis, L.; Bisogno, T.; Ligresti, A.; Bifulco, M.; Melck, D.; Di Marzo, V. Effect on cancer cell proliferation of palmitoylethanolamide, a fatty acid amide interacting with both the cannabinoid and vanilloid signalling systems. *Fundam. Clin. Pharmacol.* **2002**, *16*, 297–302. [CrossRef]

55. Sarnelli, G.; D'Alessandro, A.; Iuvone, T.; Capoccia, E.; Gigli, S.; Pesce, M.; Seguella, L.; Nobile, N.; Aprea, G.; Maione, F.; et al. Palmitoylethanolamide Modulates Inflammation-Associated Vascular Endothelial Growth Factor (VEGF) Signaling via the Akt/mTOR Pathway in a Selective Peroxisome Proliferator-Activated Receptor Alpha (PPAR-α)-Dependent Manner. *PLoS ONE* **2016**, *11*, e0156198. [CrossRef] [PubMed]

56. Friedmann Angeli, J.P.; Krysko, D.V.; Conrad, M. Ferroptosis at the crossroads of cancer-acquired drug resistance and immune evasion. *Nat. Rev. Cancer* **2019**, *19*, 405–414. [CrossRef] [PubMed]

57. Conrad, M.; Pratt, D.A. The chemical basis of ferroptosis [published correction appears in Nat Chem Biol. 2020 Jan 2]. *Nat. Chem. Biol.* **2019**, *15*, 1137–1147. [CrossRef] [PubMed]

58. Sender, R.; Fuchs, S.; Milo, R. Revised estimates for the number of human and bacteria cells in the body. *PLoS Biol.* **2016**, *14*, e1002533. [CrossRef]

59. Reed, C.F. Phospholipid exchange between plasma and erythrocytes in man and the dog. *J. Clin. Investig.* **1968**, *47*, 749–760. [CrossRef]

60. Dushianthan, A.; Cusack, R.; Koster, G.; Grocott, M.P.W.; Postle, A.D. Insight into erythrocyte phospholipid molecular flux in healthy humans and in patients with acute respiratory distress syndrome. *PLoS ONE* **2019**, *14*, e0221595. [CrossRef]

61. Ferreri, C.; Chatgilialoglu, C. *Membrane Lipidomics for Personalized Health*; Wiley: Chichester, UK, 2015.

62. Giacometti, G.; Ferreri, C.; Sansone, A.; Chatgilialoglu, C.; Marzetti, C.; Spyratou, E.; Georgakilas, A.G.; Marini, M.; Abruzzo, P.M.; Bolotta, A.; et al. High predictive values of rbc membrane-based diagnostics by biophotonics in an integrated approach for autism spectrum disorders. *Sci. Rep.* **2017**, *7*, 9854. [CrossRef]

63. Sansone, A.; Tolika, E.; Louka, M.; Sunda, V.; Deplano, S.; Melchiorre, M.; Anagnostopoulos, D.; Chatgilialoglu, C.; Formisano, C.; Di Micco, R.; et al. Hexadecenoic fatty acid isomers in human blood lipids and their relevance for the interpretation of lipidomic profiles. *PLoS ONE* **2016**, *11*, e0152378. [CrossRef]

64. Pironi, L.; Guidetti, M.; Verrastro, O.; Iacona, C.; Agostini, F.; Pazzeschi, C.; Sasdelli, A.S.; Melchiorre, M.; Ferreri, C. Functional lipidomics in patients on home parenteral nutrition: Effect of lipid emulsions. *World J. Gastroenterol.* **2017**, *23*, 4604–4614. [CrossRef] [PubMed]

65. Mikirova, N.; Riordan, H.D.; Jackson, J.A.; Wong, K.; Miranda-Massari, J.R.; Gonzalez, M.J. Erythrocyte membrane fatty acid composition in cancer patients. *P. R. Health Sci. J.* **2004**, *23*, 107–113.

66. Okunoi, M.; Hamazaki, K.; Ogura, T.; Kitade, H.; Matsuura, T.; Yoshida, R.; Hijikawa, T.; Kwon, M.; Arita, S.; Itomura, M.; et al. Abnormalities in Fatty Acids in Plasma, Erythrocytes and Adipose Tissue in Japanese Patients with Colorectal Cancer. *In Vivo* **2013**, *27*, 203–210.

67. Pala, V.; Krogh, V.; Muti, P.; Chajes, V.; Riboli, E.; Micheli, A.; Saadatian, M.; Sieri, S.; Berrino, F. Erythrocyte membrane fatty acids and subsequent breast cancer: A prospective Italian study. *J. Natl. Cancer Inst.* **2001**, *93*, 1088–1095. [CrossRef] [PubMed]

68. Coviello, G.; Tutino, V.; Notarnicola, M.; Caruso, M.G. Erythrocyte Membrane Fatty Acids Profile in Colorectal Cancer Patients: A Preliminary Study. *Anticancer Res.* **2014**, *34*, 4775–4779. [PubMed]

69. Amézaga, J.; Arranz, S.; Urruticoechea, A.; Ugartemendia, G.; Larraioz, A.; Louka, M.; Uriarte, M.; Ferreri, C.; Tueros, I. Altered red blood cell membrane fatty acid profile in cancer patients. *Nutrients* **2018**, *10*, 1853. [CrossRef] [PubMed]

70. Cottet, V.; Collin, M.; Gross, A.S.; Boutron-Ruault, M.C.; Morois, S.; Clavel-Chapelon, F.; Chajes, V. Erythrocyte membrane phospholipid fatty acid concentrations and risk of colorectal adenomas: A case-control nested in the French E3N-EPIC cohort study. *Cancer Epidemiol. Biomark. Prev.* **2013**, *22*, 1417–1427. [CrossRef] [PubMed]

71. Rahrovani, F.; Javanbakht, M.H.; Ehsani, A.H.; Esrafili, A.; Mohammadi, H.; Ghaedi, E.; Zarei, M.; Djalali, M. Erythrocyte membrane saturated fatty acids profile in newly diagnosed Basal Cell Carcinoma patients. *Clin. Nutr. Espen* **2018**, *23*, 107–111. [CrossRef] [PubMed]

72. Rahrovani, F.; Javanbakht, M.H.; Ghaedi, E.; Mohammadi, H.; Ehsani, A.H.; Esrafili, A.; Djalali, M. Erythrocyte Membrane Unsaturated (Mono and Poly) Fatty Acids Profile in Newly Diagnosed Basal Cell Carcinoma Patients. *Clin. Nutr. Res.* **2018**, *7*, 21–30. [CrossRef]

73. Kuriki, K.; Wakai, K.; Hirose, K.; Matsuo, K.; Ito, H.; Suzuki, T.; Saito, T.; Kanemitsu, Y.; Hirai, T.; Kato, T.; et al. Risk of colorectal cancer is linked to erythrocyte compositions of fatty acids as biomarkers for dietary intakes of fish, fat, and fatty acids. *Cancer Epidemiol. Biomark. Prev.* **2006**, *15*, 1791–1798. [CrossRef] [PubMed]

74. Shannon, J.; King, I.B.; Moshofsky, R.; Lampe, J.W.; Gao, D.L.; Ray, R.M.; Thomas, D.B. Erythrocyte fatty acids and breast cancer risk: A case-control study in Shanghai, China. *Am. J. Clin. Nutr.* **2007**, *85*, 1090–1097. [CrossRef] [PubMed]

75. Shannon, J.; O'Malley, J.; Mori, M.; Garzotto, M.; Palma, A.J.; King, I.B. Erythrocyte fatty acids and prostate cancer risk: A comparison of methods. *Prostaglandins Leukot. Essent. Fatty Acids* **2010**, *83*, 161–169. [CrossRef]

76. Sanchez-Rodriguez, P.; Rodriguez, M.C.; Sanchez-Yague, J. Identification of potential erythrocyte phospholipid fatty acid biomarkers of advanced lung adenocarcinoma, squamous cell lung carcinoma, and small cell lung cancer. *Tumor Biol.* **2015**, *36*, 5687–5698. [CrossRef]

77. Sanchez-Vega, F.; Mina, M.; Armenia, J.; Chatila, W.; Luna, A.; La, K.C.; Dimitriadoy, S.; Liu, D.L.; Kantheti, H.S.; Saghafinia, S.; et al. Oncogenic signaling in the cancer genome atlas. *Cell* **2018**, *173*, 321–337. [CrossRef]

78. Stromberg, L.R.; Lilley, L.M.; Mukundan, H. Advances in Lipidomics for Cancer Biomarker Discovery. In *Proteomic and Metabolomic Approaches to Biomarker Discovery*; Academic Press: Cambridge, MA, USA, 2020; pp. 421–436. [CrossRef]

79. Sansone, A.; Melchiorre, M.; Chatgilialoglu, C.; Ferreri, C. Hexadecenoic Fatty Acid Isomers: A Chemical Biology Approach for Human Plasma Biomarker Development. *Chem. Res. Toxicol.* **2013**, *26*, 1703–1709. [CrossRef]

80. Ge, L.; Gordon, J.S.; Hsuan, C.; Stenn, K.; Prouty, S.M. Identification of the Delta-6 desaturase of human sebaceous glands: Expression and enzyme activity. *J. Invest. Dermatol.* **2003**, *120*, 707–714. [CrossRef]

81. Igal, R.A. Stearoyl-CoA desaturase-1: A novel key player in the mechanisms of cell proliferation, programmed cell death and transformation to cancer. *Carcinogenesis* **2010**, *31*, 1509–1515. [CrossRef]

82. Pappas, A.; Anthonavage, M.; Gordon, J.S. Metabolic fate and selective utilization of major fatty acids in human sebaceous gland. *J. Invest. Dermatol.* **2002**, *118*, 164–171. [CrossRef]

83. Zhang, J.Y.; Kothapalli, K.S.D.; Brenna, J.T. Desaturase and elongase-limiting endogenous long-chain polyunsaturated fatty acid biosynthesis. *Curr. Opin. Clin. Nutr. Metab. Care* **2016**, *19*, 103–110. [CrossRef] [PubMed]

84. Sahebi, R.; Langari, H.; Fathinezhad, Z.; Bahari Sani, Z.; Avan, A.; Mobarhan, M.G.; Rezayi, M. Exosomes: New insights into cancer mechanisms. *J. Cell. Biochem.* **2020**, *121*, 7–16. [CrossRef]

85. Lattka, E.; Illig, T.; Koletzko, B.; Heinrich, J. Genetic variants of the FADS1 FADS2 gene cluster as related to essential fatty acid metabolism. *Curr. Opin. Lipidol.* **2010**, *21*, 64–69. [CrossRef] [PubMed]

86. Tracz-Gaszewska, Z.; Dobrzyn, P. Stearoyl-CoA Desaturase 1 as a Therapeutic Target for the Treatment of Cancer. *Cancers* **2019**, *11*, 948. [CrossRef] [PubMed]

87. Yamashita, Y.; Nishiumi, S.; Kono, S.; Takao, S.; Azuma, T.; Yoshida, M. Differences in elongation of very long chain fatty acids and fatty acid metabolism between triple-negative and hormone receptor-positive breast cancer. *BMC Cancer* **2017**, *17*, 589. [CrossRef] [PubMed]

88. Gonzalez-Bengtsson, A.; Asadi, A.; Gao, H.; Dahlman-Wright, K.; Jacobsson, A. Estrogen Enhances the Expression of the Polyunsaturated Fatty Acid Elongase Elovl2 via ERa in Breast Cancer Cells. *PLoS ONE* **2016**, *11*, e0164241. [CrossRef]

89. Tamura, K.; Makino, A.; Hullin-Matsuda, F.; Kobayashi, T.; Furihata, M.; Chung, S.; Ashida, S.; Miki, T.; Fujioka, T.; Shuin, T.; et al. Novel Lipogenic Enzyme ELOVL7 Is Involved in Prostate Cancer Growth through Saturated Long-Chain Fatty Acid Metabolism. *Cancer Res.* **2009**, *69*, 8133–8140. [CrossRef]

90. Triki, M.; Rinaldi, G.; Planque, M.; Broekaert, D.; Winkelkotte, A.M.; Maier, C.R.; Raman, S.J.; Vandekeere, A.; Van Elsen, J.; Orth, M.F.; et al. mTOR Signaling and SREBP Activity Increase FADS2 Expression and Can Activate Sapienate Biosynthesis. *Cell Rep.* **2020**, *31*, 1–9. [CrossRef]

91. Park, H.G.; Kothapalli, K.S.D.; Park, W.J.; DeAllie, C.; Liu, L.; Liang, A.; Lawrence, P.; Brenna, J.T. Palmitic acid (16:0) competes with omega-6 linoleic and omega-3 alpha-linolenic acids for FADS2 mediated Delta 6-desaturation. *BBA. Mol. Cell Biol. Lipids* **2016**, *1861*, 91–97. [CrossRef]

92. Park, H.G.; Engel, M.G.; Vogt-Lowell, K.; Lawrence, P.; Kothapalli, K.S.; Brenna, J.T. The role of fatty acid desaturase (FADS) genes in oleic acid metabolism: FADS1 Delta 7 desaturates 11-20:1 to 7,11-20:2. *Prostaglandins Leukot. Essent. Fatty Acids* **2018**, *128*, 21–25. [CrossRef]

93. Khandekar, M.J.; Cohen, P.; Spiegelman, B.M. Molecular mechanisms of cancer development in obesity. *Nat. Rev. Cancer* **2011**, *11*, 886–895. [CrossRef]

94. Ligibel, J.A.; Wollins, D. American Society of Clinical Oncology Obesity Initiative: Rationale, Progress, and Future Directions. *J. Clin. Oncol.* **2016**, *34*, 4256–4260. [CrossRef] [PubMed]

95. Gutierrez-Salmeron, M.; Chocarro-Calvo, A.; Garcia-Martinez, J.M.; De la Vieja, A.; Garcia-Jimenez, C. Epidemiological bases and molecular mechanisms linking obesity, diabetes, and cancer. *Endocrinol. Diabetes Y Nutr.* **2017**, *64*, 109–117. [CrossRef]

96. Shaw, E.; Farris, M.; McNeil, J.; Friedenreich, C. Obesity and Endometrial Cancer. In *Obesity and Cancer. Recent Results in Cancer Research*; Pischon, T., Nimptsch, K., Eds.; Springer: Cham, Switzerland, 2016; Volume 208. [CrossRef]

97. Feng, Y.H. The association between obesity and gynecological cancer. *Gynecol. Minim. Invasive Ther. Gmit* **2015**, *4*, 102–105. [CrossRef]

98. Hoyo, C.; Cook, M.B.; Kamangar, F.; Freedman, N.D.; Whiteman, D.C.; Bernstein, L.; Brown, L.M.; Risch, H.A.; Ye, W.M.; Sharp, L.; et al. Body mass index in relation to oesophageal and oesophagogastric junction adenocarcinomas: A pooled analysis from the International BEACON Consortium. *Int. J. Epidemiol.* **2012**, *41*, 1706–1718. [CrossRef] [PubMed]

99. Schlottmann, F.; Dreifuss, N.H.; Patti, M.G. Obesity and esophageal cancer: GERD, Barrett´s esophagus, and molecular carcinogenic pathways. *Exp. Rev. Gastroent. Hepatol.* **2020**, *14*, 425–433. [CrossRef]

100. Chen, Y.; Liu, L.X.; Wang, X.L.; Wang, J.H.; Yan, Z.P.; Cheng, J.M.; Gong, G.Q.; Li, G.P. Body Mass Index and Risk of Gastric Cancer: A Meta-analysis of a Population with More Than Ten Million from 24 Prospective Studies. *Cancer Epidemiol. Biomark. Prev.* **2013**, *22*, 1395–1408. [CrossRef]

101. Du, X.; Hidayat, K.; Shi, B.M. Abdominal obesity and gastroesophageal cancer risk: Systematic review and meta-analysis of prospective studies. *Biosci. Rep.* **2017**, *37*, BSR20160474. [CrossRef]

102. Olefson, S.; Moss, S.F. Obesity and related risk factors in gastric cardia adenocarcinoma. *Gastric Cancer* **2015**, *18*, 23–32. [CrossRef]

103. Saitta, C.; Pollicino, T.; Raimondo, G. Obesity and liver cancer. *Ann. Hepatol.* **2019**, *18*, 810–815. [CrossRef]

104. Rahmani, J.; Varkaneh, H.K.; Kontogiannis, V.; Ryan, P.M.; Bawadi, H.; Fatahi, S.; Zhang, Y. Waist Circumference and Risk of Liver Cancer: A Systematic Review and Meta-Analysis of over 2 Million Cohort Study Participants. *Liver Cancer* **2020**, *9*, 6–14. [CrossRef]

105. Wilson, K.M.; Cho, E. Obesity and Kidney Cancer Recent Results. *Cancer Res.* **2016**, *208*, 81–93.

106. Sanfilippo, K.M.; McTigue, K.M.; Fidler, C.J.; Neaton, J.D.; Chang, Y.F.; Fried, L.F.; Liu, S.M.; Kuller, L.H. Hypertension and Obesity and the Risk of Kidney Cancer in 2 Large Cohorts of US Men and Women. *Hypertension* **2014**, *63*, 934–941. [CrossRef] [PubMed]

107. Wang, F.R.; Xu, Y.H. Body mass index and risk of renal cell cancer: A dose-response meta-analysis of published cohort studies. *Int. J. Cancer* **2014**, *135*, 1673–1686. [CrossRef]

108. Thordardottir, M.; Lindqvist, E.K.; Lund, S.H.; Costello, R.; Burton, D.; Korde, N.; Mailankody, S.; Eiriksdottir, G.; Launer, L.J.; Gudnason, V.; et al. Obesity and risk of monoclonal gammopathy of undetermined significance and progression to multiple myeloma: A population-based study. *Blood Adv.* **2017**, *1*, 2186–2192. [CrossRef]

109. Wallin, A.; Larsson, S.C. Body mass index and risk of multiple myeloma: A meta-analysis of prospective studies. *Eur. J. Cancer* **2011**, *47*, 1606–1615. [CrossRef]

110. Bullwinkle, E.M.; Parker, M.D.; Bonan, N.F.; Falkenberg, L.G.; Davison, S.P.; DeCicco-Skinner, K.L. Adipocytes contribute to the growth and progression of multiple myeloma: Unraveling obesity related differences in adipocyte signaling. *Cancer Lett.* **2016**, *380*, 114–121. [CrossRef]

111. Niedermaier, T.; Behrens, G.; Schmid, D.; Schlecht, I.; Fischer, B.; Leitzmann, M.F. Body mass index, physical activity, and risk of adult meningioma and glioma. A meta-analysis. *Neurology* **2015**, *85*, 1342–1350. [CrossRef]

112. Rutkowski, R.; Reszec, J.; Hermanowicz, A.; Chrzanowski, R.; Lyson, T.; Mariak, Z.; Chyczewski, L. Correlation of leptin receptor expression with BMI in differential grades of human meningiomas. *Oncol. Lett.* **2016**, *11*, 2515–2519. [CrossRef]

113. Genkinger, J.M.; Spiegelman, D.; Anderson, K.E.; Bernstein, L.; Van den Brandt, P.A.; Calle, E.E.; English, D.R.; Folsom, A.R.; Freudenheim, J.L.; Fuchs, C.S.; et al. A pooled analysis of 14 cohort studies of anthropometric factors and pancreatic cancer risk. *Int. J. Cancer* **2011**, *129*, 1708–1717. [CrossRef]

114. Xu, M.; Jung, X.M.; Hines, O.J.; Eibl, G.; Chen, Y.J. Obesity and pancreatic cancer overview of epidemiology and potential prevention by weight loss. *Pancreas* **2018**, *47*, 158–162. [CrossRef]

115. Liu, P.H.; Wu, K.; Ng, K.; Zauber, A.G.; Nguyen, L.H.; Song, M.Y.; He, X.S.; Fuchs, C.S.; Ogino, S.; Willett, W.C.; et al. Association of obesity with risk of early-onset colorectal cancer among women. *JAMA Oncol.* **2019**, *5*, 37–44. [CrossRef] [PubMed]

116. Whitlock, K.; Gill, R.S.; Birch, D.W.; Karmali, S. The Association between Obesity and Colorectal Cancer. *Gastroenterol. Res. Pract.* **2012**, *2012*, 1–6. [CrossRef] [PubMed]

117. Martinez-Useros, J.; Garcia-Foncillas, J. Obesity and colorectal cancer: Molecular features of adipose tissue. *J. Transl. Med.* **2016**, *14*, 21. [CrossRef] [PubMed]

118. Campbell, P.T.; Newton, C.C.; Kitahara, C.M.; Patel, A.V.; Hartge, P.; Koshiol, J.; McGlynn, K.A.; Adami, H.O.; De Gonzalez, A.B.; Freeman, L.E.B.; et al. Body Size Indicators and Risk of Gallbladder Cancer: Pooled Analysis of Individual-Level Data from 19 Prospective Cohort Studies. *Cancer Epidemiol. Biomark. Prev.* **2017**, *26*, 597–606. [CrossRef] [PubMed]

119. Nemunaitis, J.M.; Brown-Glabeman, U.; Soares, H.; Belmonte, J.; Liem, B.; Nir, I.; Phuoc, V.; Gullapalli, R.R. Gallbladder cancer: Review of a rare orphan gastrointestinal cancer with a focus on populations of New Mexico. *BMC Cancer* **2018**, *18*, 665. [CrossRef] [PubMed]

120. Picon-Ruiz, M.; Morata-Tarifa, C.; Valle-Goffin, J.J.; Friedman, E.R.; Slingerland, J.M. Obesity and Adverse Breast Cancer Risk and Outcome: Mechanistic Insights and Strategies for Intervention. *CA Cancer J. Clin.* **2017**, *67*, 379–397. [CrossRef] [PubMed]

121. Ando, S.; Gelsomino, L.; Panza, S.; Giordano, C.; Bonofiglio, D.; Barone, I.; Catalano, S. Obesity, leptin and breast cancer: Epidemiological evidence and proposed mechanisms. *Cancers* **2019**, *11*, 62. [CrossRef]

122. Neuhouser, M.L.; Aragaki, A.K.; Prentice, R.L.; Manson, J.E.; Chlebowski, R.; Carty, C.L.; Ochs-Balcom, H.M.; Thomson, C.A.; Caan, B.J.; Tinker, L.F.; et al. Overweight, obesity, and postmenopausal invasive breast cancer risk. A secondary analysis of the Women's Health Initiative randomized clinical trials. *JAMA Oncol.* **2015**, *1*, 611–621. [CrossRef]

123. Agurs-Collins, T.; Ross, S.A.; Dunn, B.K. The many faces of obesity and its influence on breast cancer risk. *Front. Oncol.* **2019**, *9*, 765. [CrossRef]

124. Beral, V.; Hermon, C.; Peto, R.; Reeves, G.; Brinton, L.; Marchbanks, P.; Negri, E.; Ness, R.; Peeters, P.H.M.; Vessey, M.; et al. Ovarian cancer and body size: Individual participant meta-analysis including 25,157 women with ovarian cancer from 47 epidemiological studies. *PLoS Med.* **2012**, *9*, e1001200. [CrossRef]

125. Kitahara, C.M.; McCullough, M.L.; Franceschi, S.; Rinaldi, S.; Wolk, A.; Neta, G.; Adami, H.O.; Anderson, K.; Andreotti, G.; Freeman, L.E.B.; et al. Anthropometric factors and thyroid cancer risk by histological subtype: Pooled analysis of 22 prospective studies. *Thyroid* **2016**, *26*, 306–318. [CrossRef] [PubMed]

126. Zhao, S.T.; Jia, X.M.; Fan, X.J.; Zhao, L.; Pang, P.; Wang, Y.J.; Luo, Y.K.; Wang, F.L.; Yang, G.Q.; Wang, X.L.; et al. Association of obesity with the clinicopathological features of thyroid cancer in a large, operative population A retrospective case-control study. *Medicine* **2019**, *98*, e18213. [CrossRef] [PubMed]

127. Schmid, D.; Ricci, C.; Behrens, G.; Leitzmann, M.F. Adiposity and risk of thyroid cancer: A systematic review and meta-analysis. *Obes. Rev.* **2015**, *16*, 1042–1054. [CrossRef] [PubMed]

128. Dirat, B.; Bochet, L.; Dabek, M.; Daviaud, D.; Dauvillier, S.; Majed, B.; Wang, Y.Y.; Meulle, A.; Salles, B.; Le Gonidec, S.; et al. Cancer-associated adipocytes exhibit an activated phenotype and contribute to breast cancer invasion. *Cancer Res.* **2011**, *71*, 2455–2465. [CrossRef] [PubMed]

129. Dirat, B.; Bochet, L.; Dabek, M.; Daviaud, D.; Dauvillier, S.; Le Gonidec, S.; Escourrou, G.; Valet, P.; Muller, C. Cancer-associated adipocytes exhibit an activated phenotype and contribute to early breast cancer invasion in vitro and in vivo. *EJC Suppl.* **2010**, *8*, 126. [CrossRef]

130. Rybinska, I.; Agresti, R.; Trapani, A.; Tagliabue, E.; Triulzi, T. Adipocytes in breast cancer, the thick and the thin. *Cells* **2020**, *9*, 560. [CrossRef]

131. Bochet, L.; Lehuede, C.; Dauvillier, S.; Wang, Y.Y.; Dirat, B.; Laurent, V.; Dray, C.; Guiet, R.; Maridonneau-Parini, I.; Le Gonidec, S.; et al. Adipocyte-derived fibroblasts promote tumor progression and contribute to the desmoplastic reaction in breast cancer. *Cancer Res.* **2013**, *73*, 5657–5668. [CrossRef]

132. Cirri, P.; Chiarugi, P. Cancer associated fibroblasts: The dark side of the coin. *Am. J. Cancer Res.* **2011**, *1*, 482–497.

133. Nieman, K.M.; Romero, I.L.; Van Houten, B.; Lengyel, E. Adipose tissue and adipocytes support tumorigenesis and metastasis. *BBA. Mol. Cell Biol. Lipids* **2013**, *1831*, 1533–1541. [CrossRef]

134. Iyengar, N.M.; Hudis, C.A.; Dannenberg, A.J. Obesity and Cancer: Local and Systemic Mechanisms. *Annu. Rev. Med.* **2015**, *66*, 297–309. [CrossRef]

135. Wauman, J.; Tavernier, J. Leptin receptor signaling: Pathways to leptin resistance. *Front. Biosci.* **2011**, *16*, 2771–2793. [CrossRef] [PubMed]

136. Bowers, L.W.; Gung, J.; Lineberger, C.G.; Hursting, S.D. Abstract SY28-04: Obesity-associated leptin signaling promotes chemotherapy resistance in triple-negative breast cancer: The role of tumor-initiating cell enrichment. *Tumor Biol.* **2019**. [CrossRef]

137. Sanchez-Jimenez, F.; Perez-Perez, A.; De la Cruz-Merino, L.; Sanchez-Margalet, V. Obesity and breast cancer: Role of leptin. *Front. Oncol.* **2019**, *9*, 596. [CrossRef]

138. Nowak, A.; Kobierzycki, C.; Dziegiel, P. The role of leptin in pathogenesis of obesity-related cancers. *Postepy Biol. Komorki* **2015**, *42*, 309–328.

139. Park, J.; Morley, T.S.; Kim, M.; Clegg, D.J.; Scherer, P.E. Obesity and cancer-mechanisms underlying tumour progression and recurrence. *Nat. Rev. Endocrinol.* **2014**, *10*, 455–465. [CrossRef]

140. Divella, R.; De Luca, R.; Abbate, I.; Naglieri, E.; Daniele, A. Obesity and cancer: The role of adipose tissue and adipo-cytokines-induced chronic inflammation. *J. Cancer* **2016**, *7*, 2346–2359. [CrossRef]

141. Booth, A.; Magnuson, A.; Fouts, J.; Foster, M. Adipose tissue, obesity and adipokines: Role in cancer promotion. *Horm. Mol. Biol. Clin. Investig.* **2015**, *21*, 57–74. [CrossRef]

142. Tumminia, A.; Vinciguerra, F.; Parisi, M.; Graziano, M.; Sciacca, L.; Baratta, R.; Frittitta, L. Adipose tissue, obesity and adiponectin: Role in endocrine cancer risk. *Int. J. Mol. Sci.* **2019**, *20*, 2863. [CrossRef]

143. Aronis, K.N.; Siatis, K.E.; Giannopoulou, E.; Kalofonos, H.P. Adiponectin promotes autophagy and apoptosis in endometrial cancer cell lines. *Clin. Oncol.* **2019**, *4*, 1–5.

144. Zhang, L.Z.; Wen, K.; Han, X.X.; Liu, R.; Qu, Q.X. Adiponectin mediates antiproliferative and apoptotic responses in endometrial carcinoma by the AdipoRs/AMPK pathway. *Gynecol. Oncol.* **2015**, *137*, 311–320. [CrossRef]

145. Jiang, J.H.; Fan, Y.C.; Zhang, W.; Sheng, Y.L.; Liu, T.T.; Yao, M.; Gu, J.R.; Tu, H.; Gan, Y. Adiponectin suppresses human pancreatic cancer growth through attenuating the beta-catenin signaling pathway. *Int. J. Biol. Sci.* **2019**, *15*, 253–264. [CrossRef] [PubMed]

146. Grossmann, M.E.; Cleary, M.P. The balance between leptin and adiponectin in the control of carcinogenesis -focus on mammary tumorigenesis. *Biochimie* **2012**, *94*, 2164–2171. [CrossRef]

147. Pollak, M. Insulin and insulin-like growth factor signalling in neoplasia. *Nat. Rev. Cancer.* **2008**, *8*, 915–928. [CrossRef] [PubMed]

148. D'Esposito, V.; Passaretti, F.; Hammarstedt, A.; Liguoro, D.; Terracciano, D.; Molea, G.; Canta, L.; Miele, C.; Smith, U.; Beguinot, F.; et al. Adipocyte-released insulin-like growth factor-1 is regulated by glucose and fatty acids and controls breast cancer cell growth in vitro. *Diabetologia* **2012**, *55*, 2811–2822. [CrossRef]

149. Louie, S.M.; Roberts, L.S.; Nomura, D.K. Mechanisms linking obesity and cancer. *BBA. Mol. Cell Biol. Lipids* **2013**, *1831*, 1499–1508. [CrossRef] [PubMed]

150. Li, Z.; Zhang, C.; Du, J.X.; Zhao, J.; Shi, M.T.; Jin, M.W.; Liu, H. Adipocytes promote tumor progression and induce PD-L1 expression via TNF-alpha/IL-6 signaling. *Cancer Cell Int.* **2020**, *20*, 179. [CrossRef] [PubMed]

151. Duong, M.N.; Geneste, A.; Fallone, F.; Li, X.; Dumontet, C.; Muller, C. The fat and the bad: Mature adipocytes, key actors in tumor progression and resistance. *Oncotarget* **2017**, *8*, 57622–57641. [CrossRef]

152. Bolsoni-Lopes, A.; Alonso-Vale, M.I.C. Lipolysis and lipases in white adipose tissue—An update. *Arch. Endocrinol. Metab.* **2015**, *59*, 335–342. [CrossRef]

153. Ma, Y.B.; Temkin, S.M.; Hawkridge, A.M.; Guo, C.Q.; Wang, W.; Wang, X.Y.; Fan, X.J. Fatty acid oxidation: An emerging facet of metabolic transformation in cancer. *Cancer Lett.* **2018**, *435*, 92–100. [CrossRef]

154. Balaban, S.; Shearer, R.F.; Lee, L.S.; Van Geldermalsen, M.; Schreuder, M.; Shtein, H.C.; Cairns, R.; Thomas, K.C.; Fazakerley, D.J.; Grewal, T.; et al. Adipocyte lipolysis links obesity to breast cancer growth: Adipocyte-derived fatty acids drive breast cancer cell proliferation and migration. *Cancer Metab.* **2017**, *5*, 1. [CrossRef]

155. Zeng, J.; Sauter, E.R.; Li, B. FABP4: A New Player in Obesity-Associated Breast Cancer. *Trends Mol. Med.* **2020**, *26*, 437–440. [CrossRef]

156. Clement, E.; Lazar, I.; Attane, C.; Carrie, L.; Dauvillier, S.; Ducoux-Petit, M.; Esteve, D.; Menneteau, T.; Moutahir, M.; Le Gonidec, S.; et al. Adipocyte extracellular vesicles carry enzymes and fatty acids that stimulate mitochondrial metabolism and remodeling in tumor cells. *EMBO J.* **2020**, *39*, e102525. [CrossRef] [PubMed]

157. Lazar, I.; Clement, E.; Dauvillier, S.; Milhas, D.; Ducoux-Petit, M.; Le Gonidec, S.; Moro, C.; Soldan, V.; Dalle, S.; Balor, S.; et al. Adipocyte exosomes promote melanoma aggressiveness through fatty acid oxidation: A novel mechanism linking obesity and cancer. *Cancer Res.* **2016**, *76*, 4051–4057. [CrossRef] [PubMed]

158. Wang, S.H.; Xu, M.Q.; Li, X.X.; Su, X.D.; Xiao, X.; Keating, A.; Zhao, R.C. Exosomes released by hepatocarcinoma cells endow adipocytes with tumor-promoting properties. *J. Hematol. Oncol.* **2018**, *11*, 82. [CrossRef] [PubMed]

159. Park, J.; Euhus, D.M.; Scherer, P.E. Paracrine and Endocrine Effects of adipose tissue on cancer development and progression. *Endocr. Rev.* **2011**, *32*, 550–570. [CrossRef] [PubMed]

160. Castillo, J.J.; Mulkey, F.; Geyer, S.; Kolitz, J.E.; Blum, W.; Powell, B.L.; George, S.L.; Larson, R.A.; Stone, R.M. Relationship between obesity and clinical outcome in adults with acute myeloid leukemia: A pooled analysis from four CALGB (alliance) clinical trials. *Am. J. Hematol.* **2016**, *91*, 199–204. [CrossRef]

161. Cascetta, P.; Cavaliere, A.; Piro, G.; Torroni, L.; Santoro, R.; Tortora, G.; Melisi, D.; Carbone, C. Pancreatic cancer and obesity: Molecular mechanisms of cell transformation and chemoresistance. *Int. J. Mol. Sci.* **2018**, *19*, 3331. [CrossRef]

162. Mentoor, I.; Engelbrecht, A.M.; Nell, T. Fatty acids: Adiposity and breast cancer chemotherapy, a bad synergy? *Prostaglandins Leukot. Essent. Fatty Acids* **2019**, *140*, 18–33. [CrossRef]

163. Kounakis, K.; Chaniotakis, M.; Markaki, M.; Tavernarakis, N. Emerging roles of lipophagy in health and disease. *Front. Cell Dev. Biol.* **2019**, *7*, 185. [CrossRef]

164. Longo, V.D.; Fontana, L. Calorie restriction and cancer prevention: Metabolic and molecular mechanisms. *Trends Pharmacol. Sci.* **2010**, *31*, 89–98. [CrossRef]

165. Caccialanza, R.; Aprile, G.; Cereda, E.; Pedrazzoli, P. Fasting in oncology: A word of caution. *Nat. Rev. Cancer* **2019**, *19*, 177. [CrossRef] [PubMed]

166. Agostoni, C.; Bresson, J.L.; Fairweather-Tait, S.; Flynn, A.; Golly, I.; Korhonen, H.; Lagiou, P.; Lovik, M.; Marchelli, R.; Martin, A.; et al. Scientific Opinion on Dietary Reference Values for fats, including saturated fatty acids, polyunsaturated fatty acids, monounsaturated fatty acids, trans fatty acids, and cholesterol. *EFSA J.* **2010**, *8*, 1461. [CrossRef]

167. Stensrud, M.J.; Valberg, M. Inequality in genetic cancer risk suggests bad genes rather than bad luck. *Nat. Commun.* **2017**, *8*, 1165. [CrossRef]

168. Tomasetti, C.; Vogelstein, B. Cancer risk: Role of environment—Response. *Science* **2015**, *347*, 729–731. [CrossRef] [PubMed]

169. Thomas, F.; Roche, B.; Ujvari, B. Intrinsic versus extrinsic cancer risks: The debate continues. *Trends Cancer* **2016**, *2*, 68–69. [CrossRef]

170. Couzin-Frankel, J. The bad luck of cancer. *Science* **2015**, *347*, 12. [CrossRef] [PubMed]

171. Erazo-Oliveras, A.; Fuentes, N.R.; Wright, R.C.; Chapkin, R.S. Functional link between plasma membrane spatiotemporal dynamics, cancer biology, and dietary membrane-altering agents. *Cancer Metastasis Rev.* **2018**, *37*, 519–544. [CrossRef]

172. Chapkin, R.S.; DeClercq, V.; Kim, E.; Fuentes, N.R.; Fan, Y.Y. Mechanisms by Which Pleiotropic Amphiphilic n-3 PUFA Reduce Colon Cancer Risk. *Curr. Colorectal Cancer Rep.* **2014**, *10*, 442–452. [CrossRef]

173. Nicolson, G.L.; Ash, M.E. Lipid Replacement Therapy: A natural medicine approach to replacing damaged lipids in cellular membranes and organelles and restoring function. *BBA Biomembr.* **2014**, *1838*, 1657–1679. [CrossRef]

174. Peláez, R.; Pariente, A.; Pérez-Sala, Á.; Larráyoz, I.M. Sterculic Acid: The Mechanisms of Action beyond Stearoyl-CoA Desaturase Inhibition and Therapeutic Opportunities in Human Diseases. *Cells* **2020**, *9*, 140. [CrossRef]

175. Kim, K.B.; Nam, Y.A.; Kim, H.S.; Hayes, A.W.; Lee, B.M. α-Linolenic acid: Nutraceutical, pharmacological and toxicological evaluation. *Food Chem. Toxicol.* **2014**, *70*, 163–178. [CrossRef] [PubMed]

176. Lien, E.C.; Vander Heiden, M.G. A framework for examining how diet impacts tumour metabolism. *Nat. Rev. Cancer* **2019**, *19*, 651–661. [CrossRef] [PubMed]

177. Kenny, F.S.; Pinder, S.E.; Elis, I.O.; Gee, J.M.W.; Nicholson, R.I.; Bryce, R.P.; Robertson, J.F.R. Gamma linolenic acid with tamoxifen as primary therapy in breast cancer. *Int. J. Cancer* **2000**, *85*, 643–648. [CrossRef]

 *metabolites*

Review

Lipidomic-Based Advances in Diagnosis and Modulation of Immune Response to Cancer

Luis Gil-de-Gómez [1,*], **David Balgoma** [2] and **Olimpio Montero** [3]

[1]   Center for Childhood Cancer Research, Children's Hospital of Philadelphia, Colket Translational Research Center, 3501 Civic Center Blvd, PA 19104, USA

[2]   Analytical Pharmaceutical Chemistry, Department of Medicinal Chemistry, Uppsala University, Husarg. 3, 75123 Uppsala, Sweden; david.balgoma@ilk.uu.se

[3]   Spanish National Research Council (CSIC), Boecillo's Technological Park Bureau, Av. Francisco Vallés 8, 47151 Boecillo, Spain; olimpio.montero@dicyl.csic.es

*   Correspondence: gildegomel@email.chop.edu

Received: 30 June 2020; Accepted: 12 August 2020; Published: 14 August 2020

Abstract: While immunotherapies for diverse types of cancer are effective in many cases, relapse is still a lingering problem. Like tumor cells, activated immune cells have an anabolic metabolic profile, relying on glycolysis and the increased uptake and synthesis of fatty acids. In contrast, immature antigen-presenting cells, as well as anergic and exhausted T-cells have a catabolic metabolic profile that uses oxidative phosphorylation to provide energy for cellular processes. One goal for enhancing current immunotherapies is to identify metabolic pathways supporting the immune response to tumor antigens. A robust cell expansion and an active modulation via immune checkpoints and cytokine release are required for effective immunity. Lipids, as one of the main components of the cell membrane, are the key regulators of cell signaling and proliferation. Therefore, lipid metabolism reprogramming may impact proliferation and generate dysfunctional immune cells promoting tumor growth. Based on lipid-driven signatures, the discrimination between responsiveness and tolerance to tumor cells will support the development of accurate biomarkers and the identification of potential therapeutic targets. These findings may improve existing immunotherapies and ultimately prevent immune escape in patients for whom existing treatments have failed.

Keywords: immunotherapy; cancer; lipids; biomarkers; metabolism

---

1. Introduction

Following the discovery of the structure of DNA in 1953 [1], increasingly efficient technologies for the study of the whole genome (genomics) have enabled assessments of genome-based pathologies in large population cohorts [2]. However, since a broad number of factors, including environment, diet or lifestyle, are important in the etiology of diverse diseases such as cancer, a high-dimensional biological approach appears to be required [3]. A multi-omics/systems-level approach, which encompasses the combined analysis of data from genomics, RNA transcription (transcriptomics), proteins/peptides (proteomics) and metabolites (metabolomics), enables one to overlay gene information onto a complementary understanding of accrued molecular mechanisms [4]. Lipidomics represents an emerging discipline from metabolomics that connects lipid biology, technology and medicine, and that strives to build an all-inclusive atlas of the cellular/tissue lipidome [5]. In this regard, the role played by lipids in the etiology and treatment of cancer has loomed large over the last decades.

Early evidence that cancer cells undergo characteristic metabolic alterations was documented by Otto Warburg in the first half of the twentieth century. In a paradoxical process in terms of adenosine triphosphate (ATP) production, cancer cells increase the consumption of glucose to support aberrant cellular proliferation. Because proliferating tumor cells require cholesterol and other lipids, perturbations in the lipid metabolism are emerging as potential targets for therapeutic intervention in cancer [6,7]. Cancer immunotherapy has proven to have an unprecedented positive impact in clinical oncology. Increased evidence suggests that glycolytic metabolism not only rules cancer signaling but also the antitumor immune response where activated inflammatory immune cells display the same metabolic profile as tumor cells [8] (Figure 1). Multiple studies have separately reported the impact of lipids on immune cells and tumor progression. However, so far, little work has focused on reviewing how the lipid metabolism is associated with the immune response to tumors. Taking this shortfall into account, we aim to highlight the role of lipid mediators in the context of immune activation in order to explore potential biomarkers and therapeutic targets for cancer.

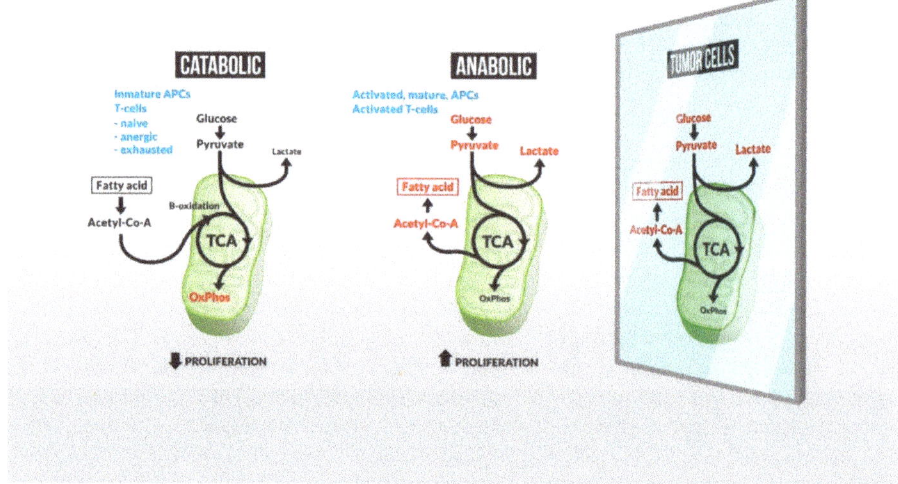

Figure 1. A metabolic shift is required by immune cells for them to respond actively to tumor cells. Inactive immune cells rely on oxidative phosphorylation (OxPhos) and fatty acid (FA) oxidation (left), while activated and responsive cells increase glucose uptake/glycolysis, resulting in an increased FA synthesis and lactate production (central panel). Lipogenesis, required for a robust cell proliferation, also characterizes tumor cell metabolism (right). Therefore, an untargeted lipid-based treatment to fuel effector immune cells may produce self-defeating effects inducing tumor cell growth. Many other lipid intermediates regulate inflammation, and exogenous lipids such as gut microbiota-derived short chain fatty acids (SCFAs) may impact the host immune response to tumor cells. Together, these findings indicate (1) the exhaustive regulation required to maintain immunity balance in the presence of tumor cells, and (2) the essential role of a large variety of lipids in this control. New precise lipidomic-based strategies may enhance therapeutic targeting and improve the capacity of existing immunotherapies to control tumor progression.

2. Lipid Metabolism Impacts Immune Activation against Tumor Progression

2.1. Lipid Interplay with Immune Regulation

Tumors impact immune cell function by supporting cancer stem cell survival, metastasis and immune evasion. The aggressiveness of tumor cells is linked to their capacity to store high levels of lipids and, in particular, cholesterol [6]. Metabolic challenges in the tumor microenvironment (TME), including hypoglycemia and hypoxia, induce changes in tumor cellular metabolism like aerobic glycolysis and fatty acid oxidation (FAO) [9]. In response, immune cells show the capacity to modulate lipid metabolism to better adapt to these special metabolic conditions.

The innate immune system is the first barrier against external stimuli, which are recognized via Toll-like receptors (TLR). TLR-dependent response, which regulates the activation of antigen-presenting cells (APC) (mainly macrophages or dendritic cells (DCs)), shifts the intracellular metabolism towards the glycolysis-fueled synthesis of fatty acid (FA) [10,11]. After the initial broad immune response, an adaptive immune response is initiated when APCs process and present antigens for recognition by certain lymphocytes such as T cells. Both phases of the immune response are characterized by a fragile equilibrium, whereas the heterogeneous groups of immune cells communicate and modulate each other via cytokine release. In this sense, cytokine production in activated DCs has been related to phospholipid remodeling to support FA demands [12]. Immune effector cells, such as T cells and macrophages, are induced by tumor-specific antigens and tumor-associated antigens. However, regulatory mechanisms of the immune system, such as immune checkpoints, make this cellular response incapable of preventing tumor progression. Immune check points are inhibitory regulators crucial for maintaining self-tolerance and controlling the duration of the immune response in order to prevent collateral tissue damage [13]. Since these key immune-regulatory molecules are used by tumor cells to promote evasion, immune checkpoint inhibitors have demonstrated their effectiveness as clinical targets for cancer immunotherapy [14]. This breakthrough is based on currently approved blocking monoclonal antibodies that inhibit cytotoxic T-lymphocyte-associated protein 4 (CTLA-4) and the programmed cell death protein PD-1/PD-L1 axis [15].

Endogenous lipid reserves provide energy to T cells but may also regulate T cell function by an immune checkpoint such as PD-1 [16]. PD-1 is a member of the cluster of differentiation 28 proteins (CD28) superfamily that delivers negative signals upon interaction with its two ligands, the PD-L1 and PD-L2 proteins. PD-1 activation impairs glucose and glutamine uptake whilst promoting FAO and catabolism of endogenous esterified fatty acids in both cytotoxic (CD8$^+$) and helper (CD4$^+$) T cells [16,17]. Another lipid pathway that targets PD-1 is regulated by the members of the peroxisome proliferator-activated receptors (PPAR) subfamily. This subfamily of nuclear receptors might be modulated by fatty acid signals derived from exogenous sources, including diet [18]. PPAR is crucial in supporting the accumulation and function of immunosuppressive regulatory T cells (Tregs) [19]. In concordance, it has been reported that PPAR-γ inhibition increases the efficiency of anti-PD-1 antibody immunotherapy, leading to the suppression of tumor progression in colon adenocarcinoma and melanoma models [20,21]; likewise, an agonist for another isomer, PPARα, is able to restore the anti-melanoma effects of tumor-infiltrating lymphocytes (TILs) by blocking the reprogramming to fatty acid catabolism in mice [22].

TILs, largely comprised of CD8$^+$ and CD4$^+$ T cells, as well as natural killer (NK) cells, are key players in tumor cell death. This particular function of both cell subtypes has been shown to be dependent on the profile of polyunsaturated fatty acids (PUFAs) in the cell membrane [23]. However, current work on how PUFA supplementation may affect TIL function in humans is often contradictory. Whereas it has been reported that the percentage of NK cells in mouse blood is reduced after dietary supplementation of docosahexaenoic acid (DHA, 22:6 n-3) and eicosapentaenoic acid (EPA, 20:5 n-3) [24], a similar previous study using EPA-rich oil in the diet did not find such differences [25]. The discovery of the G-protein-coupled receptors (GPCRs) suggests that many of the effects of dietary FAs may be receptor-mediated. This family of cell-surface free-fatty acid receptors includes the long-chain fatty acid receptors FFA1 and FFA4. Anti-inflammatory effects of omega-3 PUFAs, especially EPA and DHA, have been related directly to the expression of these FFA receptors. Hence, FFA4 knock-out mice have shown a higher proportion of pro-inflammatory macrophages than the wild type [26]. In addition, agonists of FFA receptors have been connected with the suppression of the proliferation and migration of a large variety of tumor cells [27,28].

The phenotype and maturation of T cells is also regulated by the fatty acid metabolism. Differentiation of T cells is dependent on de novo FA synthesis and uptake. In tumor tissue, the inhibition of de novo fatty acid synthase (FAS) by different targets, such as acetyl-CoA carboxylase 1, promotes Tregs but suppresses memory T cell lineage (Th17) differentiation [29]. The challenge of maintaining T cell function in a nutrient-depleted environment like the TME is resolved by other effector T cells. Unlike naïve and central memory T cells, effector memory T cells are less dependent on FA metabolism [30]. This feature plays an essential role in establishing immune equilibrium, since most effector T cells are removed after antigen elimination, whereas memory T cells remain for rapid response upon antigen re-exposure. The analysis of other molecules such as the mammalian target of rapamycin (mTOR) extends the list of lipid mediators that contribute to maintaining the immune balance. mTOR regulates Tregs differentiation, function and survival, ultimately defining the immunosuppressive profile of the TME [31]. Tregs are a dominant suppressive population that infiltrate the TME and dampen anti-tumor immune responses by inhibiting the effector T-cell function [32]. The singular metabolism of Tregs, including an increased FAO, provides them with critical advantages to survive and proliferate under hypoxia or low glucose conditions within the tumor [32,33].

The delivery and cellular distribution of PUFAs are indirectly regulated by desaturases, which perform the desaturation and elongation of essential fatty acids. However, phospholipases A_2 (PLA_2) are the main cellular regulators of PUFA release, maintaining the homeostatic levels of several free PUFAs, and in particular of those that are precursors of mediators with pro-inflammatory properties, such as arachidonic acid (AA, 20:4 n-6). In the inflammation process, AA is released by PLA_2 activity, and prostaglandin E_2 (PGE_2) is subsequently generated from arachidonic acid by the enzyme cyclooxygenase-2 (COX-2) [34,35]. One of the mechanisms that Tregs uses to suppress T cell activity is PGE_2 production, which can be reversed by COX-2 inhibitors [36]. PGE_2 is essential in homeostasis, and while its pro-inflammatory role is crucial for host cell self-preservation, its immunosuppressive effects may support tumor progression [37]. Besides directly mediating inflammation, PGE_2 might be used as an intermediate not only in the signaling between immune cells but also between immunity and tumors. Hence, PGE_2 released from DCs affects the generation and proliferation of Tregs by immunosuppressive cytokines like IL-10, whereas PGE_2 released from tumor cells is able to regulate DC maturation [37–39]. This $COX2/PGE_2$ pathway is also involved in the regulation of the immune checkpoint enzyme expression, like PD-L1, in tumor-infiltrating macrophages and other myeloid cells [40]. Moreover, a recent study suggests that the combined blockade of PD-1 and PGE_2 pathways is a promising therapeutic strategy for enhancing antitumor activity. This effect is due to an increased frequency of T cell-recognized tumor antigens, whose dysfunction is regulated by PD-1 [41].

Suppressing tumor immune surveillance may lead to the exhaustion or inactivation of pro-inflammatory immune cells and may, subsequently, promote tumor growth and metastasis. Myeloid-derived suppressor cells (MDSC) and immunosuppressive type II (M2) tumor-associated macrophages (TAMs) are fueled by the ß-oxidation of lipids, rather than glycolysis, within the TME [42]. Recent studies have shown that the phenotype of M2-like TAMs is controlled by intracellular long-chain fatty acid (LCFA) homeostasis, specifically unsaturated fatty acids like oleate [43]. Additionally, lipid metabolism provides a mechanistic explanation for TAM polarization and differentiation [44]. The upregulation of lipogenesis by sterol regulatory element-binding protein-1 (SREBP1) promotes the transcriptional response of macrophages to TLR signaling by driving the synthesis of anti-inflammatory fatty acids [45]. SREBP1 signaling also impacts tumor cells by sustaining the high energetic demands required for their growth and survival, and has been shown to be important in melanoma and prostate cancer progression [46–49]. One of the metabolic effects of SREBP1 is the regulation of the de novo lipogenesis by the upregulation of, among others, fatty acid synthase (FAS) and stearoyl-CoA desaturase-1 (SCD-1) [50,51]. Consequently, the upregulation of SREBP1 entails the upregulation of saturated and monounsaturated fatty acids, both free and in glycerolipids. Regarding macrophages, stimulation by lipopolysaccharide (LPS), a component of cell wall of gram negative bacteria, upregulates SREPB1 expression which is required for the inflammatory response [52,53]. In contrast, the activation of liver X receptors (LXRs), which also upregulate SREPB1, decreases the inflammation level in macrophages [54]. Because of this opposed effect, it is expected that the level of de novo lipogenesis in TAMs presents a complex relationship with their activation state. LXRs are regulated by oxysterols and SREPB1 by sterols in the cell environment. Consequently, not only diet but also the tumor lipid microenvironment can regulate the metabolic/pro-inflammatory status of TAMs. In addition, external palmitic acid reprograms the microglia metabolism in a way that mimics LPS treatment [55], whereas oleic acid reduces the pro-inflammatory response [56]. Furthermore, sexual hormones in the TME also play a key role in the lipid metabolism and the inflammatory state of TAMs. The androgen receptor decreases the LXR and SREBP1 activity, which decreases the de novo lipogenesis and remodels the lipid metabolism [57,58]. Interestingly, the interaction between prostate cancer cells and macrophages regulates the resistance to hormonal therapy [59]. This fact suggests an interplay in tumor growth among: (1) the activation of the androgen receptor, (2) the tumor microenvironment and (3) the LXR-mediated lipogenesis in both the tumor and TAMs. Altogether, these studies suggest that both the lipidic and hormonal microenvironment interact to reprogram the metabolic and inflammatory state of TAMs. This reprogramming is associated with therapy resistance and patient prognosis.

LXRs are major regulators of FA and cholesterol homeostasis. Cholesterol, a nonpolar lipid transported in plasma by low-density lipoproteins (LDL) and high-density lipoproteins (HDL), has been linked to the effect of IL-10 in immune regulation. The inhibition of cholesterol biosynthesis with atorvastatin or 25-hydroxycholesterol regulates IL-10 production by inducing human CD4$^+$ T cells to switch from an effector to an anti-inflammatory profile [60]. Furthermore, given the role of lipoproteins as cholesterol carriers, while they promote tumor growth by regulating T cell activation and functionality [61], recent studies have used them as anti-tumor drug delivery vehicles [62].

The impact of lipids on the immune response to cancer includes post-translational modifications. Palmitoylation has been found to be important in the context of cancer immunotherapy. This post-translational process involves the binding of palmitate (C16:0) to amino acid residues. Yao and colleagues identified palmitoyl transferase ZDHHC3, which contains a conserved Asp-His-His-Cys (DHHC) signature motif, as the main acyltransferase required for PD-L1 palmitoylation. This lipid modification stabilizes PD-L1 by blocking ubiquitination, which ultimately prevents lysosomal-driven degradation. Thus, DHHC3 targeting enhances T cell cytotoxicity against cancer cells in vitro, as well as the in vivo antitumor effect in a colon carcinoma model [63]. Other studies have related the ablation of ZDHHC3 in human mammary tumor cell xenografts to a reduced primary and lung metastasis infiltration. This effect correlates with an enhanced recruitment of macrophages and NK cells to the tumor, and its subsequent clearance [64].

2.2. Short-Chain Fatty Acids from Gut Microbiota as Effectors of the Immune System

FAs with chain lengths ranging from one to six carbon atoms are produced by trillions of harmless microorganisms that inhabit the human gastrointestinal tract. These short chain fatty acids (SCFAs) are the major end product derived from gut microbiota; very high concentrations are found in the colon [65]. The presence of SCFAs (propionic, butyric, acetic and valeric acids) regulates the intestinal microenvironment by reducing pH and impacting the microbial function and composition [66]. Besides various gut disorders, gut microbiota also play an important role in central nervous system disorders, the immune system and cancer malignancies [67]. Although the role of butyrate in fueling tumor cells proliferation has been described [68], SCFAs have been generally perceived as tumor suppressors because they induce cancer cell differentiation and apoptosis [69]. The ability of SCFAs to regulate effector immune cells is considered one of the essential mechanisms accounting for their anti-tumor properties [70]. SCFAs engage GPCRs such as FFA2 and FFA3, and act as histone deacetylases (HDACs) to regulate the activity of innate immune cells such as neutrophils, macrophages and DCs, and they also modulate antigen-specific adaptive immunity mediated by T cells and B cells [71,72].

SCFAs, particularly butyrate, directly impact the immune response to cancer through the reprogramming of the cellular metabolism. In activated CD8$^+$ T cells, butyrate increases glycolytic activity, mitochondrial mass and membrane signaling. Butyrate-stimulated CD8$^+$ T cells also show functional uncoupling of the TCA cycle from glycolysis, promoting additional sources of carbon such as glutamine and FAs [73]. An increased FA intake in butyrate-treated CD8$^+$ T cells serves to charge the TCA cycle, but triacylglycerides and phospholipids are other candidates that serve as suppliers [74,75]. The anti-inflammatory properties of SCFAs are also related to the ability of butyrate and propionate to abrogate IL-12 release from APCs, a cytokine with a primary role in effector T cell stimulation [76,77].

In contrast, both SCFAs are also associated with resistance to immune checkpoint CTLA-4 blockade and a higher proportion of Treg cells. These effects limit the clinical outcome of cancer patients treated with anti-CTLA-4 blocking monoclonal antibodies [78].

The capacity of butyrate to regulate T cell polarization and immune checkpoint blockade correlates with the diversity of commensal microbiota. In human bacterial communities, most butyrate-producing colon bacteria belong to the *Firmicutes* phylum. The equilibrium between species defines the therapeutic outcome, and a low Bacteroidetes/Firmicutes ratio has been used to identify lung cancer patients [79]. Moreover, the relative abundance of other specific bacteria, such as Bifidobacterium, increases anti-PD-L1 efficacy, promoting anti-tumor immunity [80]. Taken together, these findings point toward an alternative therapeutic strategy by targeting immune cells on a metabolic level. Augmenting the efficacy of the immune system by targeting the lipid metabolism could be useful for improving the antitumor immune response. However, as Chalmin et al. postulate, targeting the lipid metabolism may affect multiple immune populations and could have unpredictable outcomes [81]. Thus, since fatty acid oxidase is required not only for effector T cell development but also for Treg differentiation [82], its blockade limits Treg-dependent immunosuppression. Despite these drawbacks, data suggest that the capacity to define specific lipid reprogramming that correlates with disease stages will help to design new cancer treatments. The balance between immune activation and suppression is a critical feature of immunity, and lipids are able to alter this equilibrium. Therefore, targeting the lipid metabolism may be used to induce immune stimulation, which will ultimately determine the clinical success of cancer immunotherapy.

3. Lipids as Biomarkers of Immune Response to Cancer

Accurate and predictive biomarkers to diagnose early stages of disease are a critical objective of clinical and biomedical research. Lipids, among several other metabolites such as amino acids or sugars, have been described as potential predictors of systemic alterations that discriminate between healthy controls and patients [83]. Clinical success often hinges on an early diagnosis, especially in long and age-related malignances like Alzheimer's disease or cancer [84]. New technologies for the qualitative and quantitative analyses of metabolites can provide essential information on pathological conditions that can result in profound alterations in the architecture of the immune system. Identifying the metabolic profile associated with the immune response to tumor cells has emerged, parallel to immunotherapy, as a tool for obtaining an early and accurate diagnosis and for designing personalized treatments, both being essential for better clinical outcomes in cancer patients.

An increased de novo synthesis of fatty acids is required for membrane synthesis and, therefore, for the growth and proliferation of both immune and tumor cells. This makes fatty acids robust biomarker candidates. Recent studies have shown that genetic alterations observed in acute myeloid leukemia (AML) patients control lipid dynamics and metabolism [39,85]. Interestingly, patients with AML can be identified by specific lipid signatures in plasma [86] and bone marrow [87]. Whereas lipid biomarkers have been used to identify tumor progression, the relationship between a characteristic lipid profile and the immune response to cancer is still poorly understood. The major clinical advantages of immune checkpoint inhibitors have generated considerable interest in discovering biomarkers that predict the response to treatment [88]. Recent studies propose serum concentrations of very long chain fatty acids (VLCFA) as a way to identify the response to immune checkpoint inhibitors in urological cancer [42]. The rationale for this biomarker is motivated by the finding that lower serum VLCFA levels are associated with highly immunosuppressive TME with a high-VLCFA consumption rate.

As discussed previously, de novo lipogenesis is also associated in a complex manner with the metabolic/inflammatory state of TAMs. Consequently, the lipids associated with the de novo lipogenesis act as biomarkers of tumor growth and the activation of TAMs. The LXRs/SREBP1 pathway is the key player in the regulation of the de novo lipogenesis, and it is involved in tumor growth and in the inflammatory response [89]. LXRs/SREBP1 upregulation in tumor or inflammatory cells leads to an increase of saturated and monounsaturated fatty acids via the activation of FAS and SCD-1, which are incorporated into glycerolipids by acyltransferases. Consequently, the upregulation of glycerolipids with saturated and monounsaturated fatty acids acts as a biomarker for the tumor synthesis of membranes and the activation of macrophages [90,91]. In addition, LXRs/SREBP1 upregulate glycerol-3-phosphate acyltransferase 1 (GPAT-1), which has a strong preference for transferring palmitic acid to the *sn*-1 position of glycerol-3-phosphate. This leads to an enrichment of glycerolipids with palmitic acid in the *sn*-1 position of the glycerol backbone. Consequently, the triacylglycerides with palmitic acid in the external position of the glycerol act as a biomarker of LXRs/SREBP1 activation and the de novo lipogenesis [92,93]. In conclusion, these triacylglycerides have the potential to be used as biomarkers for (1) monitoring the metabolic reprogramming of TAMs in the TME, and (2) the effect or resistance to immunotherapy by evaluating the up- or downregulation of lipogenesis in the tumor.

Because SCFAs from gut microbiota have a wide-ranging impact on the host physiology, these metabolites are also increasingly studied as predictive biomarkers. SCFAs and microbiota composition have been used to determine the risk of cancer, and reduced levels of butyric acid in patients with colon cancer have been reported [94,95]. The levels of butyrate are also correlated with the responsiveness to melanoma in mice treated with antibiotics [96]. Recent results have reported a correlation between the relative abundance of certain SCFA-producing microbiota and the outcome of PD-1-based immunotherapy in melanoma patients [97]. These data correlate with those from a recent study that makes the case for the reduced serum content of SCFAs being a biomarker of refractory non-small cell lung cancer (NSCLC) [67]. According to Boticcelli et al., lower levels of SCFA are found in the fecal samples of patients with a poor prognosis treated with Nivolumab, a human PD-1-blocking antibody. Together, these results show that gut microbiota-induced immune effects are dependent on the specific cancer therapy and that certain blood lipid biomarkers are able to predict this relationship.

When cancer care is delayed, patient treatment is associated with greater clinical complications and a lower survival rate. In order to have the best chance for a successful treatment and prognosis, an early and precise diagnosis of cancer progression before and during the treatment is critical. Thus, identifying biomarkers that can monitor the tumor response in every stage of treatment has huge clinical implications. Further studies will be needed to correlate the lipid profile with the immune cell phenotype and immune checkpoint expression within the tumor. These data will help to discriminate between pro-inflammatory and immunosuppressive TME populations, resulting in more accurate biomarkers of cancer progression.

4. Active Modulation of Lipid Metabolism to Improve CAR T Cell Therapy

Chimeric antigen receptor—engineered T cell (CAR T) therapy has demonstrated its long-term clinical benefit for patients with advanced cancers [98]. CART therapy involves genetically modified patient T cells with chimeric antigen receptors that recognize specific antigens on the tumor cell surface. The antitumor efficacy of immunotherapy against hematologic cancers has been extended to other tumors [99–101]. Among diverse potential targets, such as CD19 for B-cell malignancies, GD2, a disialoganglioside glycolipid, was identified as a tumor antigen more than 30 years ago [102]. GD2 is normally present in developing brains and can be overexpressed in some tumors, with a greater recurrence in childhood cancer neuroblastoma, melanoma and diverse pediatric sarcomas [103]. However, while GD2-specific antibody therapies used in the treatment of neuroblastoma have been shown to be successful, the fatal neurotoxicity of GD2-specific CAR T cell therapy that has been observed in some studies suggests that GD2 may be a difficult target antigen for CAR T cell therapy [104].

Several studies and clinical trials reveal that CAR T cell therapy for leukemia achieved high rates of complete remission, but therapy-relapsed leukemia remains a significant source of mortality [105]. Because T cell exhaustion elevates the risk of relapse [106], additional research on how to avoid this detrimental effect is urgently needed. Differentiated effector T cells use glycolysis for proliferation, and after activation they ultimately succumb [21]. Only a small proportion of long-surviving memory T cells with OXPHOS-mediated ATP production contributes to a favorable and durable antitumor response in the TME [107]. Since Notch signaling, a conserved cellular interaction mechanism, promotes mitochondrial biogenesis and FA synthesis, recent studies have evaluated the impact of the manipulation of this metabolic pathway on the success of CAR T cells. According to Kondo et al. [108], the overexpression of Notch and its downstream gene Forkhead box M1 (FOXM1) results in enhanced anti-tumor effects as compared with conventional CART cells, suggesting a novel strategy to improve CART-based therapy [108].

The success of CAR T cell therapy in treating hematological malignancies is limited in solid tumors, where finding, entering and surviving in the tumor are extra challenges [109]. Other restrictions are driven by constraints from the on-target off-tumor toxicity of CAR T cells, where the lack of tumor specificity increases the potential risk for normal tissues to be attacked by CAR T cells [109,110]. In order to avoid these limitations, new strategies have focused on providing an anti-tumor effect with an absence of side-effects. Besides the enormous ability of PUFAs, such as AA, EPA and DHA, to regulate the immune responses, as presented above, gamma-linolenic acid (GLA, 18:3 n-6) has shown a selective effect against tumor cells [111]. According to an open-label clinical study that included 21 patients with stage IV glioma, the intra-tumor injection of GLA enhanced the sensitivity of tumor cells to chemotherapeutic drugs and radiation, producing tumor regression without harming normal cells [112]. Additionally, together with AA, EPA and DHA, GLA has been reported to regulate the antioxidant properties of glutathione peroxidase 4 (GPX4), as well as the levels of cytokines such as IL-1, IL-6 and tumor necrosis factor alpha (TNF-α) that play essential roles in inflammation [113,114]. These data suggest that the combination of PUFAs as an adjuvant may help immunotherapy block tumor progression.

Lipid metabolism has a dual impact on CAR T cell therapy. Lipids can systematically fuel tumor cells and immune cells. However, enhancing the immune response via CAR T cells presents evident advantages beyond the described obstacles. T cells can be successfully designed and prepared for the restricted metabolic conditions within the TME [115]. Identifying and reprogramming the mechanisms involved in the dysfunction of CAR T cells may help support more proliferative and ultimately successful CART cell-based therapies [109]. Therefore, metabolic targets that include the lipid metabolism may generate improved CAR T cells, so as to avoid cancer relapses related to T cell disability.

5. Conclusions

The complexity and variability of tumors still constitute a challenge for physicians and researchers. Although immunotherapy has attained ambitious milestones and improved prognoses for cancer patients, the systemic character and self-regulation capacity of immunity should be considered in order to obtain improvements. A multi-focal anti-tumor strategy, in combination with other treatments, appears to be required in order to avoid relapses; moreover, it should draw from diverse perspectives: First, a global intervention, by for instance modulating the gut microbiota, which could have positive effects on the immune cell activity; second, an early and precise diagnosis so as to achieve better clinical outcomes; and third, targeted treatments, where genetically engineered patient CAR T cells have already shown clinical benefits.

Current treatment limitations are related to an immunosuppressive TME, which modifies the T cell function in terms of differentiation and exhaustion. Combining CAR T cells with checkpoint inhibitors and the depletion of suppressive factors in the microenvironment via lipid targets may mitigate this phenomenon. Although new studies will be necessary to characterize specific metabolic pathways implicated in the immune response to tumor cells, data suggest that lipid reprogramming will be key to generating a favorable metabolic environment to avoid tumor evasion.

In conclusion, the modulation of the immune system has been extensively demonstrated to be an effective cancer treatment. However, further investigations should focus on reducing treatment limitations that ultimately lead to tumor relapse. Several studies are currently focusing on therapy improvements by facilitating energy influx to T cells, where lipids play an essential role. Targeting lipid reprogramming in the immunity setting may generate new tools to create lasting, robust and personalized therapies against cancer.

Author Contributions: L.G.-d.-G. wrote the manuscript. D.B. and O.M. contributed to writing, reviewing and editing the manuscript. All authors have read and agreed to the published version of the manuscript.

Funding: The elaboration of this manuscript received no external funding.

Acknowledgments: We thank Sue Seif and Julián Abad for their excellent figure editing assistance. We also thank Rodrigo Gier and Dinesh Dilip Thakur for improving the manuscript.

Conflicts of Interest: The authors declare no conflict of interest.

References

1. Watson, J.D.; Crick, F.H.C. Molecular Structure of Nucleic Acids: A Structure for Deoxyribose Nucleic Acid. *Nature* **1953**, *171*, 737–738. [CrossRef] [PubMed]
2. Check Hayden, E. Technology: The $1,000 genome. *Nature* **2014**, *507*, 294–295. [CrossRef] [PubMed]
3. Rattray, N.J.W.; Deziel, N.C.; Wallach, J.D.; Khan, S.A.; Vasiliou, V.; Ioannidis, J.P.A.; Johnson, C.H. Beyond genomics: Understanding exposotypes through metabolomics. *Hum. Genom.* **2018**, *12*, 4. [CrossRef] [PubMed]
4. Romero, R.; Espinoza, J.; Gotsch, F.; Kusanovic, J.; Friel, L.; Erez, O.; Mazaki-Tovi, S.; Than, N.; Hassan, S.; Tromp, G. The use of high-dimensional biology (genomics, transcriptomics, proteomics, and metabolomics) to understand the preterm parturition syndrome. *BJOG Int. J. Obstet. Gynaecol.* **2006**, *113*, 118–135. [CrossRef]
5. Lam, S.M.; Tian, H.; Shui, G. Lipidomics, en route to accurate quantitation. *Biochim. Biophys. Acta (BBA) Mol. Cell Biol. Lipids* **2017**, *1862*, 752–761. [CrossRef]
6. Mancini, R.; Noto, A.; Pisanu, M.E.; De Vitis, C.; Maugeri-Saccà, M.; Ciliberto, G. Metabolic features of cancer stem cells: The emerging role of lipid metabolism. *Oncogene* **2018**, *37*, 2367–2378. [CrossRef]
7. Mohamed, A.; Collins, J.; Jiang, H.; Molendijk, J.; Stoll, T.; Torta, F.; Wenk, M.R.; Bird, R.J.; Marlton, P.; Mollee, P.; et al. Concurrent lipidomics and proteomics on malignant plasma cells from multiple myeloma patients: Probing the lipid metabolome. *PLoS ONE* **2020**, *15*, e0227455. [CrossRef]
8. Domblides, C.; Lartigue, L.; Faustin, B. Control of the Antitumor Immune Response by Cancer Metabolism. *Cells* **2019**, *8*, 104. [CrossRef]
9. Hanahan, D.; Weinberg, R.A. Hallmarks of Cancer: The Next Generation. *Cell* **2011**, *144*, 646–674. [CrossRef]
10. Krawczyk, C.M.; Holowka, T.; Sun, J.; Blagih, J.; Amiel, E.; DeBerardinis, R.J.; Cross, J.R.; Jung, E.; Thompson, C.B.; Jones, R.G.; et al. Toll-like receptor-induced changes in glycolytic metabolism regulate dendritic cell activation. *Blood* **2010**, *115*, 4742–4749. [CrossRef]
11. Lauterbach, M.A.; Hanke, J.E.; Serefidou, M.; Mangan, M.S.J.; Kolbe, C.C.; Hess, T.; Rothe, M.; Kaiser, R.; Hoss, F.; Gehlen, J.; et al. Toll-like Receptor Signaling Rewires Macrophage Metabolism and Promotes Histone Acetylation via ATP-Citrate Lyase. *Immunity* **2019**, *51*, 997–1011. [CrossRef] [PubMed]
12. Márquez, S.; Fernández, J.J.; Mancebo, C.; Herrero-Sanchez, C.; Alonso, S.; Sandoval, T.A.; Rodríguez Prados, M.; Cubillos-Ruiz, J.R.; Montero, O.; Fernández, N.; et al. Tricarboxylic Acid Cycle Activity and Remodeling of Glycerophosphocholine Lipids Support Cytokine Induction in Response to Fungal Patterns. *Cell Rep.* **2019**, *27*, 525–536. [CrossRef]
13. Pardoll, D.M. The blockade of immune checkpoints in cancer immunotherapy. *Nat. Rev. Cancer* **2012**, *12*, 252–264. [CrossRef] [PubMed]
14. Brahmer, J.R.; Tykodi, S.S.; Chow, L.Q.M.; Hwu, W.J.; Topalian, S.L.; Hwu, P.; Drake, C.G.; Camacho, L.H.; Kauh, J.; Odunsi, K.; et al. Safety and Activity of Anti–PD-L1 Antibody in Patients with Advanced Cancer. *New Engl. J. Med.* **2012**, *366*, 2455–2465. [CrossRef]
15. Donini, C.; D'Ambrosio, L.; Grignani, G.; Aglietta, M.; Sangiolo, D. Next generation immune-checkpoints for cancer therapy. *J. Thorac. Dis.* **2018**, *10*, S1581–S1601. [CrossRef]
16. Patsoukis, N.; Bardhan, K.; Chatterjee, P.; Sari, D.; Liu, B.; Bell, L.N.; Karoly, E.D.; Freeman, G.J.; Petkova, V.; Seth, P.; et al. PD-1 alters T-cell metabolic reprogramming by inhibiting glycolysis and promoting lipolysis and fatty acid oxidation. *Nat. Commun.* **2015**, *6*, 6692. [CrossRef] [PubMed]
17. Ogando, J.; Sáez, M.E.; Santos, J.; Nuevo-Tapioles, C.; Gut, M.; Esteve-Codina, A.; Heath, S.; González-Pérez, A.; Cuezva, J.M.; Lacalle, R.A.; et al. PD-1 signaling affects cristae morphology and

leads to mitochondrial dysfunction in human CD8+ T lymphocytes. *J. Immunother. Cancer* **2019**, *7*, 151. [CrossRef] [PubMed]

18. Varga, T.; Czimmerer, Z.; Nagy, L. PPARs are a unique set of fatty acid regulated transcription factors controlling both lipid metabolism and inflammation. *Biochim. Biophys. Acta (BBA) Mol. Basis Dis.* **2011**, *1812*, 1007–1022. [CrossRef]

19. Cipolletta, D.; Feuerer, M.; Li, A.; Kamei, N.; Lee, J.; Shoelson, S.E.; Benoist, C.; Mathis, D. PPAR-γ is a major driver of the accumulation and phenotype of adipose tissue Treg cells. *Nature* **2012**, *486*, 549–553. [CrossRef]

20. Zhao, Y.; Lee, C.K.; Lin, C.H.; Gassen, R.B.; Xu, X.; Huang, Z.; Xiao, C.; Bonorino, C.; Lu, L.F.; Bui, J.D.; et al. PD-L1:CD80 Cis-Heterodimer Triggers the Co-stimulatory Receptor CD28 While Repressing the Inhibitory PD-1 and CTLA-4 Pathways. *Immunity* **2019**, *51*, 1059–1073. [CrossRef]

21. Chowdhury, P.S.; Chamoto, K.; Kumar, A.; Honjo, T. PPAR-Induced Fatty Acid Oxidation in T Cells Increases the Number of Tumor-Reactive CD8+ T Cells and Facilitates Anti-PD-1 Therapy. *Cancer Immunol. Res.* **2018**, *6*, 1375–1387. [CrossRef] [PubMed]

22. Saibil, S.D.; St. Paul, M.; Laister, R.C.; Garcia-Batres, C.R.; Israni-Winger, K.; Elford, A.R.; Grimshaw, N.; Robert-Tissot, C.; Roy, D.G.; Jones, R.G.; et al. Activation of Peroxisome Proliferator-Activated Receptors α and δ Synergizes with Inflammatory Signals to Enhance Adoptive Cell Therapy. *Cancer Res.* **2019**, *79*, 445–451. [CrossRef] [PubMed]

23. Gutiérrez, S.; Svahn, S.L.; Johansson, M.E. Effects of Omega-3 Fatty Acids on Immune Cells. *Int. J. Mol. Sci.* **2019**, *20*, 5028. [CrossRef] [PubMed]

24. Mukaro, V.R.; Costabile, M.; Murphy, K.J.; Hii, C.S.; Howe, P.R.; Ferrante, A. Leukocyte numbers and function in subjects eating n-3 enriched foods: Selective depression of natural killer cell levels. *Arthritis Res. Ther.* **2008**, *10*, R57. [CrossRef]

25. Miles, E.A.; Banerjee, T.; Wells, S.J.; Calder, P.C. Limited effect of eicosapentaenoic acid on T-lymphocyte and natural killer cell numbers and functions in healthy young males. *Nutrition* **2006**, *22*, 512–519. [CrossRef]

26. Ichimura, A.; Hirasawa, A.; Poulain-Godefroy, O.; Bonnefond, A.; Hara, T.; Yengo, L.; Kimura, I.; Leloire, A.; Liu, N.; Iida, K.; et al. Dysfunction of lipid sensor GPR120 leads to obesity in both mouse and human. *Nature* **2012**, *483*, 350–354. [CrossRef]

27. Liu, Z.; Hopkins, M.M.; Zhang, Z.; Quisenberry, C.B.; Fix, L.C.; Galvan, B.M.; Meier, K.E. Omega-3 fatty acids and other FFA4 agonists inhibit growth factor signaling in human prostate cancer cells. *J. Pharmacol. Exp. Ther.* **2015**, *352*, 380–394. [CrossRef]

28. Thirunavukkarasan, M.; Wang, C.; Rao, A.; Hind, T.; Teo, Y.R.; Siddiquee, A.A.M.; Goghari, M.A.I.; Kumar, A.P.; Herr, D.R. Short-chain fatty acid receptors inhibit invasive phenotypes in breast cancer cells. *PLoS ONE* **2017**, *12*, e0186334. [CrossRef]

29. Berod, L.; Friedrich, C.; Nandan, A.; Freitag, J.; Hagemann, S.; Harmrolfs, K.; Sandouk, A.; Hesse, C.; Castro, C.N.; Bähre, H.; et al. De novo fatty acid synthesis controls the fate between regulatory T and T helper 17 cells. *Nat. Med.* **2014**, *20*, 1327–1333. [CrossRef]

30. Ecker, C.; Guo, L.; Voicu, S.; Gil-de-Gómez, L.; Medvec, A.; Cortina, L.; Pajda, J.; Andolina, M.; Torres-Castillo, M.; Donato, J.L.; et al. Differential Reliance on Lipid Metabolism as a Salvage Pathway Underlies Functional Differences of T Cell Subsets in Poor Nutrient Environments. *Cell Rep.* **2018**, *23*, 741–755. [CrossRef]

31. Essig, K.; Hu, D.; Guimaraes, J.C.; Alterauge, D.; Edelmann, S.; Raj, T.; Kranich, J.; Behrens, G.; Heiseke, A.; Floess, S.; et al. Roquin Suppresses the PI3K-mTOR Signaling Pathway to Inhibit T Helper Cell Differentiation and Conversion of Treg to Tfr Cells. *Immunity* **2017**, *47*, 1067–1082. [CrossRef] [PubMed]

32. Angelin, A.; Gil-de-Gómez, L.; Dahiya, S.; Jiao, J.; Guo, L.; Levine, M.H.; Wang, Z.; Quinn, W.J.; Kopinski, P.K.; Wang, L.; et al. Foxp3 Reprograms T Cell Metabolism to Function in Low-Glucose, High-Lactate Environments. *Cell Metab.* **2017**, *25*, 1282–1293. [CrossRef] [PubMed]

33. Miska, J.; Lee-Chang, C.; Rashidi, A.; Muroski, M.E.; Chang, A.L.; Lopez-Rosas, A.; Zhang, P.; Panek, W.K.; Cordero, A.; Han, Y.; et al. HIF-1α Is a Metabolic Switch between Glycolytic-Driven Migration and Oxidative Phosphorylation-Driven Immunosuppression of Tregs in Glioblastoma. *Cell Rep.* **2019**, *27*, 226–237. [CrossRef] [PubMed]

34. Mizuno, R.; Kawada, K.; Sakai, Y. Prostaglandin E2/EP Signaling in the Tumor Microenvironment of Colorectal Cancer. *Int. J. Mol. Sci.* **2019**, *20*, 6254. [CrossRef]

35. Gil-de-Gómez, L.; Astudillo, A.M.; Lebrero, P.; Balboa, M.A.; Balsinde, J. Essential Role for Ethanolamine Plasmalogen Hydrolysis in Bacterial Lipopolysaccharide Priming of Macrophages for Enhanced Arachidonic Acid Release. *Front. Immunol.* **2017**, *8*, 1251. [CrossRef]

36. Li, J.; Feng, G.; Liu, J.; Rong, R.; Luo, F.; Guo, L.; Zhu, T.; Wang, G.; Chu, Y. Renal cell carcinoma may evade the immune system by converting CD4+Foxp3- T cells into CD4+CD25+Foxp3+ regulatory T cells: Role of tumor COX-2-derived PGE2. *Mol. Med. Rep.* **2010**, *3*, 959–963. [CrossRef]

37. Kalinski, P. Regulation of Immune Responses by Prostaglandin E$_2$. *J. Immunol.* **2012**, *188*, 21–28. [CrossRef]

38. Trinath, J.; Hegde, P.; Sharma, M.; Maddur, M.S.; Rabin, M.; Vallat, J.M.; Magy, L.; Balaji, K.N.; Kaveri, S.V.; Bayry, J. Intravenous immunoglobulin expands regulatory T cells via induction of cyclooxygenase-2–dependent prostaglandin E2 in human dendritic cells. *Blood* **2013**, *122*, 1419–1427. [CrossRef]

39. Loew, A.; Köhnke, T.; Rehbeil, E.; Pietzner, A.; Weylandt, K.H. A Role for Lipid Mediators in Acute Myeloid Leukemia. *Int. J. Mol. Sci.* **2019**, *20*, 2425. [CrossRef]

40. Prima, V.; Kaliberova, L.N.; Kaliberov, S.; Curiel, D.T.; Kusmartsev, S. COX2/mPGES1/PGE$_2$ pathway regulates PD-L1 expression in tumor-associated macrophages and myeloid-derived suppressor cells. *Proc. Natl. Acad. Sci. USA* **2017**, *114*, 1117–1122. [CrossRef]

41. Miao, J.; Lu, X.; Hu, Y.; Piao, C.; Wu, X.; Liu, X.; Huang, C.; Wang, Y.; Li, D.; Liu, J. Prostaglandin E2 and PD-1 mediated inhibition of antitumor CTL responses in the human tumor microenvironment. *Oncotarget* **2017**, *8*, 89802–89810. [CrossRef] [PubMed]

42. Mock, A.; Zschäbitz, S.; Kirsten, R.; Scheffler, M.; Wolf, B.; Herold-Mende, C.; Kramer, R.; Busch, E.; Jenzer, M.; Jäger, D.; et al. Serum very long-chain fatty acid-containing lipids predict response to immune checkpoint inhibitors in urological cancers. *Cancer Immunol. Immunother.* **2019**, *68*, 2005–2014. [CrossRef] [PubMed]

43. Wu, A.; Wu, Q.; Deng, Y.; Liu, Y.; Lu, J.; Liu, L.; Li, X.; Liao, C.; Zhao, B.; Song, H. Loss of VGLL4 Suppresses Tumor PD-L1 Expression and Immune Evasion. *EMBO J.* **2019**, *38*. [CrossRef] [PubMed]

44. Huang, S.C.C.; Everts, B.; Ivanova, Y.; O'Sullivan, D.; Nascimento, M.; Smith, A.M.; Beatty, W.; Love-Gregory, L.; Lam, W.Y.; O'Neill, C.M.; et al. Cell-intrinsic lysosomal lipolysis is essential for alternative activation of macrophages. *Nat. Immunol.* **2014**, *15*, 846–855. [CrossRef] [PubMed]

45. Oishi, Y.; Spann, N.J.; Link, V.M.; Muse, E.D.; Strid, T.; Edillor, C.; Kolar, M.J.; Matsuzaka, T.; Hayakawa, S.; Tao, J.; et al. SREBP1 Contributes to Resolution of Pro-inflammatory TLR4 Signaling by Reprogramming Fatty Acid Metabolism. *Cell Metab.* **2017**, *25*, 412–427. [CrossRef] [PubMed]

46. Griffiths, B.; Lewis, C.A.; Bensaad, K.; Ros, S.; Zhang, Q.; Ferber, E.C.; Konisti, S.; Peck, B.; Miess, H.; East, P.; et al. Sterol regulatory element binding protein-dependent regulation of lipid synthesis supports cell survival and tumor growth. *Cancer Metab.* **2013**, *1*, 3. [CrossRef] [PubMed]

47. Liu, C.; Chikina, M.; Deshpande, R.; Menk, A.V.; Wang, T.; Tabib, T.; Brunazzi, E.A.; Vignali, K.M.; Sun, M.; Stolz, D.B.; et al. Treg Cells Promote the SREBP1-Dependent Metabolic Fitness of Tumor-Promoting Macrophages via Repression of CD8+ T Cell-Derived Interferon-γ. *Immunity* **2019**, *51*, 381–397. [CrossRef] [PubMed]

48. Wu, S.; Näär, A.M. SREBP1-dependent de novo fatty acid synthesis gene expression is elevated in malignant melanoma and represents a cellular survival trait. *Sci. Rep.* **2019**, *9*, 10369. [CrossRef] [PubMed]

49. Li, L.; Hua, L.; Fan, H.; He, Y.; Xu, W.; Zhang, L.; Yang, J.; Deng, F.; Zeng, F. Interplay of PKD3 with SREBP1 Promotes Cell Growth via Upregulating Lipogenesis in Prostate Cancer Cells. *J. Cancer* **2019**, *10*, 6395–6404. [CrossRef]

50. Wang, B.; Tontonoz, P. Liver X receptors in lipid signalling and membrane homeostasis. *Nat. Rev. Endocrinol.* **2018**, *14*, 452–463. [CrossRef]

51. Shimano, H.; Sato, R. SREBP-regulated lipid metabolism: Convergent physiology–divergent pathophysiology. *Nat. Rev. Endocrinol.* **2017**, *13*, 710–730. [CrossRef] [PubMed]

52. Costales, P.; Castellano, J.; Revuelta-López, E.; Cal, R.; Aledo, R.; Llampayas, O.; Nasarre, L.; Juarez, C.; Badimon, L.; Llorente-Cortés, V. Lipopolysaccharide downregulates CD91/low-density lipoprotein receptor-related protein 1 expression through SREBP-1 overexpression in human macrophages. *Atherosclerosis* **2013**, *227*, 79–88. [CrossRef] [PubMed]

53. Im, S.S.; Yousef, L.; Blaschitz, C.; Liu, J.Z.; Edwards, R.A.; Young, S.G.; Raffatellu, M.; Osborne, T.F. Linking Lipid Metabolism to the Innate Immune Response in Macrophages through Sterol Regulatory Element Binding Protein-1a. *Cell Metab.* **2011**, *13*, 540–549. [CrossRef]

54. Joseph, S.B.; Castrillo, A.; Laffitte, B.A.; Mangelsdorf, D.J.; Tontonoz, P. Reciprocal regulation of inflammation and lipid metabolism by liver X receptors. *Nat. Med.* **2003**, *9*, 213–219. [CrossRef] [PubMed]

55. Chausse, B.; Kakimoto, P.A.; Caldeira-Da-Silva, C.C.; Chaves-Filho, A.B.; Yoshinaga, M.Y.; Da Silva, R.P.; Miyamoto, S.; Kowaltowski, A.J. Distinct metabolic patterns during microglial remodeling by oleate and palmitate. *Biosci. Rep.* **2019**, *39*. [CrossRef] [PubMed]

56. Oh, Y.T.; Lee, J.Y.; Lee, J.; Kim, H.; Yoon, K.S.; Choe, W.; Kang, I. Oleic acid reduces lipopolysaccharide-induced expression of iNOS and COX-2 in BV2 murine microglial cells: Possible involvement of reactive oxygen species, p38 MAPK, and IKK/NF-κB signaling pathways. *Neurosci. Lett.* **2009**, *464*, 93–97. [CrossRef] [PubMed]

57. Krycer, J.R.; Brown, A.J. Cross-talk between the androgen receptor and the liver X receptor: Implications for cholesterol homeostasis. *J. Biol. Chem.* **2011**, *286*, 20637–20647. [CrossRef]

58. Balgoma, D.; Zelleroth, S.; Grönbladh, A.; Hallberg, M.; Pettersson, C.; Hedeland, M. Anabolic androgenic steroids exert a selective remodeling of the plasma lipidome that mirrors the decrease of the de novo lipogenesis in the liver. *Metabolomics* **2020**, *16*, 12. [CrossRef]

59. Zhu, P.; Baek, S.H.; Bourk, E.M.; Ohgi, K.A.; Garcia-Bassets, I.; Sanjo, H.; Akira, S.; Kotol, P.F.; Glass, C.K.; Rosenfeld, M.G.; et al. Macrophage/Cancer Cell Interactions Mediate Hormone Resistance by a Nuclear Receptor Derepression Pathway. *Cell* **2006**, *124*, 615–629. [CrossRef]

60. Perucha, E.; Melchiotti, R.; Bibby, J.A.; Wu, W.; Frederiksen, K.S.; Roberts, C.A.; Hall, Z.; LeFriec, G.; Robertson, K.A.; Lavender, P.; et al. The cholesterol biosynthesis pathway regulates IL-10 expression in human Th1 cells. *Nat. Commun.* **2019**, *10*, 498. [CrossRef]

61. Rodrigues, N.V.; Correia, D.V.; Mensurado, S.; Nóbrega-Pereira, S.; DeBarros, A.; Kyle-Cezar, F.; Tutt, A.; Hayday, A.C.; Norell, H.; Silva-Santos, B.; et al. Low-Density Lipoprotein Uptake Inhibits the Activation and Antitumor Functions of Human Vγ9Vδ2 T Cells. *Cancer Immunol. Res.* **2018**, *6*, 448–457. [CrossRef] [PubMed]

62. Chaudhary, J.; Bower, J.; Corbin, I.R. Lipoprotein Drug Delivery Vehicles for Cancer: Rationale and Reason. *Int. J. Mol. Sci.* **2019**, *20*, 6327. [CrossRef] [PubMed]

63. Yao, H.; Lan, J.; Li, C.; Shi, H.; Brosseau, J.P.; Wang, H.; Lu, H.; Fang, C.; Zhang, Y.; Liang, L.; et al. Inhibiting PD-L1 palmitoylation enhances T-cell immune responses against tumours. *Nat. Biomed. Eng.* **2019**, *3*, 306–317. [CrossRef] [PubMed]

64. Sharma, C.; Wang, H.X.; Li, Q.; Knoblich, K.; Reisenbichler, E.S.; Richardson, A.L.; Hemler, M.E. Protein Acyltransferase DHHC3 Regulates Breast Tumor Growth, Oxidative Stress, and Senescence. *Cancer Res.* **2017**, *77*, 6880–6890. [CrossRef] [PubMed]

65. Louis, P.; Flint, H.J. Formation of propionate and butyrate by the human colonic microbiota. *Environ. Microbiol.* **2017**, *19*, 29–41. [CrossRef]

66. Kim, C.H. Immune regulation by microbiome metabolites. *Immunology* **2018**, *154*, 220–229. [CrossRef]

67. Botticelli, A.; Vernocchi, P.; Marini, F.; Quagliariello, A.; Cerbelli, B.; Reddel, S.; Del Chierico, F.; Di Pietro, F.; Giusti, R.; Tomassini, A.; et al. Gut metabolomics profiling of non-small cell lung cancer (NSCLC) patients under immunotherapy treatment. *J. Transl. Med.* **2020**, *18*, 49. [CrossRef]

68. Belcheva, A.; Irrazabal, T.; Robertson, S.J.; Streutker, C.; Maughan, H.; Rubino, S.; Moriyama, E.H.; Copeland, J.K.; Surendra, A.; Kumar, S.; et al. Gut microbial metabolism drives transformation of MSH2-deficient colon epithelial cells. *Cell* **2014**, *158*, 288–299. [CrossRef]

69. Ruemmele, F.M. Butyrate induced Caco-2 cell apoptosis is mediated via the mitochondrial pathway. *Gut* **2003**, *52*, 94–100. [CrossRef]

70. Singh, N.; Gurav, A.; Sivaprakasam, S.; Brady, E.; Padia, R.; Shi, H.; Thangaraju, M.; Prasad, P.D.; Manicassamy, S.; Munn, D.H.; et al. Activation of Gpr109a, Receptor for Niacin and the Commensal Metabolite Butyrate, Suppresses Colonic Inflammation and Carcinogenesis. *Immunity* **2014**, *40*, 128–139. [CrossRef]

71. Park, J.; Kim, M.; Kang, S.G.; Jannasch, A.H.; Cooper, B.; Patterson, J.; Kim, C.H. Short-chain fatty acids induce both effector and regulatory T cells by suppression of histone deacetylases and regulation of the mTOR–S6K pathway. *Mucosal Immunol.* **2015**, *8*, 80–93. [CrossRef] [PubMed]

72. Bilotta, A.J.; Cong, Y. Gut microbiota metabolite regulation of host defenses at mucosal surfaces: Implication in precision medicine. *Precis. Clin. Med.* **2019**, *2*, 110–119. [CrossRef]

73. Bachem, A.; Makhlouf, C.; Binger, K.J.; De Souza, D.P.; Tull, D.; Hochheiser, K.; Whitney, P.G.; Fernandez-Ruiz, D.; Dähling, S.; Kastenmüller, W.; et al. Microbiota-Derived Short-Chain Fatty Acids Promote the Memory Potential of Antigen-Activated CD8+ T Cells. *Immunity* **2019**, *51*, 285–297. [CrossRef] [PubMed]

74. O'Sullivan, D.; Van Der Windt, G.J.W.; Huang, S.C.C.; Curtis, J.D.; Chang, C.H.; Buck, M.D.; Qiu, J.; Smith, A.M.; Lam, W.Y.; DiPlato, L.M.; et al. Memory CD8+ T Cells Use Cell-Intrinsic Lipolysis to Support the Metabolic Programming Necessary for Development. *Immunity* **2014**, *41*, 75–88. [CrossRef] [PubMed]

75. Van Der Windt, G.J.W.; Everts, B.; Chang, C.H.; Curtis, J.D.; Freitas, T.C.; Amiel, E.; Pearce, E.J.; Pearce, E.L. Mitochondrial Respiratory Capacity Is a Critical Regulator of CD8+ T Cell Memory Development. *Immunity* **2012**, *36*, 68–78. [CrossRef] [PubMed]

76. Nastasi, C.; Candela, M.; Bonefeld, C.M.; Geisler, C.; Hansen, M.; Krejsgaard, T.; Biagi, E.; Andersen, M.H.; Brigidi, P.; Ødum, N.; et al. The effect of short-chain fatty acids on human monocyte-derived dendritic cells. *Sci. Rep.* **2015**, *5*, 16148. [CrossRef]

77. Nastasi, C.; Fredholm, S.; Willerslev-Olsen, A.; Hansen, M.; Bonefeld, C.M.; Geisler, C.; Andersen, M.H.; Ødum, N.; Woetmann, A. Butyrate and propionate inhibit antigen-specific CD8+ T cell activation by suppressing IL-12 production by antigen-presenting cells. *Sci. Rep.* **2017**, *7*, 14516. [CrossRef]

78. Coutzac, C.; Jouniaux, J.-M.; Paci, A.; Schmidt, J.; Mallardo, D.; Seck, A.; Asvatourian, V.; Cassard, L.; Saulnier, P.; Lacroix, L.; et al. Systemic short chain fatty acids limit antitumor effect of CTLA-4 blockade in hosts with cancer. *Nat. Commun.* **2020**, *11*, 2168. [CrossRef]

79. Zhang, W.Q.; Zhao, S.K.; Luo, J.W.; Dong, X.P.; Hao, Y.T.; Li, H.; Shan, L.; Zhou, Y.; Shi, H.B.; Zhang, Z.Y.; et al. Alterations of fecal bacterial communities in patients with lung cancer. *Am. J. Transl. Res.* **2018**, *10*, 3171–3185.

80. Sivan, A.; Corrales, L.; Hubert, N.; Williams, J.B.; Aquino-Michaels, K.; Earley, Z.M.; Benyamin, F.W.; Man Lei, Y.; Jabri, B.; Alegre, M.L.; et al. Commensal Bifidobacterium promotes antitumor immunity and facilitates anti-PD-L1 efficacy. *Science* **2015**, *350*, 1084–1089. [CrossRef]

81. Chalmin, F.; Bruchard, M.; Vegran, F.; Ghiringhelli, F. Regulation of T cell antitumor immune response by tumor induced metabolic stress. *Cell Stress* **2019**, *3*, 9–18. [CrossRef] [PubMed]

82. Michalek, R.D.; Gerriets, V.A.; Jacobs, S.R.; Macintyre, A.N.; MacIver, N.J.; Mason, E.F.; Sullivan, S.A.; Nichols, A.G.; Rathmell, J.C. Cutting Edge: Distinct Glycolytic and Lipid Oxidative Metabolic Programs Are Essential for Effector and Regulatory CD4 [+] T Cell Subsets. *J. Immunol.* **2011**, *186*, 3299–3303. [CrossRef] [PubMed]

83. Santaren, I.D.; Bazinet, R.P.; Liu, Z.; Johnston, L.W.; Sievenpiper, J.L.; Giacca, A.; Retnakaran, R.; Harris, S.B.; Zinman, B.; Hanley, A.J. The Distribution of Fatty Acid Biomarkers of Dairy Intake across Serum Lipid Fractions: The Prospective Metabolism and Islet Cell Evaluation (PROMISE) Cohort. *Lipids* **2019**, *54*, 617–627. [CrossRef] [PubMed]

84. Olazarán, J.; Gil-de-Gómez, L.; Rodríguez-Martín, A.; Valentí-Soler, M.; Frades-Payo, B.; Marín-Muñoz, J.; Antúnez, C.; Frank-García, A.; Acedo-Jiménez, C.; Morlán-Gracia, L.; et al. A Blood-Based, 7-Metabolite Signature for the Early Diagnosis of Alzheimer's Disease. *J. Alzheimer's Dis.* **2015**, *45*, 1157–1173. [CrossRef] [PubMed]

85. Stuani, L.; Riols, F.; Millard, P.; Sabatier, M.; Batut, A.; Saland, E.; Viars, F.; Tonini, L.; Zaghdoudi, S.; Linares, L.; et al. Stable Isotope Labeling Highlights Enhanced Fatty Acid and Lipid Metabolism in Human Acute Myeloid Leukemia. *Int. J. Mol. Sci.* **2018**, *19*, 3325. [CrossRef] [PubMed]

86. Pabst, T.; Kortz, L.; Fiedler, G.M.; Ceglarek, U.; Idle, J.R.; Beyoğlu, D. The plasma lipidome in acute myeloid leukemia at diagnosis in relation to clinical disease features. *BBA Clin.* **2017**, *7*, 105–114. [CrossRef]

87. Stefanko, A.; Thiede, C.; Ehninger, G.; Simons, K.; Grzybek, M. Lipidomic approach for stratification of acute myeloid leukemia patients. *PLoS ONE* **2017**, *12*, e0168781. [CrossRef]

88. Lavoie, J.M.; Black, P.C.; Eigl, B.J. Predictive Biomarkers for Checkpoint Blockade in Urothelial Cancer: A Systematic Review. *J. Urol.* **2019**, *202*, 49–56. [CrossRef]

89. Wang, B.; Tontonoz, P. Phospholipid Remodeling in Physiology and Disease. *Annu. Rev. Physiol.* **2019**, *81*, 165–188. [CrossRef]

90. Yang, L.; Bai, Y.; Han, X.; Shi, Y.; Liu, H. Plasma Lipidomic Analysis to Identify Novel Biomarkers for Hepatocellular Carcinoma. *J. Anal. Test.* **2017**, *1*, 223–232. [CrossRef]

91. Lee, J.W.; Mok, H.J.; Lee, D.Y.; Park, S.C.; Kim, G.S.; Lee, S.E.; Lee, Y.S.; Kim, K.P.; Kim, H.D. UPLC-QqQ/MS-Based Lipidomics Approach To Characterize Lipid Alterations in Inflammatory Macrophages. *J. Proteome Res.* **2017**, *16*, 1460–1469. [CrossRef]

92. Balgoma, D.; Pettersson, C.; Hedeland, M. Common Fatty Markers in Diseases with Dysregulated Lipogenesis. *Trends Endocrinol. Metab.* **2019**, *30*, 283–285. [CrossRef] [PubMed]

93. Balgoma, D.; Guitton, Y.; Evans, J.J.; Le Bizec, B.; Dervilly-Pinel, G.; Meynier, A. Modeling the fragmentation patterns of triacylglycerides in mass spectrometry allows the quantification of the regioisomers with a minimal number of standards. *Anal. Chim. Acta* **2019**, *1057*, 60–69. [CrossRef] [PubMed]

94. O'Keefe, S.J.D.; Ou, J.; Aufreiter, S.; O'Connor, D.; Sharma, S.; Sepulveda, J.; Fukuwatari, T.; Shibata, K.; Mawhinney, T. Products of the colonic microbiota mediate the effects of diet on colon cancer risk. *J. Nutr.* **2009**, *139*, 2044–2048. [CrossRef] [PubMed]

95. Chen, H.M.; Yu, Y.N.; Wang, J.L.; Lin, Y.W.; Kong, X.; Yang, C.Q.; Yang, L.; Liu, Z.J.; Yuan, Y.Z.; Liu, F.; et al. Decreased dietary fiber intake and structural alteration of gut microbiota in patients with advanced colorectal adenoma. *Am. J. Clin. Nutr.* **2013**, *97*, 1044–1052. [CrossRef] [PubMed]

96. Uribe-Herranz, M.; Rafail, S.; Beghi, S.; Gil-de-Gómez, L.; Verginadis, I.; Bittinger, K.; Pustylnikov, S.; Pierini, S.; Perales-Linares, R.; Blair, I.A.; et al. Gut microbiota modulate dendritic cell antigen presentation and radiotherapy-induced antitumor immune response. *J. Clin. Investig.* **2020**, *130*, 466–479. [CrossRef]

97. Matson, V.; Fessler, J.; Bao, R.; Chongsuwat, T.; Zha, Y.; Alegre, M.L.; Luke, J.J.; Gajewski, T.F. The commensal microbiome is associated with anti–PD-1 efficacy in metastatic melanoma patients. *Science* **2018**, *359*, 104–108. [CrossRef]

98. Milone, M.C.; Fish, J.D.; Carpenito, C.; Carroll, R.G.; Binder, G.K.; Teachey, D.; Samanta, M.; Lakhal, M.; Gloss, B.; Danet-Desnoyers, G.; et al. Chimeric receptors containing CD137 signal transduction domains mediate enhanced survival of T cells and increased antileukemic efficacy in vivo. *Mol. Ther.* **2009**, *17*, 1453–1464. [CrossRef]

99. George, P.; Dasyam, N.; Giunti, G.; Mester, B.; Bauer, E.; Andrews, B.; Perera, T.; Ostapowicz, T.; Frampton, C.; Li, P.; et al. Third-generation anti-CD19 chimeric antigen receptor T-cells incorporating a TLR2 domain for relapsed or refractory B-cell lymphoma: A phase I clinical trial protocol (ENABLE). *BMJ Open* **2020**, *10*, e034629. [CrossRef]

100. June, C.H.; O'Connor, R.S.; Kawalekar, O.U.; Ghassemi, S.; Milone, M.C. CAR T cell immunotherapy for human cancer. *Science* **2018**, *359*, 1361–1365. [CrossRef]

101. Wang, D.; Starr, R.; Chang, W.C.; Aguilar, B.; Alizadeh, D.; Wright, S.L.; Yang, X.; Brito, A.; Sarkissian, A.; Ostberg, J.R.; et al. Chlorotoxin-directed CAR T cells for specific and effective targeting of glioblastoma. *Sci. Transl. Med.* **2020**, *12*. [CrossRef] [PubMed]

102. Schulz, G.; Cheresh, D.A.; Varki, N.M.; Yu, A.; Staffileno, L.K.; Reisfeld, R.A. Detection of ganglioside GD2 in tumor tissues and sera of neuroblastoma patients. *Cancer Res.* **1984**, *44*, 5914–5920. [PubMed]

103. Dobrenkov, K.; Ostrovnaya, I.; Gu, J.; Cheung, I.Y.; Cheung, N.K.V. Oncotargets GD2 and GD3 are highly expressed in sarcomas of children, adolescents, and young adults. *Pediatric Blood Cancer* **2016**, *63*, 1780–1785. [CrossRef] [PubMed]

104. Richman, S.A.; Nunez-Cruz, S.; Moghimi, B.; Li, L.Z.; Gershenson, Z.T.; Mourelatos, Z.; Barrett, D.M.; Grupp, S.A.; Milone, M.C. High-Affinity GD2-Specific CAR T Cells Induce Fatal Encephalitis in a Preclinical Neuroblastoma Model. *Cancer Immunol. Res.* **2018**, *6*, 36–46. [CrossRef]

105. Tasian, S.K.; Loh, M.L.; Hunger, S.P. Philadelphia chromosome–like acute lymphoblastic leukemia. *Blood* **2017**, *130*, 2064–2072. [CrossRef]

106. Finney, O.C.; Brakke, H.; Rawlings-Rhea, S.; Hicks, R.; Doolittle, D.; Lopez, M.; Futrell, B.; Orentas, R.J.; Li, D.; Gardner, R.; et al. CD19 CAR T cell product and disease attributes predict leukemia remission durability. *J. Clin. Investig.* **2019**, *129*, 2123–2132. [CrossRef]

107. Konjar, Š.; Veldhoen, M. Dynamic Metabolic State of Tissue Resident CD8 T Cells. *Front. Immunol.* **2019**, *10*, 1683. [CrossRef]

108. Kondo, T.; Ando, M.; Nagai, N.; Tomisato, W.; Srirat, T.; Liu, B.; Mise-Omata, S.; Ikeda, M.; Chikuma, S.; Nishimasu, H.; et al. The NOTCH–FOXM1 Axis Plays a Key Role in Mitochondrial Biogenesis in the Induction of Human Stem Cell Memory–like CAR-T Cells. *Cancer Res.* **2020**, *80*, 471–483. [CrossRef]

109. Martinez, M.; Moon, E.K. CAR T Cells for Solid Tumors: New Strategies for Finding, Infiltrating, and Surviving in the Tumor Microenvironment. *Front. Immunol.* **2019**, *10*, 128. [CrossRef]

110. Fucà, G.; Reppel, L.; Landoni, E.; Savoldo, B.; Dotti, G. Enhancing Chimeric Antigen Receptor T-Cell Efficacy in Solid Tumors. *Clin. Cancer Res.* **2020**, *26*, 2444–2451. [CrossRef]

111. Vartak, S.; Robbins, M.E.C.; Spector, A.A. Polyunsaturated fatty acids increase the sensitivity of 36B10 rat astrocytoma cells to radiation-induced cell kill. *Lipids* **1997**, *32*, 283–292. [CrossRef] [PubMed]

112. Das, U.N. Gamma-linolenic acid therapy of human glioma-a review of in vitro, in vivo, and clinical studies. *Med. Sci. Monit.* **2007**, *13*, RA119–RA131. [PubMed]

113. Belavgeni, A.; Bornstein, S.R.; Von Mässenhausen, A.; Tonnus, W.; Stumpf, J.; Meyer, C.; Othmar, E.; Latk, M.; Kanczkowski, W.; Kroiss, M.; et al. Exquisite sensitivity of adrenocortical carcinomas to induction of ferroptosis. *Proc. Natl. Acad. Sci. USA* **2019**, *116*, 22269–22274. [CrossRef] [PubMed]

114. Das, U.N. Can Bioactive Lipids Augment Anti-cancer Action of Immunotherapy and Prevent Cytokine Storm? *Arch. Med. Res.* **2019**, *50*, 342–349. [CrossRef]

115. Van Bruggen, J.A.C.; Martens, A.W.J.; Fraietta, J.A.; Hofland, T.; Tonino, S.H.; Eldering, E.; Levin, M.D.; Siska, P.J.; Endstra, S.; Rathmell, J.C.; et al. Chronic lymphocytic leukemia cells impair mitochondrial fitness in CD8+ T cells and impede CAR T-cell efficacy. *Blood* **2019**, *134*, 44–58. [CrossRef]

Review

The Cardiac Lipidome in Models of Cardiovascular Disease

Mateusz M. Tomczyk [1,2,3] and **Vernon W. Dolinsky** [1,2,3,*]

[1] Diabetes Research Envisioned and Accomplished in Manitoba (DREAM) Theme of the Children's Hospital Research Institute of Manitoba, 715 McDermot Avenue, Winnipeg, MB R3E 3P4, Canada; tomczyk3@myumanitoba.ca

[2] Department of Pharmacology and Therapeutics, University of Manitoba, Winnipeg, MB R3E 0T6, Canada

[3] Rady Faculty of Health Science, College of Medicine, University of Manitoba, Winnipeg, MB R3T 2N2, Canada

* Correspondence: vdolinsky@chrim.ca

Received: 22 April 2020; Accepted: 11 June 2020; Published: 17 June 2020

Abstract: Cardiovascular disease (CVD) is the leading cause of death worldwide. There are numerous factors involved in the development of CVD. Among these, lipids have an important role in maintaining the myocardial cell structure as well as cardiac function. Fatty acids (FA) are utilized for energy, but also contribute to the pathogenesis of CVD and heart failure. Advances in mass spectrometry methods have enabled the comprehensive analysis of a plethora of lipid species from a single sample comprised of a heterogeneous population of lipid molecules. Determining cardiac lipid alterations in different models of CVD identifies novel biomarkers as well as reveals molecular mechanisms that underlie disease development and progression. This information could inform the development of novel therapeutics in the treatment of CVD. Herein, we provide a review of recent studies of cardiac lipid profiles in myocardial infarction, obesity, and diabetic and dilated cardiomyopathy models of CVD by methods of mass spectrometry analysis.

Keywords: cardiovascular disease; heart failure; myocardial infarction; obesity; diabetic cardiomyopathy; dilated cardiomyopathy; lipids; lipidomics; mass spectrometry

---

1. Introduction

Cardiovascular disease (CVD) is the leading cause of death worldwide [1]. CVD encompasses stroke, cardiomyopathy, coronary artery disease (CAD), and other disorders that can lead to myocardial infarctions and heart failure. The pathophysiological processes in each of these diseases can differ, but lipids play a significant role in every model of CVD [2]. Lipid molecules are important structural components of cardiomyocyte plasma and organelle membranes. For example, a specific phospholipid molecular species composition is necessary for the assembly of the electron transport chain in the mitochondria [3]. In addition, fats are the primary fuels utilized for cardiac energy production [4–6]. Therefore, lipids have a direct role in cardiovascular function. On the other hand, regarding lipid excess experienced when diets high in fat are consumed, hyperlipidemia and hypercholesterolemia can result, which puts patients at risk for developing atherosclerosis and cardiometabolic disease [7,8].

The development of mass spectrometry (MS) technology such as high performance liquid or gas-chromatography MS (HPLC-MS/GC-MS) separation techniques and ionization methods such as electrospray ionization MS (ESI-MS), and matrix assisted laser desorption/ionization MS (MALDI-MS) have enabled the detailed analysis of chemically complex lipids from biological tissues, which contain heterogeneous pools of lipid species [9]. These MS methods are increasingly utilized to analyze multiple lipid species from a single sample in a methodology termed lipidomics. This has been an important advance for identifying potential biomarkers of disease. A number of studies have analyzed the changes in serum lipid profiles of patients with CVD [10–16]. However, information about lipid profiles in cardiac tissues in models of CVD is more limited and has not been reviewed. The following review will focus on recent lipidomic research findings about lipid profiles in cardiac tissue in experimental models of CVD that has contributed novel information about lipid biomarkers for myocardial infarction, obesity, and diabetic and dilated cardiomyopathies by MS methods. Furthermore, this review will focus on current and novel therapies that alter cardiac lipid profiles.

1.1. Importance of Lipids in the Development of Cardiovascular Disease

Lipids are a class of amphipathic molecules that are characterized as being insoluble in water [17]. They consist of a wide array of structures including some that are depicted in Table 1. Lipids play an important role in CVD development. Beyond their well recognized structural function in lipid bilayers [18], lipids can also act as signalling molecules and secondary messenger molecules such as those involved in G protein coupled receptor signalling [19,20]. An excess of deleterious lipid species can also contribute to CVD progression [7]. Obesity is a growing epidemic and patients characterized as obese are at risk for developing cardiovascular complications that could be linked to 3.4 million deaths worldwide in 2010 [21]. High fat and high cholesterol diets found in Western countries contribute to the development of cardiovascular risk factors such as hyperlipidemia and hypercholesterolemia [8]. High levels of these circulating lipids can lead to the accumulation of lipid plaques in arterial walls, which are also known as atherosclerosis [22]. Specifically, high quantities of low-density lipoprotein (LDL) increase the likelihood of LDL translocating from the arterial lumen to the endothelial intima [23]. LDL oxidation results in the release of cytokines, which signal uptake of the modified lipoproteins by macrophages [24]. Macrophage-filled particles or foam cells can efflux cholesterol out of the arterial wall into the blood stream or undergo apoptosis, which results in fatty streaks [25]. Fatty streaks are then converted to fibrous plaques, which can block arterial blood flow. Furthermore, macrophages release growth factors, which initiates smooth muscle cell proliferation from the media across the internal elastic membrane and into the intima. This results in further bulging and blockage of blood flow [25]. CAD is a result of atherosclerotic plaques that occur in the micro vessels, which supply blood to the heart. When these arteries are blocked, it results in ischemic injury as a result of hypoxic conditions [26]. In this environment, the heart relies on anaerobic respiration. Further cardiovascular compensation and influx of blood flow can result in reperfusion injury since sudden increases in oxygen leads to increased reactive oxygen species (ROS) and calcium flux, which causes cardiomyocyte damage (e.g., myocardial infarction (MI)) and death [26].

High density lipoprotein (HDL) and LDL cholesterol are standard measurements for patients at risk for the development of myocardial infarctions and CVD in the clinic [27]. Troponin I and creatine kinase are used as markers for cardiac damage [28]. However, new biomarkers for earlier disease diagnosis are needed to prevent CVD progression. Recent advances in MS technology have enabled the determination of lipid quantities and composition in serum as well as in myocardial tissues. In order to analyze the large datasets that accompany lipidomic analyses, researchers must apply consistent computational and statistical approaches. The lipidomics standard initiative, launched in 2018, aims to overcome challenges presented by working with lipidomic data [29,30]. Specifically, using software for lipid annotation, overreporting and using arbitrary units rather than concentrations when reporting lipid species. For example, according to this initiative, when quantifying lipids from tissue internal standards must be added prior to lipid extraction, standards should not be present in samples and tissue samples should be normalized to wet weight or protein [29]. Standardization is critical in order to determine clinical reference values, which can bring the lipid biomarkers identified at the lab bench to clinical use at the bedside for patient diagnosis.

Table 1. Lipid Classes and Examples of General Structures.

Lipid Class		Examples of General Structure
Fatty Acyl Lipids	Fatty Acids (FA)	Ex. Saturated FA: Palmitic acid - C16:0 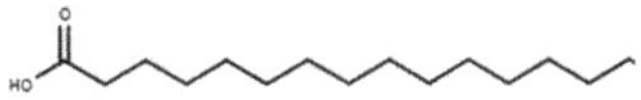 Ex. Unsaturated FA: Arachidonic acid 20:4(5,8,11,14) 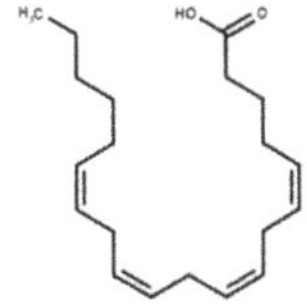
Glycerolipids	Diacylglycerol (DG) Triacylglycerol (TG)	Ex. Diacylglyceride 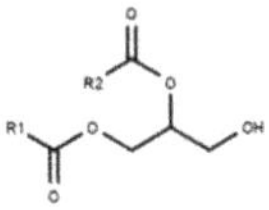 Ex.Triacylglyceride 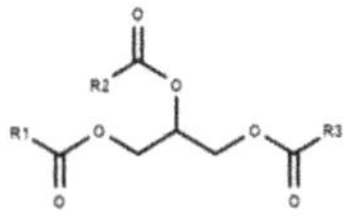
Glycero-phospholipids	Phosphatidylcholine (PC) Phosphatidylethanolamine (PE) Phosphatidylserine (PS) Phosphatidylinositol (PI) Phosphatidylglycerol (PG) Cardiolipin (CL)	Ex. Glycerophospholipid 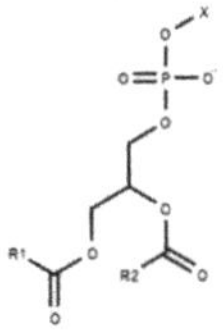 Ex. Cardiolipin 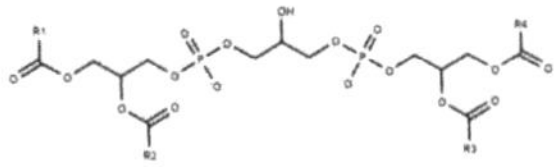
Sphingolipids	Sphingosine Sphingomyelin (SM) Ceramide (CER)	Ex. Sphingosine 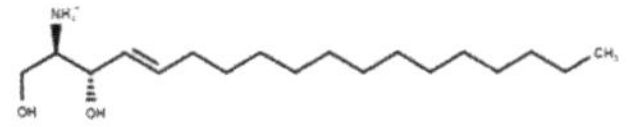 Ex. Ceramide 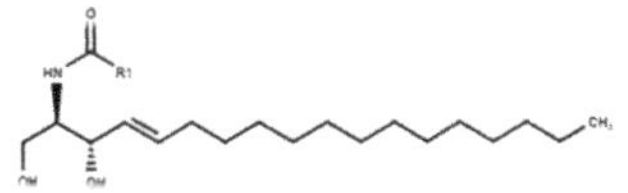
Sterol Lipids	Cholesterol Cholesterol Ester (CE)	Ex. Cholesterol 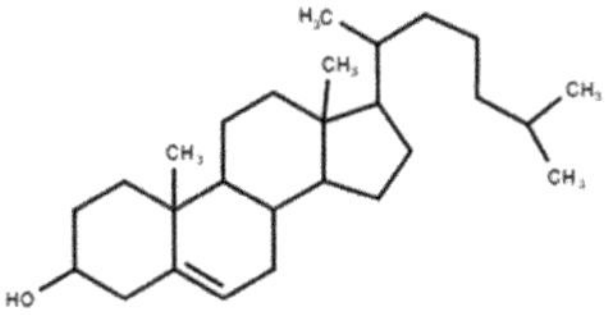 Ex. Cholesterol Ester 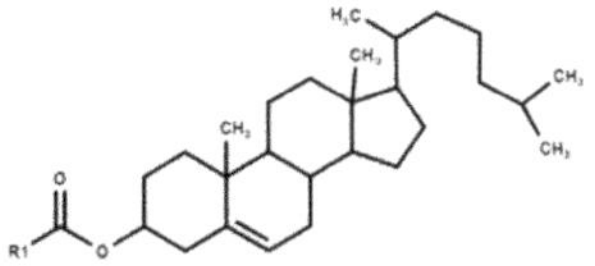

Table is limited to lipids discussed in the review. Structures are examples or general structures. R1, R2, R3, and R4 indicate unspecified fatty acid groups. X specifies phosphatidyl head group. Structures made with MarvinSketch Version 20.11 [31].

1.2. Cardiac Lipid Composition

The heart is composed of numerous cell and tissue types. Cardiomyocytes account for the largest percentage (30–40%) of cells within the heart, which occupy ~70–85% of heart volume [32,33]. Studies investigating the lipid composition in the heart began in the 1950s. Gray and colleagues performed the first studies to isolate and determine the composition of phospholipids from the ox heart by chromatography separation [34]. These studies established that lipids, which constitute cardiac tissues include free fatty acids (FA), triglycerides (TG), diglycerides (DG), cardiolipin (CL) phosphatidylethanolamine (PE), phosphatidylserine (PS), and phosphatidylcholine (PC) [35]. A recent large scale lipidomic study in rats has been able to decipher tissue-specific lipid composition [36]. These data reinforce the classical cardiac lipid composition and compare the lipid composition of major tissue types. The heart exhibited a high abundance of PC, PE, PS, phosphatidylinositol (PI), phosphatidylglycerol (PG), and CL species [36]. PG is a precursor for CL synthesis and was enriched in the heart when compared to other tissues, which may be indicative of the high mitochondrial content in cardiac tissue [36]. Mitochondria are closely linked to cardiac function since cardiomyocyte contractility require an abundance of ATP production.

Cardiac muscle contains high numbers of mitochondria in order to produce sufficient amounts of ATP to supply the heart with abundant energy needed for the mechanical action of pumping blood throughout the body. CL is abundant in the heart and is a major phospholipid of the inner mitochondrial membrane [37]. It is a unique phospholipid since it composed of two glycerol phosphatidyl moieties. This means CL is comprised of four fatty acyl molecules (Table 1) [38]. Tetra-linoleic acid is the predominant form of CL in the mature heart [38]. It is responsible for the mitochondrial structure and the function of inner mitochondrial membrane proteins. For example, it is required for the efficient transfer of electrons and the formation of super complexes in the respiratory chain [39]. Therefore, the cardiac lipid composition and the lipids in the mitochondria play an important role in cardiac energy production and, as a result, are implicated in cardiac function.

1.3. Cardiac Lipid Utilization

FA and glucose are the major fuel sources of the heart [5,40]. Specifically, the utilization of FA through beta-oxidation and subsequent oxidation-reduction reactions within the tricarboxylic (TCA) cycle are responsible for the majority of ATP production in the heart [41]. FA are transported into the cardiomyocyte through the plasma membrane by the FA binding protein. CD36 and FA transport proteins (FATPs) [42,43]. The FA are acylated by some transporters (e.g., FATPs) or through acyl-Coenzyme A synthetase. Carnitine palmitoyltransferase I (CPT-I) then converts the acyl-CoA derivatives into long-chain acylcarnitine molecules on the outer side of the outer mitochondrial membrane. The acyl-carnitine molecules are transported through the inner membrane space and then across the inner mitochondrial membrane by carnitine-acylcarnitine translocase [42]. On the inner membrane, CPT-II is responsible for transferring the acyl residue from carnitine back onto a CoA molecule. The FA acyl-CoA molecules can then enter beta-oxidation for the conversion of FA into acetyl-CoA, which enters the TCA cycle [42]. The TCA cycle produces nicotinamide adenine dinucleotide (NADH) and flavin adenine dinucleotide (FADH$_2$) molecules, which can be utilized by the electron transport chain for the production of ATP [44].

In CVD including MI and pathological cardiac hypertrophy, cardiac metabolism changes from a state of primarily relying on FA and glucose through oxidative phosphorylation to utilizing anaerobic energy production such as glycolysis [41,45,46]. Glycolysis is an inefficient means of energy production in the failing heart [41]. Inefficient ATP production can lead to increased ROS and further oxidation of phospholipids and cardiotoxicity [47,48]. Lipid molecules play an important role in cardiac energy production, the plasma membrane, and organelle composition as well as the progression and pathogenesis of CVD such as the development of atherosclerotic plaques. Mass spectrometry methods are now being used in order to determine lipid levels and examine lipid composition in models of CVD.

2. Models of Cardiovascular Disease

One of the first MS studies performed in cardiac tissue was reported in 1968 by Funasaki and Gilbertson who isolated and identified cholesteryl alkyl ethers from bovine cardiac muscle [49]. Advancements in MS methods [50] have enabled researchers to use these technologies to investigate how lipid species are being altered in the cardiac lipidome (the entire lipid composition of the heart) and how it is altered in different models of CVD.

2.1. Cardiac Lipid Profiles in Experimental Myocardial Infarction Models

MI is the loss of blood flow that leads to myocardial damage [26]. As described above, atherosclerotic plaques are the main cause for MIs. Hsueh et al. were the first to identify that ischemia stimulates fatty acid release in rabbit heart tissue by chromatography separation [51]. The following study identified increases in arachidonic acid in ischemic canine myocardium using high pressure LC separation [52]. Researchers also began using MS methods to examine molecular changes that occur post-MI. The first study to utilize MS technology to examine lipids after a MI was in 1979, where Epps and colleagues identified an increase in N-acylethanolamine 24 h after a canine heart was subjected to ligation of the left descending artery [53]. Since then, numerous studies have examined the effect of MIs on cardiac lipids. For example, in a recent paper utilizing an ischemia and/or starvation model in the H9c2 rat cardiomyocyte cell line, differences in lipid levels including PC (34:1), PC (36:2), lyso-phosphatidylcholine lysoPC (16:0), lysoPC (18:1), lysoPC (18:0), PE (34:1), PS (36:1), PI (36:2), PI (38:3), PI (38:5), sphingomyelin (SM) (34:1), CL (68:4), CL (72:5), and CL (74:7) were observed [54]. Increases in lyso-PCs and decreases in CL were observed in ischemic/starvation conditions compared to controls [54]. Similar results were reported by Nam and colleagues who performed metabolomic and lipidomic analysis of rat hearts, which followed ligation of the left anterior descending coronary artery by ultra-HPLC-MS. In this animal model, gradual increases of free FA, ceramides, PE (40:6), lysoPE(16:0), PC (30:0), PC (32:1), lysoPC (16:0), lysoPC (o-16:0), lysoPC (o-18:0), PG (40:8), PG(42:9), lysoPG (18:2), PS (38:4), SM, and mono and triacylglycerides were observed [55]. Alterations in acylcarnitines, adenine, S-adenosyl methionine, adenosine monophosphate, NAD+, and succinic acid were reported, which suggested a disruption in lipid metabolism. Additional lipidomic studies have also described increased de novo ceramide synthesis and accumulation of long chain ceramides in human serum myocardial tissue [56]. Using an animal model of ischemic left ventricular dysfunction and the serine palmitoyl transferase (SPT) inhibitor, myriocin, this study also showed that reduced ceramide accumulation (C16, C24:1, C24) prevented ventricular remodelling post-MI [56].

A study using a rat MI model reported increases in lysoPC (16:0), lysoPC (18:0), lysoPC (18:2), lysoPC (18:1), lysoPC (20:4), and lysoPE (18:0) in heart tissue by MALDI-MS imaging technology similar to that of the cell and animal studies discussed previously [57]. However, in a comprehensive lipidomic analysis of post-MI cardiac mouse tissue by Halade and colleagues, strong increases in lysophospholipids were not observed [58]. This discrepancy could be due to species' specific differences but warrants further investigation. MALDI imaging MS technology has been used to identify spatial distribution of lipids within cardiac tissue such as the even distribution of tetralinoleic acid CL within healthy heart sections [54,55]. The study by Halade et al. is unique as it identified FA substrates such as arachidonic, docosahexaenoic, eicosapentaenoic acid, and their bioactive lipid mediators (e.g., hydroxydocosahexaenoic acid, hydroxyeicosapentaenoic acid) in infarct LV post MI tissue using MALDI-imaging [53]. The study also reported an increase in PC (36:4), PC (40:8), PC (40:6), and oxidized PC (O-32:0), PC (O-34:0), and PC (O-42:2), which could be indicative of oxidative stress as a result of ischemic conditions that occur during MIs. More specialized studies have focused on the lipid composition of organelles' membranes such as the nucleus. Williams and colleagues employed ESI-MS methods and identified the loss of choline and ethanolamine glycerophospholipids in the nuclear membrane from ischemic and reperfused rat myocardial tissue [59].

Novel studies are now using transgenic and knockout (KO) animal models to decipher pathways, which contribute metabolic signalling in CVD. In a follow-up study, researchers used lipoxygenase

(LOX$^{-/-}$) deficient mice to study its effect on ischemic heart failure [60]. LOX enzymes are a class of FA metabolizing enzymes and, therefore, play an important role in regulating bioactive lipid mediators and FA utilization during myocardial injury [60]. Hearts from LOX deficient mice displayed both increases and decreases in certain PC and lysoPCs species with specific fatty acyl compositions [60]. Additionally, they exhibited increases in sphingolipids in comparison to wild-type mice. Ultimately, the LOX$^{-/-}$ mice showed altered lipidomic and metabolomic profiles and exhibited delayed heart failure progression and improved survival. However, additional studies are needed to decipher important protein, enzyme, and lipid targets that contribute to lipidome alterations during CVD progression and how these alterations can be prevented.

Lipidomic studies of cardiac tissue post MI allow for a greater understanding of how the cardiac lipidome is altered as well as the changes of the fatty acyl chains within these lipid classes. This is notable because it can provide insights into the molecular mechanisms of the pathogenesis of CVD. Understanding these changes may help the development of better therapies to prevent and treat MIs and could be translated to other models of CVD.

2.2. Cardiac Lipid Profiles in Animal Models of Obesity

Weight gain puts patients at risk for developing dyslipidemia and lipotoxicity [61]. In addition to hypertension and diabetes, patients that are characterized as obese are at twice the risk for developing cardiovascular complications [62]. Thus, analysis of the lipidome provides a wealth of information about mechanisms of disease progression. In a study comparing standard, high fat, or high fat/high sucrose (western) diets in rat hearts, researchers identified increases in C16, C18, C20, and C24 ceramides in the western diet group by HPLC-ESI-MS methods [63]. As expected, increased cardiac TG levels were also observed [63]. In heart tissue of mice fed a high fat diet, increases in polyunsaturated fatty acyl chains were observed in ceramides, glycosphingolipids, and sphingomyelins whereas decreases in monounsaturated fatty acyl chains were observed in phospholipids and sphingomyelins [64]. In another study, researchers determined that feeding mice a diet enriched in polyunsaturated fatty acids (arachidonic acid, eicosapentaenoic acid, or docosahexaenoic acid supplemented) for two weeks decreased cardiac phospholipids containing linoleic acid when compared to control mice on a fish meal free diet [65]. This study also went on to describe differences in the oxylipin profiles of tissues in a targeted lipidomics approach.

Peroxisome proliferator-activated receptor-gamma coactivator 1β (PGC1β) is a transcriptional co-activator, which has a role in regulating mitochondrial biogenesis genes and is thought to have a role in the development of obesity and diabetes [66,67]. McCombie et al. utilized a PGC1β KO mouse model to investigate lipidomic changes induced by a high fat diet [68]. In this study, LC-MS lipidomics of cardiac tissue revealed alterations in polar lipid composition and increases in TG. The preceding study focused on results from a combined dataset of male and female mice. However, they did report larger differences in male datasets in KO mice fed a high fat diet when compared to females using partial least squares discriminant analysis (PLS-DA) models, which indicated the importance of performing sex-specific lipidomic studies.

Using cardiac specific diacylglycerol O-acyltransferase 1 (DGAT1) transgenic mice as a model of cardiac steatosis, LC-MS analysis of myocardial tissue revealed no changes in ceramides [62]. In contrast, exposure of these mice to angiotensin II resulted in increased ceramide levels [62]. Increased ceramide ratios (C16:0/24:0) in plasma have been associated with increased cardiac remodelling and cardiac dysfunction in a human study, which examined 2652 Framingham Offspring Study participants [69]. Therefore, activation of the renin-angiotensin system exacerbates the risk of cardiac lipid remodelling. This could be a rationale for investigating whether angiotensin converting enzyme (ACE) inhibitors prevent increased ceramides in models of CVD.

More comprehensive models are now being developed where diets are coupled with models of CVD and aging. A recent study investigated the effect of high-unsaturated fatty acid diet (HUFA) on rats subjected to supra-valvar aortic stenosis (SVAS). The study reported decreases in unsaturated (oleic and linoleic) free fatty acids as well as diacylglycerol and triacylglycerol molecules in SVAS heart tissue. However, the HUFA diet did not restore these lipids to normal levels [70]. In one model, mice on a PUFA diet had impaired wound healing post-MI [71]. The study specifically observed an increase in plasma arachidonic acid by LC-MS analysis, which implicated a high PUFA diet that increases pro-inflammatory lipid metabolites capable of affecting post-MI tissue [71]. In another study combining obesity and MI, researchers examined mitochondrial lipid species from cardiac tissue [72]. Decreases in PCs, PEs, and increases in TGs, lysoPCs, and lysoPEs were reported in mitochondrial lipids from total cardiac tissue of rats subject to MI. In contrast, MI rats fed high fat diets did not exhibit such drastic changes in cardiac mitochondrial glycerophospholipids [72]. This study also reported decreases in total CL. Under closer inspection, the researchers reported decreases in CL (18:2) but increases in CL (20:4, 22:6) in high fat fed groups post MI. This interesting finding suggests that CL composition was altered from a form enriched with linoleic acid to one that is increased in arachidonic acid [72]. The study went on to show an association between levels of fibrosis with cardiac lipids such as TG, CL, ceramide, and several plasma microRNA (miRNA) species including 194-5p, 301a-3p, 144-5p, and 15b-5p [72]. These findings could be significant since miRNA play an important role in transcriptional regulation. Studies such as these could bridge the gap between lipidomic alterations and epigenetic regulation. Complex models of obesity and MI are more representative of cardiac lipid changes that occur in patients in the clinic. Furthermore, they can be used to more accurately decipher the molecular pathways and epigenetic changes that occur in these diseased states.

2.3. Cardiac Lipid Profiles in Diabetic Cardiomyopathy Models

Diabetic cardiomyopathy is characterized by structural and functional changes that occur in the myocardium as a result of diabetes mellitus [73]. Specifically, these changes occur without the presence of CAD or hypertension but are a direct result of diabetes [73]. Hypertrophy or thickening of ventricular walls is a characteristic of diabetic cardiomyopathy and leads to diastolic dysfunction typically with conserved systolic function [74,75]. Ultimately, these structural and functional changes can lead to heart failure. The first study to use MS technology (by ESI-MS) was performed by Han and colleagues who examined alterations in the lipid profile of the diabetic myocardium [76]. Utilizing a rat model and a single injection of streptozotocin, they identified alterations to ethanolamine glycerophospholipids. Specifically, a 24% increase in PE and a 44% increase in plasmenylethanolamine could be restored by insulin treatment [76]. The study identified a 60% decrease in TG, which was not prevented by insulin treatment. Furthermore, a 44% increase in PI and small increases in PG and PS molecular species were also observed. No changes in CL were identified in the heart tissue from this streptozotocin-induced diabetes model. Reminiscent of the PUFA MI model, this group also identified a predominance of PC molecules with arachidonic acid FA moieties in their lipid fractions. However, no statistically significant differences between the diabetic and control rats were observed [76]. The same group followed up with a separate study to examine the acylcarnitine species from cardiac tissue in the streptozotocin-induced diabetes rat model using ESI-MS approaches. They identified a four-fold increase in long-chain acylcarnitines (16:0, 18:2, 18:1, 20:4) in diabetic myocardium compared to controls that could be partially or fully reversed with insulin treatment [77]. These data suggest that impaired FA transport or β-oxidation of FA leads to the accumulation of acylcarnitine species in diabetic cardiomyopathy. The same group also performed a study in streptozotocin-induced diabetic mice using a shotgun MS approach. Unlike rats, CL depletion as well the CL precursor PG was observed in diabetic myocardium mice (7.2 nmol/mg to 3.1 nmol/mg in diabetic hearts) [78]. These findings suggest that CL depletion occurs through ineffective FA utilization, which leads to lipotoxicity that manifests into diabetic cardiomyopathy.

Using leptin receptor deficient mice as a model of diabetic cardiomyopathy increases in TGs and DGs as determined by MS analysis, which were observed in cardiac tissue. There were increases in C14:1, C16:1, C16:0, C18:1, and C20:4 free fatty acid molecular subspecies in the leptin receptor deficient mice compared to controls [79]. Similar to the MI models reviewed above, leptin receptor deficient mice also exhibited increases in ceramides, SM, PC, lysoPC, and PE. In a more recent paper using UPLC/QTOF/MS with ESI positive and negative modes to distinguish acyl chains at the sn-1 and sn-2 positions in myocardial tissue revealed down regulation of PC (22:6/18:2), PC (22:6/18:1), PC (20:4/16:1), PC (16:1/18:3), PE (20:4/18:2), PE (20:4/16:0) and an increase in PC (20:2/18:2), PC (18:0/16:0), PC (20:4/18:0) in a streptozotocin-induced rat model [80]. This study revealed that diabetic cardiomyopathy also induced differences in the fatty acyl chain composition of glycerolipids such as PC and PE. Reporting these changes allows for a more reliable comparison of changes in lipid profiles between models of CVD such as MI and other cardiomyopathies. The power of MS technology is now allowing researchers to decipher how lipid species are being affected at a global scale and at the chemical level. However, more investigation needs to occur in other cardiomyopathy models.

2.4. Lipid Profiles in Cardiac Hypertrophy

Cardiac hypertrophy is characterized by ventricular wall thickening, which can be accompanied by both reduced systolic and diastolic function [81]. Pathological causes for cardiac hypertrophy include hypertension and valvular disease [82]. A common experimental model of cardiac hypertrophy is transverse aortic constriction (TAC) in which surgical ligation of the transverse aorta leads to a pressure-overload induced hypertrophy [83]. A recent study used this model of TAC to investigate molecular changes in mice with cardiac restricted acyl-coenzyme A synthetase-1 overexpression (ACSL1) [84]. ACSL1 is responsible for mediating the activation of long-chain fatty acids to acyl-CoA substrates, which can undergo further β-oxidation for energy production within the heart [85]. LC-ESI-MS/MS of TAC heart tissue revealed increases in ceremide levels (C16, C24:1, C24), which were not observed in the ACSL1 overexpressing hearts subject to TAC. ACSL1 TAC hearts exhibited increases in C20 and C22 ceramides. The authors suggest that ACSL1 overexpression could, therefore, mitigate TAC-induced cardiac hypertrophy through mitochondrial oxidative metabolism.

2.5. Lipid Profiles in Dilated Cardiomyopathy

Dilated cardiomyopathy is characterized by an enlarged ventricle, ventricular wall thinning, reduced ejection fraction, and decreased cardiac output [86]. In contrast to diabetic cardiomyopathy, it is characterized by both systolic and diastolic dysfunction. It can be caused by genetic mutations (e.g., Tafazzin, β-Myosin heavy chain, α-Tropomyosin, Cardiac troponin T, Lamin/C (LMNA)) or chemical toxicity (e.g., anthracycline chemotherapeutics). However, it is often idiopathic [87]. One study performed lipidomic analysis of serum from control individuals and patients with dilated cardiomyopathy as a result of an LMNA mutation. In the serum, changes in PC (38:5e, 38:2) and TGs were identified [10]. Sparagna and colleagues performed a study examining CL in human left ventricular tissue samples and in the spontaneously hypertensive heart failure (SHHF) rat model, which exhibits idiopathic dilated cardiomyopathy (IDC) [88]. The human tissue in this study was isolated from the left ventricle of explanted hearts of patients diagnosed with IDC (*n* = 10) and exhibited decreases in tetra-linoleic CL [88]. Similar decreases in tetra-linoleoyl CL in subsarcolemmal and interfibrillar cardiac mitochondria isolated from 5-month rats and 15-month rats were observed in parallel with increases in CL species with oleic and arachidonic acid side chains. Additionally, a positive relationship with decreased tetra-linoleic CL and impaired cytochrome oxidase activity was observed [88]. This shows the importance that cardiac tissue lipid composition plays in mitochondrial function. In the same study, a rat model of heart failure using SHHF rats subject to thoracic aortic banding (TAB) surgery was also used and decreased tetra-linoleic CL. Increased CL species containing oleic and arachidonic acid side chains were observed in the rat heart tissue. While, in a follow-up study, LC-MS/MS analysis of cardiac tissue explants from eight human patients with dilated cardiomyopathy revealed lower levels of linoleic acid and also reported similar increases in arachidonic and docosahexaenoic acid phospholipid species [89]. These elevated polyunsaturated fatty acid product/precursor ratios suggested that delta-6-desaturase enzyme activity was elevated in dilated cardiomyopathy. Notably, inhibition of the delta-6-desaturase enzyme (with SC-26196 for four weeks) reversed these changes in polyunsaturated fatty acid composition in two different rat models of heart failure (SHHF and TAC). Inhibition of delta-6-desaturase also attenuated elevations in pathogenic eicosanoids and lipid peroxides and normalized the CL fatty acyl chain composition in the rat heart [89]. Another study performed LC-ESI-MS in left ventricular tissue from pediatric patients with IDC and reported similar decreases in total and tetra linoleic CL. The authors do, however, report a unique pediatric cardiac CL profile attributed to differences in the expression of CL biosynthesis genes with age [90].

Doxorubicin (DOX) is an anthracycline chemotherapeutic used in treating pediatric leukemias and lymphomas, but its utility is limited since high dosages of DOX put patients at risk for developing a dilated cardiomyopathy [91,92]. In animal models, DOX is frequently used to induce dilated cardiomyopathy. There has been a modest number of published studies examining cardiac tissue profiles by MS methods in DOX models of dilated cardiomyopathy. In one study, male and female rats were injected with 2 mg/kg of DOX weekly for seven weeks and lipidomic analysis was performed [93]. This study uncovered sex-specific differences in the cardiac lipid profile with response to DOX treatment. Male rats exhibited decreased phospholipid content in cardiac tissue after DOX treatment. Specifically, sex-specific fatty acid composition of PE and PC were different in males and females prior to and after DOX treatment. Furthermore, analysis of CL species revealed no sex differences, but DOX treatment induced a decrease in the most abundant tetra-linoleic CL and an increase in every other CL species [93]. In another study, rats were injected with 2.5 mg/kg of DOX for two weeks and MS analysis of ceramides revealed an increase in C16 and C18 ceramide levels in heart tissue [63]. These two studies illustrate some of the similarities of cardiac lipid profile alterations to other models of CVD discussed above including the depletion of CL in models of diabetic cardiomyopathy and obesity and the increase in ceramides seen in MI models.

2.6. Similarities in Cardiac Lipid Profiles in Models of Cardiovascular Disease

CVD encompasses a wide range of cardiac diseases that have different underlying causes. Cardiac lipid profiles in these models share many similarities. Specifically, most models (whether they be MI, obesity, diabetes, or dilated cardiomyopathy) show increases in ceramide, sphingomyelin, and lyso-phospholipids in cardiac tissue, which suggests these may be molecular markers of disease progression. Increases in ceramides have also been linked to increases in apoptosis in a variety of models including neonatal rat cardiomyocytes [94–96]. Furthermore, ceramides have been shown to modulate lipotoxic cardiomyopathy in mice through interactions with proteins involved in cardiac contractility, apoptosis, and lipogenesis (myosin chaperone, annexin, and fatty acid synthase) [97]. Other similarities in the findings from lipidomic studies in different models of CVD was increases in arachidonic acid fatty acid acyl chains in the failing heart. Specifically, in most models of CVD, when measured, there appears to be decreases in tetralinoleoyl CL species and increases in other forms of CL such as those containing arachidonic. Since CL is so closely linked to the electron transport chain, changes in lipid composition of CL could be related to disrupted oxidative phosphorylation super-complex formation and, thus, decreases in cardiac energy production.

Where the lipidomic studies differ is in changes to phospholipid fatty acyl composition. Specifically, phospholipid fatty acid molecules are shown to be increased and decreased in different models of CVD. This could be an indication of different alterations to FA metabolism, which may be present in different models (e.g., diabetic vs. dilated cardiomyopathies). Other differences include changes in glycerolipids. For example, obesity and diabetic models frequently cite increases in TG and DG lipid species. Specifically, DG lipid accumulation has been linked to impaired insulin-stimulated glucose oxidation in the heart [98] and incomplete oxidation of fatty acids in skeletal muscle, which leads to insulin resistance and mitochondrial dysfunction [99]. In contrast, MI and dilated cardiomyopathy models do not exhibit altered DG species or did not report them altogether. The lipidomic studies discussed are summarized in Table 2.

Metabolites **2020**, *10*, 254

Table 2. Summary of lipidomics studies in models of cardiovascular disease by mass spectrometry.

Lipid Species columns are grouped as: Glycolipids (TG, DG); PC; PE; PI; Phospholipids (PS, PG); LysoPL; OxPL; CL; Sphingolipids (CER, SM).

CVD Model	Animal/Cell Species	N Number	Other	TG	DG	PC	PE	PI	PS	PG	LysoPL	OxPL	CL	CER	SM	Mass Spectrometry (MS) Technology	Reference
MI and IR Models																	
Starvation/Ischemic	H9c2	3	FFA↕			↕	↑	↕	↕		↕		↕		↓	HPLC-MS/MS	[54]
LAD CA Ligation	Rat	8	FFA/AC↑	↑		↑	-		↑	↑	↑			↑	↑	UPLC-QTOF-MS	[55]
Ischemic CM	Patient (Serum/Tissue)	15–64												↕		LC-MS	[56]
LAD CA Ligation	Mice (Serum/Tissue)	4–20												↕		LC-MS	[56]
LAD CA Ligation	Rat	5				↕					↑					MALDI-MS	[57]
LAD CA Ligation	Mice	6	UFA/SFA↑			↑						↑				MALDI-MSI and LC-MS/MS	[58]
IR Injury (15 min)	Rat	6				↓	↓									ESI-MS/MS	[59]
LAD CA Ligation	LOX−/− Mice	37–49	AC ↕					↕PL			↕				↑	LC-MS/MS	[60]
Obesity Models																	
HF Diet or HFHS Diet	Rat	6												↑		LC-MS	[63]
HF Diet	Mice	10		↑	↑			↕PL						↑	↑	GC-MS	[64]
PUFA Diet	Mice	5	TC↓/FA↕	↑						↕						GC-MS	[65]
HF Diet	PGC1β−/− Mice	5–10		↑		↕	↕									GC-MS LC-MS	[68]
Cardiac Steatosis	DGAT1 Mice	6–9		↑										↕		UPLC-QTOF-MS	[62]
Mixed Models																	
SVAS HUFA	Mice	11–14	UFA↓/↑SA	↕	↕											GC-MS	[70]
HF Diet/Aging/ LAD CA Ligation	Mice (plasma)	3–8	AA ↑													LC-MS/MS	[71]
HF Diet/LAD CA Ligation	Rat	8–10		↑		↓MI	↓MI				↕ MI		↕	↕		UPLC-QTOF-MS	[72]
Diabetic CM Models																	
Streptozotocin Injection	Rat	6		↑		-	↓	↑	↑	↑			-	-		ESI-MS	[76]
Streptozotocin Injection	Rat	4–6	AC↑													ESI-MS	[77]
Streptozotocin Injection	Mice	7		↑							↓		↓			ESI-MS	[78]
Genetic	LepR^db/db Mice	5–6	FFA↑	↑	↑	↑	↑				↑			↑	↑	ESI-MS	[79]
HF Diet and Streptozotocin	Rat	11–12				↕	↑									UPLC-QTOF-MS	[80]

Table 2. *Cont.*

CVD Model	Animal/Cell Species	N Number	Other	Glycolipids TG	DG	PC	PE	PI	Phospholipids PS	PG	LysoPL	OxPL	CL	Sphingolipids CER	SM	Mass Spectrometry (MS) Technology	Reference
Hypertrophy Models																	
TAC	ACL1 Mice	3–17		↓										↕		ESI-MS/MS	[84]
Dilated CM Models (SHHF Rats as Validation)																	
IDCM	Patient (Serum)	8–11		↓		↓			↓							UPLC-MS	[10]
SHHF/TAB	Rat	4											↕			LC-ESI-MS	[88]
IDCM	Patient (LV Tissue)	10–11											↕			LC-ESI-MS	[88]
SSHF	Rat	4–10	↕AA													LC-ESI-MS	[89]
IDCM	Human (LV Tissue)	8	↕AA													LC-ESI-MS	[89]
IDCM	Pediatric (LV Tissue)	20–44											↓			LC-ESI-MS	[90]
DOX (2mg/kg Weekly 7X)	Rat	4				↕	↕			↕			↕			LC-MS/MS	[93]
DOX/HFHS Diet (15 mg/kg CD)	Rat	6												↑		LC-MS	[63]

CVD: Cardiovascular Disease. MI: Myocardial infarction. IR: Reperfusion injury. LAD CA: Left anterior descending coronary artery. HF: High fat. PUFA: polyunsaturated fatty acid diet. SVAS: supra-valvar aortic stenosis. HUFA: high unsaturated fatty acid diet. TAC: transverse aortic constriction. IDCM: idiopathic dilated cardiomyopathy. SHHF: Spontaneously hypertensive heart failure. FFA: Free fatty acids. TG: Triglyceride. DG: Diglycerides. PC: Phosphatidylcholine. PE: Phosphatidylethanolamine. PI: Phosphatidylinositol. PS: Phosphatidylserine. PG: Phosphatidylglycerol. LysoPL: Lyso-phospholipids. OxPL: Oxidized phospholipids. CL: Cardiolipin. CER: Ceramides. SM: sphingomyelin. AC: Acylcarnitine. AA: Arachidonic Acid. ↑ increase. ↓ decrease. ↕ increase or decrease depending on FA chain. -: No change. Blank: Not reported.

3. The Effect of Current and Novel Therapies on Cardiac Lipid Profiles

Extensive efforts have gone into examining how cardiac lipid profiles are altered in different models of CVD. The next area of lipidomic research reviewed focuses on understanding how current therapeutics used in treating cardiovascular and lipid disorders affect cardiac lipids.

3.1. The Effect of Non-Pharmacological Interventions on Cardiac Lipid Profiles

Non-pharmacological interventions such as diet and lifestyle changes are often the front line to prevent CVD in patients who are at risk [100]. A recent study utilized the power of MS technology to examine how cardiac lipid profiles are altered in models of exercise and CVD [101]. Specifically, they were interested in examining the differences between physiological hypertrophy that occurs as a compensatory mechanism in response to exercise and the pathological hypertrophy that occurs during CVD. Using a swim model of exercise and a four-week model of pressure overload TAC, LC-MS/MS technology was utilized to perform lipidomic analysis of cardiac tissue. A total of 104 lipid species were significantly altered in swimming mice compared to controls, and 100 lipid species in the severe TAC model. Lipid concentrations in this study were determined by internal standards and normalized to levels of PC rather than protein concentrations or tissue weight. In these models, differences between PC lipids were not observed. However, phospholipids such as alkylphosphatidylcholine (PC(O)), alkylkphosphatidylethanolamine (PE(O)), and phosphatidyl-ethanolamine plasmalogens (PE(P)) were decreased in the hearts of exercised mice and unchanged in the TAC mice. Furthermore, sphingolipids were decreased in cardiac tissue from the exercise model and increased in the TAC model of CVDs. This study suggests that differences in cardiac sphingolipid levels could distinguish between physiological and pathological hypertrophy, which are indicative of damage to cardiomyocyte cell membranes. Identification of how non-pharmacological interventions affect myocardial lipids is important since it may provide information on the actionable mechanism of classic and novel therapeutics used in treating CVD.

3.2. The Effect of Commonly Prescribed CVD Medications on Cardiac Lipids

There is a modest amount of literature that focuses on how common drugs (e.g., statins, fenofibrates) used to treat cardiovascular and lipid disorders affect the cardiac tissue lipidome. Statins prevent cholesterol synthesis by inhibiting 3-hydroxy-3-methyl-glutaryl–CoA reductase and, in turn, reduce circulating levels of LDL. There are several studies that examine serum lipidomics in patients treated with statins [27,102–105]. They report decreased plasma TGs and circulating sphingomyelins in patients treated with statins. However, to date, no study has used MS technology to examine the effect of statin therapy on the cardiac tissue lipidome in obesity models.

Other commonly used therapeutics in treating CVD such as atherosclerosis are fibric acid derivatives. Drugs such as gemfibrozil, fenofibrate, and clofibrate lower TG and LDL levels by increasing lipoprotein lipase activity and inhibiting synthesis of very low-density lipoprotein by activating peroxisome proliferator activated receptor α (PPARα)[46,106]. The Fibrate Intervention and Event Lowering in Diabetes (FIELD) study identified that patients treated with fenofibrates did not benefit the primary endpoint of coronary heart disease events [107]. A substudy of the FIELD assessed serum from patients treated with fenofibrates and identified decreases in lysoPCs and increases in SM. Consistent with the paucity of research surrounding the cardiac lipidome in response to statin therapy in models of CVD, there is also a lack of studies that examine how these tissues are affected by other classic drugs used in treating CVD such as fibric acid derivates. Future studies should aim to examine how drugs already used in treating CVD affect cardiac lipids.

3.3. The Effect of Natural Health Products and Novel Drugs on Cardiac Lipid Profiles

Resveratrol is a polyphenolic molecule derived from plants shown to improve myocardial lipid oxidation and cardiac function in rats [108]. We have shown that, in the spontaneously hypertensive rat model of cardiac hypertrophy as well as the Wistar control rat strain, resveratrol attenuates pathological cardiac hypertrophy and using mass spectroscopy increases total CL mass as well as the tetra-linoleic CL species [109]. Therefore, resveratrol-induced increases in cardiac CL could be linked to improved mitochondrial function. Berberine is a naturally occurring alkaloid extracted from various plants and used in traditional Chinese medicine. It is also available at health food markets [110]. A recent study examined the effect of berberine on myocardial lipid profiles in a high fat, high sucrose diet and a streptozotocin-induced rat model of diabetic cardiomyopathy [110]. Berberine partially reversed alterations to PC (16:0/20:4), PC (18:0/18:2), PC (18:0/18:2), PC (18:0/22:5), PC (20:4/0:0), PC (20:4/18:0), PC (20:4/20:2), PE (18:2/0:0), and SM (d18:0/16:0) in diabetic heart tissue. Berberine also decreased SM, which is a lipid species often reported as upregulated in other models of CVD (obesity, dilated cardiomyopathy). Resveratrol and berberine are thought to have antioxidant capabilities as indicated by decreased ROS levels [111,112]. However, when compared to placebos in clinical trails, antioxidants have had little success in treating CVD [113]. This could be due to improper timing or dosing. Other concerns regarding natural health products such as resveratrol or berberine is their lack of specificity. These compounds have multiple targets, which means they can have a broad impact on metabolism. However, initial studies suggest that some natural health products could be broadly protective in CVD by modifying lipid profiles. This result merits further investigation [114–116]. Therefore, it may be efficacious to investigate novel drugs that have specific protein or lipid targets.

Another novel therapeutic that is gaining attention in treating CVD is elamipretide (a.k.a. Bendavia, MTP-131 and SS-31). Elamipretide is a cell permeable tetrapeptide, which is targeted to the mitochondria by binding directly to CL and reducing ROS formation while increasing mitochondrial function [117]. It has been shown to have cardioprotective effects in animal models of atherosclerotic renovascular disease [118], ischemic-reperfusion injury [119], myocardial infarction [120], hypertension [121], DOX-induced cardiomyopathy models [122], and improvement in mitochondrial function in failing human myocardium [123]. One study has employed MS approaches to examine how elamipretide alters lipids. Specifically, they examined a decrease in tetra-linoleic CL in explanted failing heart tissue from pediatric and adult patients. Treatment with elamipretide prevented changes in CL when compared to untreated controls [123]. The study reports coupling of oxidative phosphorylation supercomplex activity as the mechanism of action. However, more comprehensive lipidomics studies are needed to assess the effect of elamipretide on the entire cardiac lipidome.

4. Conclusions

Lipidomic analysis by MS technology is an expanding field of research. Lipids play an important role in cardiac structure, function, and disease progression. Utilizing this sensitive technique to determine changes that occur in cardiac lipid profiles in models of CVD (MI, obesity, diabetic, or dilated cardiomyopathies, etc.) is important in understanding the pathology behind each disease. Furthermore, performing lipidomic studies in experimental models of CVD holds the promise of increasing our understanding of how novel therapeutics affect the heart. New challenges facing the ever-growing field of lipidomics will include data standardization to generate comparable and reproducible results. Future cardiac lipidomic studies should also focus on sorted cell populations from cardiac tissue to address heterogenous cell populations found in cardiac tissue during CVD. Ultimately, the intention of utilizing MS approaches will be to integrate lipidomics data with other -omics technology to get a better understanding of how the cardiovascular system is affected in its entirety in disease models.

Author Contributions: Conceptualization: M.M.T. and V.W.D. Writing—original draft preparation: M.M.T. Writing—review and editing: M.M.T. and V.W.D. All authors have read and agreed to the published version of the manuscript.

Funding: This research received no external funding.

Acknowledgments: Thank you to Grant Hatch for his expert insights during the paper writing process. M.M.T. is the recipient of a Research Manitoba Studentship in Partnership with CHRIM. V.W.D. is the Allen Rouse-Manitoba Medical Services Foundation Basic Scientist.

Conflicts of Interest: The authors declare no conflict of interest.

References

1. World Health Organization. Available online: https://www.who.int/health-topics/cardiovascular-diseases/#tab=tab_1 (accessed on 17 March 2020).
2. Ference, B.A.; Graham, I.; Tokgozoglu, L.; Catapano, A.L. Impact of Lipids on Cardiovascular Health: JACC Health Promotion Series. *J. Am. Coll. Cardiol.* **2018**, *72*, 1141–1156. [CrossRef] [PubMed]
3. Schenkel, L.C.; Bakovic, M. Formation and Regulation of Mitochondrial Membranes. *Int. J. Cell Biol.* **2014**, *2014*, 709828. [CrossRef] [PubMed]
4. Alberts, B.; Johnson, A.; Lewis, J.; Raff, M.; Roberts, K.; Walter, P. *The Lipid Bilayer*; Garland Science: New York, NY, USA, 2002. ISBN 0-8153-3218-1.
5. Bing, R.J.; Siegel, A.; Ungar, I.; Gilbert, M. Metabolism of the human heart. II. Studies on fat, ketone and amino acid metabolism. *Am. J. Med.* **1954**, *16*, 504–515. [CrossRef]
6. Van Der Vusse, G.J.; Van Bilsen, M.; Glatz, J.F.C. Cardiac fatty acid uptake and transport in health and disease. *Cardiovasc. Res.* **2000**, *45*, 279–293. [CrossRef]
7. Nelson, R.H. Hyperlipidemia as a Risk Factor for Cardiovascular Disease. *Prim. Care Clin. Off. Pract.* **2013**, *40*, 195–211. [CrossRef]
8. Leutner, M.; Göbl, C.; Wielandner, A.; Howorka, E.; Prünner, M.; Bozkurt, L.; Harreiter, J.; Prosch, H.; Schlager, O.; Charwat-Resl, S.; et al. Cardiometabolic Risk in Hyperlipidemic Men and Women. *Int. J. Endocrinol.* **2016**, *2016*, 2647865. [CrossRef]
9. Köfeler, H.C.; Fauland, A.; Rechberger, G.N.; Trötzmüller, M. Mass spectrometry based lipidomics: An overview of technological platforms. *Metabolites* **2012**, *2*, 19–38. [CrossRef]
10. Sysi-Aho, M.; Koikkalainen, J.; Seppänen-Laakso, T.; Kaartinen, M.; Kuusisto, J.; Peuhkurinen, K.; Kärkkäinen, S.; Antila, M.; Lauerma, K.; Reissell, E.; et al. Serum Lipidomics Meets Cardiac Magnetic Resonance Imaging: Profiling of Subjects at Risk of Dilated Cardiomyopathy. *PLoS ONE* **2011**, *6*, e15744. [CrossRef]
11. Syme, C.; Czajkowski, S.; Shin, J.; Abrahamowicz, M.; Leonard, G.; Perron, M.; Richer, L.; Veillette, S.; Gaudet, D.; Strug, L.; et al. Glycerophosphocholine Metabolites and Cardiovascular Disease Risk Factors in Adolescents: A Cohort Study. *Circulation* **2016**, *134*, 1629–1636. [CrossRef]
12. Poss, A.M.; Maschek, J.A.; Cox, J.E.; Hauner, B.J.; Hopkins, P.N.; Hunt, S.C.; Holland, W.L.; Summers, S.A.; Playdon, M.C. Machine learning reveals serum sphingolipids as cholesterol-independent biomarkers of coronary artery disease. *J. Clin. Investig.* **2020**, *130*, 1363–1376. [CrossRef]
13. Anjos, S.; Feiteira, E.; Cerveira, F.; Melo, T.; Reboredo, A.; Colombo, S.; Dantas, R.; Costa, E.; Moreira, A.; Santos, S.; et al. Lipidomics Reveals Similar Changes in Serum Phospholipid Signatures of Overweight and Obese Pediatric Subjects. *J. Proteome Res.* **2019**, *18*, 3174–3183. [CrossRef]
14. Zalloua, P.; Kadar, H.; Hariri, E.; Abi Farraj, L.; Brial, F.; Hedjazi, L.; Le Lay, A.; Colleu, A.; Dubus, J.; Touboul, D.; et al. Untargeted Mass Spectrometry Lipidomics identifies correlation between serum sphingomyelins and plasma cholesterol. *Lipids Health Dis.* **2019**, *18*, 38. [CrossRef] [PubMed]
15. Kohno, S.; Keenan, A.L.; Ntambi, J.M.; Miyazaki, M. Lipidomic insight into cardiovascular diseases. *Biochem. Biophys. Res. Commun.* **2018**, *504*, 590–595. [CrossRef] [PubMed]
16. Brown, J.M.; Hazen, S.L. Seeking a unique lipid signature predicting cardiovascular disease risk. *Circulation* **2014**, *129*, 1799–1803. [CrossRef] [PubMed]
17. Fahy, E.; Cotter, D.; Sud, M.; Subramaniam, S. Lipid classification, structures and tools. *Biochim. Biophys. Acta Mol. Cell Biol. Lipids* **2011**, *1811*, 637–647. [CrossRef] [PubMed]
18. Katz, A.M.; Messineo, F.C. Lipid-membrane interactions and the pathogenesis of ischemic damage in the myocardium. *Circ. Res.* **1981**, *48*, 1–16. [CrossRef] [PubMed]
19. Uhlenbrock, K.; Gassenhuber, H.; Kostenis, E. Sphingosine 1-phosphate is a ligand of the human gpr3, gpr6 and gpr12 family of constitutively active G protein-coupled receptors. *Cell. Signal.* **2002**, *14*, 941–953. [CrossRef]
20. Ohanian, J.; Liu, G.; Ohanian, V.; Heagerty, A.M. Lipid second messengers derived from glycerolipids and sphingolipids, and their role in smooth muscle function. *Acta Physiol. Scand.* **1998**, *164*, 533–548. [CrossRef]

21. Ng, M.; Fleming, T.; Robinson, M.; Thomson, B.; Graetz, N.; Margono, C.; Mullany, E.C.; Biryukov, S.; Abbafati, C.; Abera, S.F.; et al. Global, regional, and national prevalence of overweight and obesity in children and adults during 1980-2013: A systematic analysis for the Global Burden of Disease Study 2013. *Lancet* **2014**, *384*, 766–781. [CrossRef]
22. Page, I.H.; Stare, F.J.; Corcoran, A.C.; Pollack, H.; Wilkinson, C.F. Atherosclerosis and the Fat Content of the Diet. *Circulation* **1957**, *16*, 163–178. [CrossRef]
23. Rafieian-Kopaei, M.; Setorki, M.; Doudi, M.; Baradaran, A.; Nasri, H. Atherosclerosis: Process, indicators, risk factors and new hopes. *Int. J. Prev. Med.* **2014**, *5*, 927–946. [PubMed]
24. Singh, R.B.; Mengi, S.A.; Xu, Y.J.; Arneja, A.S.; Dhalla, N.S. Pathogenesis of atherosclerosis: A multifactorial process. *Exp. Clin. Cardiol.* **2002**, *7*, 40–53. [PubMed]
25. Bentzon, J.F.; Otsuka, F.; Virmani, R.; Falk, E. Mechanisms of plaque formation and rupture. *Circ. Res.* **2014**, *114*, 1852–1866. [CrossRef]
26. Hausenloy, D.J.; Yellon, D.M. Myocardial ischemia-reperfusion injury: A neglected therapeutic target. *J. Clin. Investig.* **2013**, *123*, 92–100. [CrossRef] [PubMed]
27. Arsenault, B.J.; Boekholdt, S.M.; Kastelein, J.J.P. Lipid parameters for measuring risk of cardiovascular disease. *Nat. Rev. Cardiol.* **2011**, *8*, 197–206. [CrossRef]
28. Bodor, G.S. Biochemical Markers of Myocardial Damage. *J. Int. Fed. Clin. Chem. Lab. Med.* **2016**, *27*, 95–111.
29. Lipidomics-Standards-Initiative (LSI). Available online: https://lipidomics-standards-initiative.org/ (accessed on 18 March 2020).
30. Liebisch, G.; Ahrends, R.; Arita, M.; Arita, M.; Bowden, J.A.; Ejsing, C.S.; Griffiths, W.J.; Holčapek, M.; Köfeler, H.; Mitchell, T.W.; et al. Lipidomics needs more standardization. *Nat. Metab.* **2019**, *1*, 745–747.
31. ChemAxon. Marvin Was Used for Drawing and Displaying Chemical Structures. In *Marvin V20.11.*; Available online: https://chemaxon.com/products/marvin/download (accessed on 16 April 2020).
32. Pinto, A.R.; Ilinykh, A.; Ivey, M.J.; Kuwabara, J.T.; D'antoni, M.L.; Debuque, R.; Chandran, A.; Wang, L.; Arora, K.; Rosenthal, N.A.; et al. Revisiting cardiac cellular composition. *Circ. Res.* **2016**, *118*, 400–409. [CrossRef]
33. Zhou, P.; Pu, W.T. Recounting cardiac cellular composition. *Circ. Res.* **2016**, *118*, 368–370. [CrossRef]
34. Gray, G.M.; Macfarlane, M.G. Separation and composition of the phospholipids of ox heart. *Biochem. J.* **1958**, *70*, 409–425. [CrossRef]
35. Wheeldon, L.W.; Schumert, Z.; Turner, D.A. Lipid composition of heart muscle homogenate. *J. Lipid Res.* **1965**, *6*, 481–489. [PubMed]
36. Pradas, I.; Huynh, K.; Cabré, R.; Ayala, V.; Meikle, P.J.; Jové, M.; Pamplona, R. Lipidomics reveals a tissue-specific fingerprint. *Front. Physiol.* **2018**, *9*, 1165. [CrossRef]
37. Crowe, M.O.; Walker, A. The ultraviolet absorption spectra and other physical data for cardiolipin, a new phospholipid, and lecithin isolated from beef heart. *J. Opt. Soc. Am.* **1945**, *35*, 800. [PubMed]
38. Schlame, M.; Rua, D.; Greenberg, M.L. The biosynthesis and functional role of cardiolipin. *Prog. Lipid Res.* **2000**, *39*, 257–288. [CrossRef]
39. Paradies, G.; Paradies, V.; De Benedictis, V.; Ruggiero, F.M.; Petrosillo, G. Functional role of cardiolipin in mitochondrial bioenergetics. *Biochim. Biophys. Acta Bioenerg.* **2014**, *1837*, 408–417. [CrossRef] [PubMed]
40. Depre, C.; Vanoverschelde, J.L.J.; Taegtmeyer, H. Glucose for the heart. *Circulation* **1999**, *99*, 578–588. [CrossRef] [PubMed]
41. Stanley, W.C.; Recchia, F.A.; Lopaschuk, G.D. Myocardial substrate metabolism in the normal and failing heart. *Physiol. Rev.* **2005**, *85*, 1093–1129. [CrossRef] [PubMed]
42. Brivet, M.; Boutron, A.; Slama, A.; Costa, C.; Thuillier, L.; Demaugre, F.; Rabier, D.; Saudubray, J.M.; Bonnefont, J.P. Defects in activation and transport of fatty acids. *J. Inherit. Metab. Dis.* **1999**, *22*, 428–441. [CrossRef]
43. Son, N.H.; Basu, D.; Samovski, D.; Pietka, T.A.; Peche, V.S.; Willecke, F.; Fang, X.; Yu, S.Q.; Scerbo, D.; Chang, H.R.; et al. Endothelial cell CD36 optimizes tissue fatty acid uptake. *J. Clin. Investig.* **2018**, *128*, 4329–4342. [CrossRef]
44. Bolisetty, S.; Jaimes, E. Mitochondria and Reactive Oxygen Species: Physiology and Pathophysiology. *Int. J. Mol. Sci.* **2013**, *14*, 6306–6344. [CrossRef]
45. Allard, M.F. Energy substrate metabolism in cardiac hypertrophy. *Curr. Hypertens. Rep.* **2004**, *6*, 430–435. [CrossRef] [PubMed]

46. Taegtmeyer, H.; Young, M.E.; Lopaschuk, G.D.; Abel, E.D.; Brunengraber, H.; Darley-Usmar, V.; Des Rosiers, C.; Gerszten, R.; Glatz, J.F.; Griffin, J.L.; et al. Assessing Cardiac Metabolism. *Circ. Res.* **2016**, *118*, 1659–1701. [CrossRef] [PubMed]

47. Peoples, J.N.; Saraf, A.; Ghazal, N.; Pham, T.T.; Kwong, J.Q. Mitochondrial dysfunction and oxidative stress in heart disease. *Exp. Mol. Med.* **2019**, *51*, 1–13. [CrossRef] [PubMed]

48. Allen, D.; Hasanally, D.; Ravandi, A. Role of oxidized phospholipids in cardiovascular pathology. *Clin. Lipidol.* **2013**, *8*, 205–215. [CrossRef]

49. Funasaki, H.; Gilbertson, J.R. Isolation and identification of cholesteryl alkyl ethers from bovine cardiac muscle. *J. Lipid Res.* **1968**, *9*, 766–768.

50. Han, X.; Gross, R.W. Global analyses of cellular lipidomes directly from crude extracts of biological samples by ESI mass spectrometry: A bridge to lipidomics. *J. Lipid Res.* **2003**, *44*, 1071–1079. [CrossRef]

51. Hsueh, W.; Isakson, P.C.; Needleman, P. Hormone selective lipase activation in the isolated rabbit heart. *Prostaglandins* **1977**, *13*, 1073–1091. [CrossRef]

52. Chien, K.R.; Han, A.; Sen, A.; Buja, L.M.; Willerson, J.T. Accumulation of unesterified arachidonic acid in ischemic canine myocardium. Relationship to a phosphatidylcholine deacylation-reacylation cycle and the depletion of membrane phospholipids. *Circ. Res.* **1984**, *54*, 313–322. [CrossRef]

53. Epps, D.E.; Schmid, P.C.; Natarajan, V.; Schmid, H.H.O. N-acylethanolamine accumulation in infarcted myocardium. *Biochem. Biophys. Res. Commun.* **1979**, *90*, 628–633. [CrossRef]

54. Sousa, B.; Melo, T.; Campos, A.; Moreira, A.S.P.; Maciel, E.; Domingues, P.; Carvalho, R.P.; Rodrigues, T.R.; Girão, H.; Domingues, M.R.M. Alteration in Phospholipidome Profile of Myoblast H9c2 Cell Line in a Model of Myocardium Starvation and Ischemia. *J. Cell. Physiol.* **2016**, *231*, 2266–2274. [CrossRef]

55. Nam, M.; Jung, Y.; Ryu, D.H.; Hwang, G.S. A metabolomics-driven approach reveals metabolic responses and mechanisms in the rat heart following myocardial infarction. *Int. J. Cardiol.* **2017**, *227*, 239–246. [CrossRef] [PubMed]

56. Ji, R.; Akashi, H.; Drosatos, K.; Liao, X.; Jiang, H.; Kennel, P.J.; Brunjes, D.L.; Castillero, E.; Zhang, X.; Deng, L.Y.; et al. Increased de novo ceramide synthesis and accumulation in failing myocardium. *JCI Insight* **2017**, *2*, e82922. [CrossRef] [PubMed]

57. Menger, R.F.; Stutts, W.L.; Anbukumar, D.S.; Bowden, J.A.; Ford, D.A.; Yost, R.A. MALDI mass spectrometric imaging of cardiac tissue following myocardial infarction in a rat coronary artery ligation model. *Anal. Chem.* **2012**, *84*, 1117–1125. [CrossRef] [PubMed]

58. Halade, G.V.; Dorbane, A.; Ingle, K.A.; Kain, V.; Schmitter, J.M.; Rhourri-Frih, B. Comprehensive targeted and non-targeted lipidomics analyses in failing and non-failing heart. *Anal. Bioanal. Chem.* **2018**, *410*, 1965–1976. [CrossRef] [PubMed]

59. Williams, S.D.; Hsu, F.F.; Ford, D.A. Electrospray ionization mass spectrometry analyses of nuclear membrane phospholipid loss after reperfusion of ischemic myocardium. *J. Lipid Res.* **2000**, *41*, 1585–1595. [PubMed]

60. Halade, G.V.; Kain, V.; Tourki, B.; Jadapalli, J.K. Lipoxygenase drives lipidomic and metabolic reprogramming in ischemic heart failure. *Metabolism* **2019**, *96*, 22–32. [CrossRef]

61. Lim, H.-Y.; Bodmer, R. Phospholipid homeostasis and lipotoxic cardiomyopathy. *Fly (Austin)* **2011**, *5*, 234–236. [CrossRef]

62. Glenn, D.J.; Cardema, M.C.; Ni, W.; Zhang, Y.; Yeghiazarians, Y.; Grapov, D.; Fiehn, O.; Gardner, D.G. Cardiac steatosis potentiates angiotensin II effects in the heart. *Am. J. Physiol. Circ. Physiol.* **2015**, *308*, H339–H350. [CrossRef]

63. Butler, T.J.; Ashford, D.; Seymour, A.M. Western diet increases cardiac ceramide content in healthy and hypertrophied hearts. *Nutr. Metab. Cardiovasc. Dis.* **2017**, *27*, 991–998. [CrossRef]

64. Pakiet, A.; Jakubiak, A.; Mierzejewska, P.; Zwara, A.; Liakh, I.; Sledzinski, T.; Mika, A. The Effect of a High-Fat Diet on the Fatty Acid Composition in the Hearts of Mice. *Nutrients* **2020**, *12*, 824. [CrossRef]

65. Naoe, S.; Tsugawa, H.; Takahashi, M.; Ikeda, K.; Arita, M. Characterization of Lipid Profiles after Dietary Intake of Polyunsaturated Fatty Acids Using Integrated Untargeted and Targeted Lipidomics. *Metabolites* **2019**, *9*, 241. [CrossRef] [PubMed]

66. Jornayvaz, F.R.; Shulman, G.I. Regulation of mitochondrial biogenesis. *Essays Biochem.* **2010**, *47*, 69–84.

67. Mootha, V.K.; Lindgren, C.M.; Eriksson, K.F.; Subramanian, A.; Sihag, S.; Lehar, J.; Puigserver, P.; Carlsson, E.; Ridderstråle, M.; Laurila, E.; et al. PGC-1α-responsive genes involved in oxidative phosphorylation are coordinately downregulated in human diabetes. *Nat. Genet.* **2003**, *34*, 267–273. [CrossRef] [PubMed]

68. McCombie, G.; Medina-Gomez, G.; Lelliott, C.J.; Vidal-Puig, A.; Griffin, J.L. Metabolomic and Lipidomic Analysis of the Heart of Peroxisome Proliferator-Activated Receptor-γ Coactivator 1-β Knock Out Mice on a High Fat Diet. *Metabolites* **2012**, *2*, 366–381. [CrossRef] [PubMed]

69. Nwabuo, C.C.; Duncan, M.; Xanthakis, V.; Peterson, L.R.; Mitchell, G.F.; McManus, D.; Cheng, S.; Vasan, R.S. Association of Circulating Ceramides With Cardiac Structure and Function in the Community: The Framingham Heart Study. *J. Am. Heart Assoc.* **2019**, *8*, e013050. [CrossRef]

70. Casquel De Tomasi, L.; Salomé Campos, D.H.; Grippa Sant'Ana, P.; Okoshi, K.; Padovani, C.R.; Masahiro Murata, G.; Nguyen, S.; Kolwicz, S.C.; Cicogna, A.C. Pathological hypertrophy and cardiac dysfunction are linked to aberrant endogenous unsaturated fatty acid metabolism. *PLoS ONE* **2018**, *13*, e0193553. [CrossRef]

71. Lopez, E.F.; Kabarowski, J.H.; Ingle, K.A.; Kain, V.; Barnes, S.; Crossman, D.K.; Lindsey, M.L.; Halade, G.V. Obesity superimposed on aging magnifies inflammation and delays the resolving response after myocardial infarction. *Am. J. Physiol. Hear. Circ. Physiol.* **2015**, *308*, H269–H280. [CrossRef]

72. Marín-Royo, G.; Ortega-Hernández, A.; Martínez-Martínez, E.; Jurado-López, R.; Luaces, M.; Islas, F.; Gómez-Garre, D.; Delgado-Valero, B.; Lagunas, E.; Ramchandani, B.; et al. The Impact of Cardiac Lipotoxicity on Cardiac Function and Mirnas Signature in Obese and Non-Obese Rats with Myocardial Infarction. *Sci. Rep.* **2019**, *9*, 1–11. [CrossRef]

73. Jia, G.; Hill, M.A.; Sowers, J.R. Diabetic cardiomyopathy: An update of mechanisms contributing to this clinical entity. *Circ. Res.* **2018**, *122*, 624–638. [CrossRef]

74. Boudina, S.; Abel, E.D. Diabetic cardiomyopathy, causes and effects. *Rev. Endocr. Metab. Disord.* **2010**, *11*, 31–39. [CrossRef]

75. Ritchie, R.H.; Abel, E.D. Basic Mechanisms of Diabetic Heart Disease. *Circ. Res.* **2020**, *126*, 1501–1525. [CrossRef] [PubMed]

76. Han, X.; Abendschein, D.R.; Kelley, J.G.; Gross, R.W. Diabetes-induced changes in specific lipid molecular species in rat myocardium. *Biochem. J.* **2000**, *352*, 79–89. [CrossRef] [PubMed]

77. Su, X.; Han, X.; Mancuso, D.J.; Abendschein, D.R.; Gross, R.W. Accumulation of long-chain acylcarnitine and 3-hydroxy acylcarnitine molecular species in diabetic myocardium: Identification of alterations in mitochondrial fatty acid processing in diabetic myocardium by shotgun lipidomics. *Biochemistry* **2005**, *44*, 5234–5245. [CrossRef] [PubMed]

78. Han, X.; Yang, J.; Cheng, H.; Yang, K.; Abendschein, D.R.; Gross, R.W. Shotgun lipidomics identifies cardiolipin depletion in diabetic myocardium linking altered substrate utilization with mitochondrial dysfunction. *Biochemistry* **2005**, *44*, 16684–16694. [CrossRef]

79. DeMarco, V.G.; Ford, D.A.; Henriksen, E.J.; Aroor, A.R.; Johnson, M.S.; Habibi, J.; Ma, L.; Yang, M.; Albert, C.J.; Lally, J.W.; et al. Obesity-related alterations in cardiac lipid profile and nondipping blood pressure pattern during transition to diastolic dysfunction in male db/db mice. *Endocrinology* **2013**, *154*, 159–171. [CrossRef]

80. Dong, S.; Zhang, R.; Liang, Y.; Shi, J.; Li, J.; Shang, F.; Mao, X.; Sun, J. Changes of myocardial lipidomics profiling in a rat model of diabetic cardiomyopathy using UPLC/Q-TOF/MS analysis. *Diabetol. Metab. Syndr.* **2017**, *9*, 56. [CrossRef]

81. Gradman, A.H.; Alfayoumi, F. From Left Ventricular Hypertrophy to Congestive Heart Failure: Management of Hypertensive Heart Disease. *Prog. Cardiovasc. Dis.* **2006**, *48*, 326–341. [CrossRef]

82. Frey, N.; Katus, H.A.; Olson, E.N.; Hill, J.A. Hypertrophy of the heart: A new therapeutic target? *Circulation* **2004**, *109*, 1580–1589. [CrossRef] [PubMed]

83. Dealmeida, A.C.; Van Oort, R.J.; Wehrens, X.H. Transverse Aortic Constriction in Mice. *J. Vis. Exp.* **2010**, *1729*. [CrossRef]

84. Goldenberg, J.R.; Carley, A.N.; Ji, R.; Zhang, X.; Fasano, M.; Schulze, P.C.; Lewandowski, E.D. Preservation of Acyl Coenzyme A Attenuates Pathological and Metabolic Cardiac Remodeling Through Selective Lipid Trafficking. *Circulation* **2019**, *139*, 2765–2777. [CrossRef]

85. Roche, C.M.; Blanch, H.W.; Clark, D.S.; Louise Glass, N. Physiological role of acyl coenzyme a synthetase homologs in lipid metabolism in Neurospora crassa. *Eukaryot. Cell* **2013**, *12*, 1244–1257. [CrossRef] [PubMed]

86. Dadson, K.; Hauck, L.; Billia, F. Molecular mechanisms in cardiomyopathy. *Clin. Sci.* **2017**, *131*, 1375–1392. [CrossRef]

87. McNally, E.M.; Mestroni, L. Dilated cardiomyopathy: Genetic determinants and mechanisms. *Circ. Res.* **2017**, *121*, 731–748. [CrossRef] [PubMed]

88. Sparagna, G.C.; Chicco, A.J.; Murphy, R.C.; Bristow, M.R.; Johnson, C.A.; Rees, M.L.; Maxey, M.L.; McCune, S.A.; Moore, R.L. Loss of cardiac tetralinoleoyl cardiolipin in human and experimental heart failure. *J. Lipid Res.* **2007**, *48*, 1559–1570. [CrossRef]

89. Le, C.H.; Mulligan, C.M.; Routh, M.A.; Bouma, G.J.; Frye, M.A.; Jeckel, K.M.; Sparagna, G.C.; Lynch, J.M.; Moore, R.L.; McCune, S.A.; et al. Delta-6-desaturase links polyunsaturated fatty acid metabolism with phospholipid remodeling and disease progression in heart failure. *Circ. Hear. Fail.* **2014**, *7*, 172–183. [CrossRef]

90. Chatfield, K.C.; Sparagna, G.C.; Sucharov, C.C.; Miyamoto, S.D.; Grudis, J.E.; Sobus, R.D.; Hijmans, J.; Stauffer, B.L. Dysregulation of cardiolipin biosynthesis in pediatric heart failure. *J. Mol. Cell. Cardiol.* **2014**, *74*, 251–259. [CrossRef] [PubMed]

91. Kremer, L.C.M.; Van Dalen, E.C.; Offringa, M.; Voûte, P.A. Frequency and risk factors of anthracycline-induced clinical heart failure in children: A systematic review. *Ann. Oncol.* **2002**, *13*, 503–512. [CrossRef]

92. Pinder, M.C.; Duan, Z.; Goodwin, J.S.; Hortobagyi, G.N.; Giordano, S.H. Congestive heart failure in older women treated with adjuvant anthracycline chemotherapy for breast cancer. *J. Clin. Oncol.* **2007**, *25*, 3808–3815. [CrossRef]

93. Moulin, M.; Solgadi, A.; Veksler, V.; Garnier, A.; Ventura-Clapier, R.; Chaminade, P. Sex-specific cardiac cardiolipin remodelling after doxorubicin treatment. *Biol. Sex. Differ.* **2015**, *6*, 20. [CrossRef]

94. Bielawska, A.E.; Shapiro, J.P.; Jiang, L.; Melkonyan, H.S.; Piot, C.; Wolfe, C.L.; Tomei, L.D.; Hannun, Y.A.; Umansky, S.R. Ceramide is involved in triggering of cardiomyocyte apoptosis induced by ischemia and reperfusion. *Am. J. Pathol.* **1997**, *151*, 1257–1263.

95. Delpy, E.; Hatem, S.N.; Andrieu, N.; De Vaumas, C.; Henaff, M.; Rücker-Martin, C.; Jaffrézou, J.P.; Laurent, G.; Levade, T.; Mercadier, J.J. Doxorubicin induces slow ceramide accumulation and late apoptosis in cultured adult rat ventricular myocytes. *Cardiovasc. Res.* **1999**, *43*, 398–407. [CrossRef]

96. Tohyama, J.; Oya, Y.; Ezoe, T.; Vanier, M.T.; Nakayasu, H.; Fujita, N.; Suzuki, K. Ceramide accumulation is associated with increased apoptotic cell death in cultured fibroblasts of sphingolipid activator protein-deficient mouse but not in fibroblasts of patients with Farber disease. *J. Inherit. Metab. Dis.* **1999**, *22*, 649–662. [CrossRef] [PubMed]

97. Walls, S.M.; Cammarato, A.; Chatfield, D.A.; Ocorr, K.; Harris, G.L.; Correspondence, R.B. Ceramide-Protein Interactions Modulate Ceramide-Associated Lipotoxic Cardiomyopathy. *Cell Rep.* **2018**, *22*, 2702–2715. [CrossRef]

98. Zhang, L.; Ussher, J.R.; Oka, T.; Cadete, V.J.J.; Wagg, C.; Lopaschuk, G.D. Cardiac diacylglycerol accumulation in high fat-fed mice is associated with impaired insulin-stimulated glucose oxidation. *Cardiovasc. Res.* **2011**, *89*, 148–156. [CrossRef] [PubMed]

99. Koves, T.R.; Ussher, J.R.; Noland, R.C.; Slentz, D.; Mosedale, M.; Ilkayeva, O.; Bain, J.; Stevens, R.; Dyck, J.R.B.; Newgard, C.B.; et al. Mitochondrial Overload and Incomplete Fatty Acid Oxidation Contribute to Skeletal Muscle Insulin Resistance. *Cell Metab.* **2008**, *7*, 45–56. [CrossRef]

100. Myers, J. Exercise and Cardiovascular Health Jonathan Myers. *Health Care (Don. Mills).* **2003**, *107*, 1–4.

101. Tham, Y.K.; Bernardo, B.C.; Huynh, K.; Giles, C.; Meikle, P.J.; Mcmullen Correspondence, J.R. Lipidomic Profiles of the Heart and Circulation in Response to Exercise versus Cardiac Pathology: A Resource of Potential Biomarkers and Drug Targets. *Cell Rep.* **2018**, *24*, 2757–2772. [CrossRef] [PubMed]

102. Chen, F.; Maridakis, V.; O'Neill, E.A.; Hubbard, B.K.; Strack, A.; Beals, C.; Herman, G.A.; Wong, P. The effects of simvastatin treatment on plasma lipid-related biomarkers in men with dyslipidaemia. *Biomarkers* **2011**, *16*, 321–333. [CrossRef]

103. Meikle, P.J.; Wong, G.; Tan, R.; Giral, P.; Robillard, P.; Orsoni, A.; Hounslow, N.; Magliano, D.J.; Shaw, J.E.; Curran, J.E.; et al. Statin action favors normalization of the plasma lipidome in the atherogenic mixed dyslipidemia of MetS: Potential relevance to statin-associated dysglycemia. *J. Lipid Res.* **2015**, *56*, 2381–2392. [CrossRef]

104. Bergheanu, S.C.; Reijmers, T.; Zwinderman, A.H.; Bobeldijk, I.; Ramaker, R.; Liem, A.H.; van der Greef, J.; Hankemeier, T.; Jukema, J.W. Lipidomic approach to evaluate rosuvastatin and atorvastatin at various dosages: Investigating differential effects among statins. *Curr. Med. Res. Opin.* **2008**, *24*, 2477–2487. [CrossRef]

105. Orsoni, A.; Thérond, P.; Tan, R.; Giral, P.; Robillard, P.; Kontush, A.; Meikle, P.J.; Chapman, M.J. Statin action enriches HDL3 in polyunsaturated phospholipids and plasmalogens and reduces LDL-derived phospholipid hydroperoxides in atherogenic mixed dyslipidemia. *J. Lipid Res.* **2016**, *57*, 2073–2087. [CrossRef] [PubMed]

106. Yetukuri, L.; Huopaniemi, I.; Koivuniemi, A.; Maranghi, M.; Hiukka, A.; Nygren, H.; Kaski, S.; Taskinen, M.R.; Vattulainen, I.; Jauhiainen, M.; et al. High density lipoprotein structural changes and drug response in lipidomic profiles following the long-term fenofibrate therapy in the FIELD substudy. *PLoS ONE* **2011**, *6*, e23589. [CrossRef]

107. Keech, A.; Simes, R.J.; Barter, P.; Best, J.; Scott, R.; Taskinen, M.R.; Forder, P.; Pillai, A.; Davis, T.; Glasziou, P.; et al. The FIELD study investigators Effects of long-term fenofibrate therapy on cardiovascular events in 9795 people with type 2 diabetes mellitus (the FIELD study): Randomised controlled trial. *Lancet* **2005**, *366*, 1849–1861. [PubMed]

108. Dolinsky, V.W.; Jones, K.E.; Sidhu, R.S.; Haykowsky, M.; Czubryt, M.P.; Gordon, T.; Dyck, J.R.B. Improvements in skeletal muscle strength and cardiac function induced by resveratrol during exercise training contribute to enhanced exercise performance in rats. *J. Physiol.* **2012**, *590*, 2783–2799. [CrossRef] [PubMed]

109. Dolinsky, V.W.; Cole, L.K.; Sparagna, G.C.; Hatch, G.M. Cardiac mitochondrial energy metabolism in heart failure: Role of cardiolipin and sirtuins. *Biochim. Biophys. Acta Mol. Cell Biol. Lipids* **2016**, *1861*, 1544–1554. [CrossRef]

110. Dong, S.; Zhang, S.; Chen, Z.; Zhang, R.; Tian, L.; Cheng, L.; Shang, F.; Sun, J. Berberine Could Ameliorate Cardiac Dysfunction via Interfering Myocardial Lipidomic Profiles in the Rat Model of Diabetic Cardiomyopathy. *Front. Physiol.* **2018**, *9*, 1042. [CrossRef]

111. Gao, R.Y.; Mukhopadhyay, P.; Mohanraj, R.; Wang, H.; Horváth, B.; Yin, S.; Pacher, P. Resveratrol attenuates azidothymidine-induced cardiotoxicity by decreasing mitochondrial reactive oxygen species generation in human cardiomyocytes. *Mol. Med. Rep.* **2011**, *4*, 151–155.

112. Qi, M.Y.; Feng, Y.; Dai, D.Z.; Li, N.; Cheng, Y.S.; Dai, Y. CPU86017, a berberine derivative, attenuates cardiac failure through normalizing calcium leakage and downregulated phospholamban and exerting antioxidant activity. *Acta Pharmacol. Sin.* **2010**, *31*, 165–174. [CrossRef]

113. Steinhubl, S.R. Why Have Antioxidants Failed in Clinical Trials? *Am. J. Cardiol.* **2008**, *101*, S14–S19. [CrossRef]

114. Britton, R.G.; Kovoor, C.; Brown, K. Direct molecular targets of resveratrol: Identifying key interactions to unlock complex mechanisms. *Ann. N. Y. Acad. Sci.* **2015**, *1348*, 124–133. [CrossRef]

115. Kulkarni, S.S.; Cantó, C. The molecular targets of resveratrol. *Biochim. Biophys. Acta Mol. Basis Dis.* **2015**, *1852*, 1114–1123. [CrossRef]

116. Feng, X.; Sureda, A.; Jafari, S.; Memariani, Z.; Tewari, D.; Annunziata, G.; Barrea, L.; Hassan, S.T.S.; Smejkal, K.; Malaník, M.; et al. Berberine in cardiovascular and metabolic diseases: From mechanisms to therapeutics. *Theranostics* **2019**, *9*, 1923–1951. [CrossRef] [PubMed]

117. Birk, A.V.; Liu, S.; Soong, Y.; Mills, W.; Singh, P.; Warren, J.D.; Seshan, S.V.; Pardee, J.D.; Szeto, H.H. The mitochondrial-targeted compound SS-31 re-energizes ischemic mitochondria by interacting with cardiolipin. *J. Am. Soc. Nephrol.* **2013**, *24*, 1250–1261. [CrossRef] [PubMed]

118. Eirin, A.; Ebrahimi, B.; Zhang, X.; Zhu, X.Y.; Woollard, J.R.; He, Q.; Textor, S.C.; Lerman, A.; Lerman, L.O. Mitochondrial protection restores renal function in swine atherosclerotic renovascular disease. *Cardiovasc. Res.* **2014**, *103*, 461–472. [CrossRef] [PubMed]

119. Lee, F.Y.; Shao, P.L.; Wallace, C.G.; Chua, S.; Sung, P.H.; Ko, S.F.; Chai, H.T.; Chung, S.Y.; Chen, K.H.; Lu, H.I.; et al. Combined therapy with SS31 and mitochondria mitigates myocardial ischemia-reperfusion injury in rats. *Int. J. Mol. Sci.* **2018**, *19*, 2782. [CrossRef]

120. Cho, J.; Won, K.; Wu, D.; Soong, Y.; Liu, S.; Szeto, H.H.; Hong, M.K. Potent mitochondria-targeted peptides reduce myocardial infarction in rats. *Coron. Artery Dis.* **2007**, *18*, 215–220. [CrossRef]

121. Dai, D.F.; Chen, T.; Szeto, H.; Nieves-Cintrón, M.; Kutyavin, V.; Santana, L.F.; Rabinovitch, P.S. Mitochondrial targeted antioxidant peptide ameliorates hypertensive cardiomyopathy. *J. Am. Coll. Cardiol.* **2011**, *58*, 73–82. [CrossRef]

122. Min, K.; Kwon, O.-S.; Smuder, A.J.; Wiggs, M.P.; Sollanek, K.J.; Christou, D.D.; Yoo, J.-K.; Hwang, M.-H.; Szeto, H.H.; Kavazis, A.N.; et al. Increased mitochondrial emission of reactive oxygen species and calpain activation are required for doxorubicin-induced cardiac and skeletal muscle myopathy. *J. Physiol.* **2015**, *593*, 2017–2036. [CrossRef]

123. Chatfield, K.C.; Sparagna, G.C.; Chau, S.; Phillips, E.K.; Ambardekar, A.V.; Aftab, M.; Mitchell, M.B.; Sucharov, C.C.; Miyamoto, S.D.; Stauffer, B.L.; et al. Elamipretide Improves Mitochondrial Function in the Failing Human Heart. *JACC Basic Trans. Sci.* **2019**, *4*, 147–157. [CrossRef]

MDPI
St. Alban-Anlage 66
4052 Basel
Switzerland
Tel. +41 61 683 77 34
Fax +41 61 302 89 18
www.mdpi.com

Metabolites Editorial Office
E-mail: metabolites@mdpi.com
www.mdpi.com/journal/metabolites